# CAMBRIDGE LIBRARY COLLECTION

*Books of enduring scholarly value*

## Botany and Horticulture

Until the nineteenth century, the investigation of natural phenomena, plants and animals was considered either the preserve of elite scholars or a pastime for the leisured upper classes. As increasing academic rigour and systematisation was brought to the study of 'natural history', its subdisciplines were adopted into university curricula, and learned societies (such as the Royal Horticultural Society, founded in 1804) were established to support research in these areas. A related development was strong enthusiasm for exotic garden plants, which resulted in plant collecting expeditions to every corner of the globe, sometimes with tragic consequences. This series includes accounts of some of those expeditions, detailed reference works on the flora of different regions, and practical advice for amateur and professional gardeners.

## The Trees of Great Britain and Ireland

Although without formal scientific training, Henry John Elwes (1846–1922) devoted his life to natural history. He had studied birds, butterflies and moths, but later turned his attention to collecting and growing plants. Embarking on his most ambitious project in 1903, he recruited the Irish dendrologist Augustine Henry (1857–1930) to collaborate with him on this well-illustrated work. Privately printed in seven volumes between 1906 and 1913, it covers the varieties, distribution, history and cultivation of tree species in the British Isles. The strictly botanical parts were written by Henry, while Elwes drew on his extensive knowledge of native and non-native species to give details of where remarkable examples could be found. Each volume contains photographic plates as well as drawings of leaves and buds to aid identification. The species covered in Volume 4 (1909) include fir, chestnut, ash and birch.

Cambridge University Press has long been a pioneer in the reissuing of out-of-print titles from its own backlist, producing digital reprints of books that are still sought after by scholars and students but could not be reprinted economically using traditional technology. The Cambridge Library Collection extends this activity to a wider range of books which are still of importance to researchers and professionals, either for the source material they contain, or as landmarks in the history of their academic discipline.

Drawing from the world-renowned collections in the Cambridge University Library and other partner libraries, and guided by the advice of experts in each subject area, Cambridge University Press is using state-of-the-art scanning machines in its own Printing House to capture the content of each book selected for inclusion. The files are processed to give a consistently clear, crisp image, and the books finished to the high quality standard for which the Press is recognised around the world. The latest print-on-demand technology ensures that the books will remain available indefinitely, and that orders for single or multiple copies can quickly be supplied.

The Cambridge Library Collection brings back to life books of enduring scholarly value (including out-of-copyright works originally issued by other publishers) across a wide range of disciplines in the humanities and social sciences and in science and technology.

# The Trees
# of Great Britain
# and Ireland

VOLUME 4

HENRY JOHN ELWES
AUGUSTINE HENRY

CAMBRIDGE
UNIVERSITY PRESS

# CAMBRIDGE
## UNIVERSITY PRESS

University Printing House, Cambridge, CB2 8BS, United Kingdom

Published in the United States of America by Cambridge University Press, New York

Cambridge University Press is part of the University of Cambridge.
It furthers the University's mission by disseminating knowledge in the pursuit of
education, learning and research at the highest international levels of excellence.

www.cambridge.org
Information on this title: www.cambridge.org/9781108069359

© in this compilation Cambridge University Press 2014

This edition first published 1909
This digitally printed version 2014

ISBN 978-1-108-06935-9 Paperback

# THE TREES OF GREAT BRITAIN AND IRELAND

NATIVE SCOTS PINE AT INVERGARRY

*From a Drawing by Miss Ruth Brand*

# The Trees

*of*

# Great Britain

## & Ireland

BY

## Henry John Elwes, F.R.S.

AND

## Augustine Henry, M.A.

VOLUME IV

Edinburgh: Privately Printed

MCMIX

# CONTENTS

iii

# Contents

# ILLUSTRATIONS

vii

# ABIES

*Abies*, Linnæus, *Gen. Pl.* 294 (in part) (1737); Bentham et Hooker, *Gen. Pl.* iii. 441 (1880); Masters, *Journ. Linn. Soc. (Bot.)* xxx. 34 (1893); Hickel, *Bull. Soc. Dendr. France*, 1907, pp. 5, 41, and 82; 1908, pp. 5 and 179.

*Picea*, D. Don, in Loudon, *Arb. et Frut. Brit.* iv. 2293 (1838).

EVERGREEN trees belonging to the order Coniferæ; bark containing numerous resin-vesicles; branches whorled. Buds, with numerous imbricated scales, with or without resin, usually two to five at the ends of the branchlets, the central bud terminal and largest, the others surrounding it in a circle on upright shoots, whilst on lateral branchlets those on the upper side are not developed; buds also occur rarely and few in number in the axils of the leaves on the branchlets below. Branchlets of one kind, usually smooth, but in certain species grooved, with raised pulvini; each season's shoot[1] marked by a sheath at the base, composed of the persistent bud-scales of the previous spring.

Leaves on fertile and barren branchlets, often different in length and thickness and in the nature of the apex; arising from the branchlets in spiral order, radially disposed on vertical shoots, but variously arranged according to the species on lateral branchlets; persisting for many years and giving the tree a dense mass of foliage; leaving as they fall circular scars on the branchlets; sessile, but usually narrowed just above the expanded circular base; linear, flattened and thin in most species, quadrangular in section in a few species; ventral surface always with two greyish or white stomatic bands, one on each side of the raised green midrib; dorsal surface with or without stomata, which when present are either in continuous lines, as in the quadrangular-leaved species, or are confined to near the tip of the leaf in the middle line, as in some flat-leaved species; apex acute, acuminate, or obtuse, notched or entire, spine-pointed in one or two species; resin-canals[2] two, constant in position for each species in the leaves on lateral branchlets, but in some species[3] differing in position in the leaves on the upright or fertile branchlets, either *median*,

---

[1] In *A. bracteata*, all the bud-scales usually fall off, leaving ring-like scars at the base of the shoot.

[2] The position of the resin-canals is easily seen on examining a thin section with a lens; and can often be made out by squeezing the leaf, after it is cut across, when the resin will be observed exuding from the two canals.

[3] In *A. pectinata*, *A. cephalonica*, and *A. Nordmanniana*, the resin-canals are marginal in the leaves of lateral branches, and are median in the leaves of cone-bearing branches. Cf. Guinier and Maire, in *Bull. Soc. Bot. France*, lv. 189 (1908).

when situated in the substance of the leaf about equidistant between its upper and lower surfaces, or *marginal* or *sub-epidermal*, when placed in the lower part of the leaf close to the epidermis; fibro-vascular bundle simple in some species, divided into two parts in other species.

Flowers monœcious, the two sexes on separate branchlets; male flowers usually abundant and on the lower side of the branchlets over the upper half of the tree; female cones on the upper side of the branchlets, usually only near the top of the tree, but in some species borne all over the upper half of the tree. Staminate flowers,[1] solitary in the axils of the leaves of the preceding year's shoot; stamens spirally crowded on a central axis, anthers surmounted by a knob-like projection and dehiscing transversely.    Female cones,[1] arising as short shoots, composed of numerous imbricated fan-shaped ovuliferous scales, and an equal number of much longer mucronate bracts; ovules inverted, two on each scale.

Mature cones erect on the branchlets, composed of closely imbricated woody scales, more or less fan-shaped with short stalks.    Bracts adnate to the outer surface of the scales at the base; either concealed between the scales or with their tips exserted and then often reflexed over the margin of the scale next below; dilated at the apex, entire or two-lobed, prolonged into a triangular mucro.    Seeds two on the inner surface of each scale, winged, and with resin-vesicles.    The cones ripen in one season; and the scales, bracts, and seeds fall away from the central spindle-like axis of the cone, which persists for a long time on the tree.    The seedling has four to ten cotyledons, stomatiferous on their upper surface.

The species of Abies are distinguishable from all other conifers by the circular base of the leaves, which on falling leave circular scars on the branchlets.

The species of Abies have been variously divided into sections by different authors, but no satisfactory arrangement has yet been made out.    Mayr proposed three sections based on the colour of the cones; but, as Sargent[2] points out, colour is not a constant character in several species.    The cones are of value in the discrimination of the species, by taking into account their age, general appearance, and characters as a whole; but the scales are often very variable in shape in the same species, and the bracts, while more constant in form, often show considerable variation in their length.    It is most convenient, in practice, especially as cones are in most cases not available for examination, to group the species, according to the characters of the buds, branchlets, and foliage, which are, as a rule, very constant in the same species.    Hickel[3] proposes three sections, based on the characters of the branchlets and buds; but his division is artificial, as it separates species closely allied by the characters of their cones.

Some notes on the genus Abies, for which we are indebted to Mr. J. D. Crozier, forester to H. R. Baird, Esq. of Durris, Kincardineshire, are inserted. Mr. Crozier's long experience in the east of Scotland gives a special value to his opinion on their respective qualities for planting in Scotland, which our own

---

[1] Both the staminate flowers and the young female cones are surrounded at the base by involucres of bud-scales.

[2] *Silva N. Amer.* xii. 97, *adnot.* (1898).    Sargent proposes three sections, based on the characters of the leaves.

[3] *Bull. Soc. Dendr. France,* 1907, p. 11.

could not have, though in almost every case he confirms the conclusions at which we had already arrived.

About thirty species are known, of which twenty-six have been introduced and are distinguished below. The silver firs are natives of the temperate parts of the northern hemisphere, usually occurring in mountainous regions; attaining high elevations towards the south, as in Guatemala, Algeria, Himalayas, and Formosa; and descending to low levels in the extreme north, as Alaska, Labrador, and Siberia.

The following table is based upon characters taken from the foliage, buds, and shoots of lateral branches, occurring on the lower part of the tree. As regards the leaves, their arrangement upon the branchlets, the position of the resin-canals, and whether the apex is entire or bifid must be noted. The presence of stomata on the upper surface of the leaf is peculiar to certain species. The young shoots are either smooth or deeply grooved with prominent pulvini; and are glabrous in some species, pubescent in others, the pubescence when present being either confined to the grooves or spread over the whole branchlet. The buds vary in size and shape and also in the quantity of resin, which in some cases is so slight that they may be described as non-resinous; whilst in other species the scales are covered with or deeply immersed in resin.

Certain species are distinguishable at a glance by some prominent character. *A. bracteata* has a bud entirely different from that of any other species. *A. Pinsapo*, with its short, thick, rigid leaves, standing out radially from the shoot, is unmistakable. *A. cephalonica*, with a more imperfect radial arrangement, is distinguished by its long flattened leaves ending in a single sharp cartilaginous point. *A. firma* is peculiar in its remarkably broad very coriaceous leaves, which end in two sharp unequal points. *A. grandis* has the leaves quite pectinate in the horizontal plane, those of the upper rank about half the size of those below. *A. Mariesii* is distinguished by the shoot being densely covered with a ferruginous tomentum. *A. brachyphylla* and *A. Webbiana* have deeply-furrowed shoots with prominent pulvini, which become more marked in the second year; and the bark begins to scale very early on the branches and trunk of the tree. *A. nobilis* and *A. magnifica* are peculiar in the upper median leaves curving up from the shoot after being appressed to it for some distance. *A. Pindrow* has long pale green leaves very irregularly arranged.

I. *Leaves radially arranged on the branchlets; apex of the leaf not bifid.*

1. *Abies Pinsapo*, Boissier. Spain. See p. 732.
   Leaves rigid, short, less than ¾ inch long, thick, acute at the apex; resin-canals median. Shoots glabrous. Buds resinous.
2. *Abies cephalonica*, Loudon. Greece. See p. 739.
   Leaves thin, flattened, about 1 inch long, ending in a sharp cartilaginous point; resin-canals marginal. Shoots glabrous. Buds resinous.
   In var. *Apollinis*, the radial arrangement is imperfect, and the leaves end in a short point.

II. *Leaves on the lateral branches pectinate in arrangement; the two lateral sets either in one plane, or with their upper ranks directed upwards as well as outwards, showing a V-shaped depression, as seen from above, between the two sets.*

#### * Resin-canals marginal.[1]

3. *Abies bracteata*, Nuttall.   California.   See p. 796.

Leaves long, 2 inches or more, rigid, ending in a spine-like point.   Shoots glabrous.   Buds peculiar in the genus, elongated, fusiform, membranous, non-resinous.

4. *Abies grandis*, Lindley.   Western N. America.   See p. 773.

Leaves all in one plane, those in the upper rank about half the length of those below, up to 2 inches long, bifid at the apex; upper surface grooved and without stomata.   Shoots minutely pubescent.   Buds small, resinous.

5. *Abies Lowiana*, Murray.   California.   See p. 779.

Leaves in a V-shaped arrangement, $1\frac{1}{2}$ to $2\frac{1}{2}$ inches long, bifid at the apex; upper surface grooved and with eight lines of stomata.   Shoots and buds as in *A. grandis*.

6. *Abies firma*, Siebold and Zuccarini.   Japan.   See p. 762.

Leaves in a V-shaped arrangement, rigid, very coriaceous, broad, up to $1\frac{1}{2}$ inch long, ending in two sharp cartilaginous points.   Shoots pubescent in the furrows between the slightly raised pulvini.   Buds small, ovoid, only slightly resinous.

7. *Abies homolepis*, Siebold and Zuccarini.   Japan.   See p. 764.

Leaves in arrangement and appearance like *A. firma*; but shorter, less coriaceous, narrower, and whiter beneath.   Shoots with prominent pulvini, glabrous.   Buds ovoid, resinous, larger than in *A. firma*.

8. *Abies pectinata*, De Candolle.   Europe.   See p. 720.

Leaves pectinate in one plane or tending to a V-shaped arrangement, about an inch long, slightly bifid at the apex.   Shoot grey, with short pubescence. Buds ovoid, non-resinous.

9. *Abies Webbiana*, Lindley.   Himalayas.   See p. 750.

Leaves V-shaped in arrangement, up to $2\frac{1}{2}$ inches long, bifid, silvery white beneath.   Shoots with prominent pulvini and deep grooves, with a reddish pubescence confined to the grooves.   Buds large, globose, resinous.

#### ** Resin-canals median.[2]

10. *Abies balsamea*, Miller.   Eastern N. America.   See p. 803.

Leaves slender, scarcely 1 inch long, bifid at the apex, with six to eight lines of stomata in each band on the lower surface.   Shoots, smooth, grey, with scattered short erect grey pubescence.   Buds globose, resinous.

11. *Abies Fraseri*, Poiret.   Alleghany Mountains.   See p. 806.

Leaves as in *A. balsamea*, but shorter and whiter beneath, with eight to

---

[1] *A. cilicica* and *A. numidica*, with weak shoots, come in this section.   See Nos. 22 and 23.

[2] *Abies lasiocarpa*, Nuttall, often has the leaves more or less pectinate, and might be sought for here.   See No. 26.

twelve lines of stomata in each band beneath. Shoots smooth, yellowish, with dense reddish curved or twisted pubescence. Buds globose, resinous.

12. *Abies brachyphylla*, Maximowicz.[1] Japan. See p. 765.

Leaves in a V-shaped arrangement, short, scarcely exceeding $\frac{7}{8}$ inch, slightly bifid, white beneath. Shoots glabrous, with prominent pulvini and deep grooves. Buds conical, resinous.

III. *Leaves on lateral branches not pectinate above, but densely crowded, those in the middle line directed forwards in imbricated ranks, their bases not being appressed to the branchlet. On the lower side of the shoot the leaves are in two lateral sets.*

*\* Resin-canals marginal.*[2]

13. *Abies Nordmanniana*, Spach.[3] Caucasus, Northern Asia Minor. See p. 746.

Leaves up to $1\frac{1}{4}$ inch long, with rounded bifid apex. Shoots smooth, with short scattered erect pubescence. Buds ovoid, brown, non-resinous.

14. *Abies amabilis*, Forbes. Western N. America. See p. 782.

Leaves in arrangement and size like those of *A. Nordmanniana*, but much darker shining green, and with a truncate bifid apex; they emit a fragrant odour when bruised. Shoots smooth, with short wavy pubescence. Buds small, globose, resinous.

15. *Abies religiosa*, Schlechtendal. Mexico, Guatemala. See p. 808.

Leaves about 1 inch long, gradually narrowing from the middle to the usually entire apex, which is occasionally slightly emarginate. Shoots with prominent pulvini and dense minute erect pubescence. Buds shortly cylindrical, resinous.

The median upper leaves are much less numerous than in the two preceding species.

16. *Abies Mariesii*, Masters. Japan, Formosa. See p. 771.

Leaves shorter and broader than in *Abies Veitchii*, widest in their upper third, with a rounded and bifid apex. Shoot densely covered with a ferruginous tomentum. Buds small, globose, resinous.

*\*\* Resin-canals median.*

17. *Abies Veitchii*, Lindley. Japan. See p. 768.

Leaves up to 1 inch long, truncate and bifid at the apex, uniform in width, very white beneath, with nine to ten lines of stomata in each band. Shoots smooth, covered with dense short erect pubescence. Buds small, globose, resinous.

The upper median leaves, pointing forwards, stand off from the shoot at a wider angle than in *A. Nordmanniana*.

[1] *Abies umbellata*, Mayr, is said to be very similar in foliage to this species. See the description of this species, p. 768.
[2] *A. numidica* with strong shoots, is distinguished from all these species by the leaves of the upper side being directed backwards. See No. 23.
[3] *A. cilicica*, with strong shoots, resembles a weak *A. Nordmanniana*. See No. 22.

18. *Abies sachalinensis*, Masters.    Saghalien, Yezo, Kurile Isles.    See p. 760.

Leaves long and slender, up to $1\frac{3}{4}$ inch, uniform in width, with a rounded and bifid apex, white beneath, seven to eight lines in each stomatic band. Shoots with prominent pulvini, and a dense short pubescence confined to the grooves.    Buds small, globose, resinous.

19. *Abies sibirica*, Ledebour.    N. E. Russia, Siberia, Turkestan.    See p. 758.

Leaves long and slender, up to $1\frac{1}{2}$ inch, uniform in width; apex rounded and either slightly bifid or entire; four to five lines in each stomatic band beneath.    Shoots ashy grey, quite smooth, with a scattered minute pubescence.    Buds small, globose, resinous.

IV. *Leaves on lateral branches not pectinate above; those in the middle line covering the branchlet, and curving upwards after being appressed to the shoot for some distance at their base.    The leaves are in two lateral sets on the lower side of the branchlet.    Resin-canals marginal.*

20. *Abies nobilis*, Lindley.    Washington, Oregon, California.    See p. 786.

Leaves above closely appressed by their bases to the branchlet, which they completely conceal; about 1 inch long, entire at the apex, flattened, grooved on the upper surface in the middle line; stomata usually present on both surfaces.    Shoots with a dense, short brown pubescence.    Terminal buds girt at the base by a ring of acute or subulately-pointed pubescent scales.

21. *Abies magnifica*, Murray.    Oregon, California.    See p. 792.

Leaves above appressed at their bases, for a short distance only, to the branchlet, which they do not completely conceal; longer than in *A. nobilis*, up to $1\frac{3}{4}$ inch, entire at the apex, quadrangular in section, not grooved on the upper surface; stomata always present on both surfaces.    Shoots and buds as in *A. nobilis*.

V. *Leaves on lateral branches arranged in two ways, which are often observable on the same tree, and depend upon the vigour of the shoots.*

22. *Abies cilicica*, Carrière.    Asia Minor.    See p. 744.

Leaves either (A) pectinate above with a V-shaped depression between the lateral sets, or (B) with the median leaves above crowded and covering the branchlet, as in *A. Nordmanniana*.    The leaves are slender, up to $1\frac{1}{4}$ inch long, not conspicuously white below, slightly bifid at the rounded or acute apex; resin-canals marginal.    Shoots smooth, with scattered short erect pubescence.    Buds small, ovoid, non-resinous.

Vigorous shoots of this species resemble a weak *A. Nordmanniana*; but with the leaves shorter, more slender, and less white beneath, the buds being much smaller.

23. *A. numidica*, De Lannoy.    Algeria.    See p. 737.

Leaves either (A) pectinate above with a V-shaped depression; or (B) crowded and covering the upper side of the branchlet, but different from all other species in the median leaves above, in that case, being directed backwards and not forwards. Leaves short, up to ¾ inch long, broad, rounded at the entire or slightly bifid apex; in most cases with four to six broken lines of stomata on their upper surface near the tip; resin-canals marginal. Shoots brown, shining, glabrous. Buds large, ovoid, non-resinous.

VI. *Leaves irregularly arranged; those on the lower side of the branches not truly pectinate.*

24. *Abies Pindrow*, Spach. W. Himalayas. See p. 755.

Leaves all directed more or less forwards; those above irregularly and imperfectly covering the branchlet; those below mostly pectinate, but with some directed downwards and forwards. Leaves soft, pale green, up to 2½ inches long, bifid at the apex with two sharp cartilaginous points; resin-canals marginal. Shoots grey, glabrous. Buds large, globose, resinous.

25. *Abies concolor*, Lindley and Gordon. Colorado, Utah, Arizona, New Mexico, Northern Mexico, Southern California. See p. 777.

Leaves imperfectly pectinate both above and below, some in the middle line being always directed forwards and not laterally outwards; up to 2 to 3 inches long; apex entire; upper surface convex and not grooved, bearing fifteen to sixteen lines of stomata; resin-canals marginal. Shoots smooth, olive-green, glabrous. Bud large, conical, resinous.

26. *Abies lasiocarpa*, Nuttall. Western N. America. See p. 800.

Leaves either (A) in an imperfect pectinate arrangement, or (B) with most of the leaves directed upwards, those in the middle line above crowded, and standing edgeways; 1½ inches long, narrow, usually entire, with conspicuous lines of stomata on the upper surface, especially in its anterior half. Resin-canals median. Shoots smooth, with a moderately dense, short wavy pubescence. Buds small, conical, resinous.

Four species, *A. Delavayi*, Franchet;[1] *A. Fargesii*, Franchet;[2] *A. squamata*, Masters;[3] and *A. recurvata*, Masters;[4] occur in the mountains of western China and are not included in the above list. The two first species are reported by Masters to have been introduced by Wilson; but, on inquiry, we find that only one species of Abies from China is now growing in the Coombe Wood nursery. It is probably *A. Fargesii*; but, as the plants are still very young, we are uncertain of this identification, and think it best to leave this species undescribed for the present.

(A. H.)

[1] *Journ. de Bot.* 1899, p. 255; Masters, *Gard. Chron.* xxxix. 212, fig. 82 (1906).
[2] *Journ. de Bot.* 1899, p. 256; Masters, *Gard. Chron.* xxxix. 212, fig. 83 (1906).
[3] *Gard. Chron.* xxxix. 299, fig. 121 (1906), and *Journ. Linn. Soc.* (*Bot.*), xxxvii. 423 (1906).
[4] *Journ. Linn. Soc.* (*Bot.*), xxxvii. 423 (1906).

## ABIES PECTINATA, COMMON SILVER FIR

*Abies pectinata*, De Candolle, in Lamarck, *Flore Franç.* iii. 276 (1805); Willkomm, *Forstliche Flora*,
    112 (1887); Mathieu, *Flore Forestière*, 525 (1897); Kent, Veitch's *Man. Coniferæ*, 530 (1900).
*Abies alba*,[1] Miller, *Dict.* ed. 8, No. 1 (1768); Kirchner, *Lebengesch. Blütenpfl. Mitteleuropas*, i. 78
    (1904).
*Abies vulgaris*, Poiret, in Lamarck, *Dict.* vi. 514 (1804).
*Abies Picea*, Lindley, *Penny Cycl.* i. 29 (not Miller) (1833).
*Pinus Picea*, Linnæus, *Sp. Pl.* 1001 (1753).
*Pinus Abies*, Du Roi, *Obs. Bot.* 39 (1771).
*Pinus pectinata*, Lamarck, *Fl. Franç.* ii. 202 (1778).
*Picea pectinata*, Loudon, *Arb. et Frut. Brit.* iv. 2329 (1838).

A tree attaining under favourable conditions about 150 feet in height and 20 feet or more in girth. Bark on young trees, smooth, greyish; ultimately fissuring and becoming rough and scaly. Buds small, ovoid, non-resinous; scales few, brownish, rounded at the apex. Young shoots grey, smooth, with a scattered short erect pubescence, which is retained in the second year.

Leaves on lateral branches pectinately arranged in two lateral sets; those below the longest and directed outwards and slightly forwards in the horizontal plane; those above directed upwards and outwards, forming between the two sets a shallow V-shaped depression. Leaves about 1 inch long, $\frac{1}{12}$ inch broad, linear, flattened, narrowed at the base, tapering slightly to the rounded, bifid apex; upper surface dark green, shining, with a continuous median groove and without stomata; lower surface with two white bands of stomata, each of seven to eight lines; resin-canals marginal.

On leading shoots the leaves are radially arranged, and differ considerably from those on lateral branches; they are thicker, with median resin-canals, acute and not bifid at the apex, and often show lines of stomata on their upper surface towards the tip. Leaves on cone-bearing branches are nearly all directed upwards, very sharp-pointed, and almost tetragonal in section.

Trees, standing in an isolated position, usually begin to flower at about thirty years old; when crowded in dense forests, much later, usually not before sixty years old.

Staminate flowers, surrounded at the base by numerous imbricated scales, cylindrical, about 1 inch long, with greenish-yellow stamens. Female cones, appearing in August of the previous year as large rounded buds, enclosed in brown scales, and situated just behind the apex of the shoot; in spring, when developed, erect, cone-shaped, about 1 inch long, surrounded at the base by fringed scales; bracts numerous, imbricated, denticulate, ending in long, acuminate points, and completely concealing the much smaller ovate, rounded ovuliferous scales.

---

[1] *Abies alba*, the oldest name under the correct genus, was never in use until lately, when it has been resuscitated by Sargent and some continental botanists. This is one of the cases where adhesion to strict priority would lead to great confusion; and hence we have adopted the name *Abies pectinata*, by which the tree is generally known.

Cones on short stout stalks, cylindrical, slightly narrowed at both ends, obtuse at the apex, about 6 inches long, 2 inches in diameter, greenish when growing, dull brown when mature, with the points of the bracts exserted and reflexed. Scales tomentose externally, fan-shaped, about 1 inch broad and long; upper margin slightly uneven; lateral margins denticulate, each usually with a sinus, below the slight wings on the outer side of the scale; claw clavate. Bract with an oblong claw, extending up three-quarters the height of the scale, and expanding above into a lozenge-shaped denticulate lamina, which ends in a sharp long triangular mucro. Seed with wing about an inch long; wing about twice as long as the body of the seed.

## SEEDLING

Seed sown in spring germinates in three or four weeks. The cotyledons, usually five in number, are at first enveloped, as with a cap, by the albumen of the seed; but speedily casting this off, they spread radially in a whorl at the summit of the short caulicle, and remain green on the plant for several years; about an inch in length, linear, obtuse at the apex, flat beneath, and slightly ridged on the upper surface, which shows two whitish bands of stomata. In the first year only a single whorl of true leaves, arising immediately above the cotyledons and alternating with them, is produced. Primary leaves short, acute, or obtuse, but not emarginate at the apex, and with the stomatic bands on the lower surface. A terminal bud closes the first season's growth, the plant scarcely attaining two inches high. In the second year ordinary leaves, arranged spirally on the stem, are produced. The growth of the plant in the first two or three years is mainly concentrated in the root, which descends deep into the soil, the increase in height of the stem above ground being trifling. The stem branches in the third or fourth year, and produces annually for some years one or two lateral branches, making no great growth in height, reaching in the ninth year an average of two feet. About the tenth year normal verticillate branching begins; and from this onwards the plant makes rapid growth.

## VARIETIES

Dr. Klein gives in *Vegetationsbilder* illustrations of some remarkable forms[1] which the silver fir assumes at high elevations in Central Europe, and which he calls "Wettertanne" or "Schirmtanne." These trees have lost their main leader through lightning, wind, or otherwise, and have developed immense side branches which spread and then ascend, sometimes forming a candelabra-like shape. The finest of this type known to him is at St. Cerques in Switzerland, and measures at breast height no less than 7.40 metres in girth, about the same as the largest of the Roseneath[2] trees.

Other varieties, distinguished by their peculiar habit, occur in the wild state.

[1] These forms are also described by Dr. Christ in *Garden and Forest*, ix. 273 (1896).

[2] One of the trees at Roseneath, Dumbartonshire, has a similar growth of erect branches, like leaders from some of the horizontal limbs. This is figured, from a photograph by Vernon Heath, in *Gard. Chron.* xxii. 8, fig. 1 (1884). At Powerscourt there is also a large tree, 13 feet 3 inches in girth, with branches prostrate on the ground and sending up several upright stems.

Var. *pendula*,[1] with weeping branches, has been found in the Vosges and in East Friesland.

Var. *virgata*,[2] found in Alsace and Bohemia, has long pendulous branches, only giving off branchlets near their apices, and densely covered with leaves.

Var. *pyramidalis*.[3] This form, which in habit resembles the cypress or a Lombardy poplar, was found growing wild in the department of Isère in France. A very fine example, about 35 feet high in 1904, is growing in the arboretum of Segrez.

Var. *columnaris*,[4] very slender in habit, with numerous short branches, all of equal length, and with leaves shorter and broader than in the type.

Var. *tortuosa*, a dwarf form, with twisted branches, and bent, irregularly-arranged leaves.

Var. *brevifolia*, another dwarf form, distinguished by its short broad leaves.

Remarkable variations in the cones have also been observed. A tree, discovered by Purkyne[5] in Bohemia, bore cones, umbonate at the apex, and with short and non-reflexed bracts. Beissner[6] mentions a tree, growing in the park at Wörlitz near Dessau, which produced cones a foot in length.

### DISTRIBUTION

The common silver fir is a native of the mountainous regions of central and southern Europe. The northern limit of its area of distribution begins in the western Pyrenees about lat. 43° in the neighbourhood of Roncesvalles in Navarre; and crossing the chain it extends along its northern slope as far as St. Béat; from here it bends northwards to the mountains of Auvergne, whence it is continued in a north-easterly direction through Burgundy and French Lorraine, crossing the eastern slope of the Vosges about the latitude of Strasburg. From here it curves for some distance westward, and reaching Luxemburg, is continued through Trier and Bonn to southern Westphalia. Across the rest of Germany, according to Drude, who gives a map of the distribution of the species, the northern limit extends as an irregular line about lat. 51°, which touches Hersfeld, Eisenach, the northern edge of the Thuringian forest, Glauchau, Rochlitz, Dresden, Bautzen, and Görlitz; and ends in the southern point of the province of Posen. Around Spremberg to the north of the limit just traced, it is found wild in a small isolated territory.

The eastern limit, beginning in Posen, extends through Poland along the River Wartha to Kolo, crosses to Warsaw, and descending through Galicia west of Lemberg, reaches the Carpathians in Bukowina; and is continued along the mountains of Transylvania to Orsova on the Danube.

The southern limit is not clearly known as regards the Balkan peninsula, as the silver fir, which occurs in the mountains of Roumelia, Macedonia, and Thrace,

---

[1] Kottmeier found peculiar weeping silver firs in the Friedeberg forest, near Wittmund in East Friesland, in 1882. Cf. Wittmack's *Gartenzeitung*, 1882, p. 406, and Conwentz, *Seltene Waldbäume in Westpreussen*, 161 (1895).

[2] Caspary, in Hempel's *Oesterr. Forstzeitung*, 1883, p. 43.

[3] Carrière, *Conif.* 280.

[4] Carrière, *Rev. Hort.* 1859, p. 39.

[5] Willkomm, *Forstliche Flora*, 118 (1887).

[6] *Nadelholzkunde*, 433 (1891).

supposed to be *A. pectinata*, is more probably a form of *A. Apollinis*. In Italy the common silver fir reaches its most southerly point on the Nebroden and Madonia Mountains of Sicily at lat. 38°. From here the limit follows the Apennines up through Italy, crosses into Corsica, and from there passes into Spain, where it extends from Monseny, near the Mediterranean coast in lat. 41° 25′, parallel to the Pyrenees, through the mountains of Catalonia and northern Aragon to Navarre. In Spain the silver fir also occurs westwards on a few points of the northern littoral in the Basque provinces and Asturias.

Within the extensive territory just delimited, the silver fir is very irregularly distributed, being totally absent in many parts, as on the plains and lower mountains of southern Europe. In the eastern part of its area it occurs only as isolated trees or in small groups in the beech and spruce forests; whereas, in the western part, as in France and in parts of Germany, it forms forests of great extent, either pure or in which it is the dominant species.

In France the largest forests of the silver fir are in the Vosges and in the Jura. Important forests also occur in the eastern parts of the Pyrenees, the Cevennes, the mountains of Auvergne, and the Alps of Dauphiné. It is rare on the hills of Burgundy, and does not occur in the Ardennes. There are small woods of this species on some of the hills in Normandy, which are, however, supposed to be planted and not indigenous. The great forest of the Vosges[1] is about 50 miles long by 5 to 10 miles in width, and contains about 200,000 acres, situated mainly between 1100 and 3300 feet elevation. This forest consists chiefly of silver fir, though, in some parts, there is a considerable mixture of beech, spruce, and common pine. The most productive woods are on siliceous soil, and only contain 10 per cent of beech and pine; their mean annual production being about 100 cubic feet per acre, the volume of timber standing on each acre averaging 4500 cubic feet.

In the Jura there are even richer and more homogeneous forests than in the Vosges, being according to Huffel the finest in Europe. Here the soil is limestone. One of these forests, which covers Mount La Joux, between 2100 and 3000 feet altitude, contains 10,600 acres, and consists of about 90 per cent silver fir and 10 per cent of spruce. The annual yield per acre is 170 cubic feet of timber. The total volume of standing timber, including only trees over 2 feet in girth, is 6000 cubic feet per acre. The net revenue is thirty-two shillings an acre. There are several other forests equally valuable in this region.

One of the finest silver firs[1] in France, a tree called "Le Président," is growing in the forest of La Joux. It is 163 feet high, with a clean stem of 93 feet, and a girth of 15 feet; and contains 1600 cubic feet of timber. In the forest[2] of Gérardmer, in the Vosges, there are two fine trees. One, the *Beau Sapin*, has a height of 144 feet and a girth of 13 feet 8 inches; it contains 777 cubic feet of timber, and is valued at £16. The other, the *Géant Sapin*, has a height of 157 feet and a girth of 14 feet 5 inches; it contains 1095 cubic feet of timber, and is valued at £27. In the Pyrenees the silver fir occurs between 4500 and 6500 feet elevation, and trees

[1] See Huffel, *Économie Forestière*, i. 349, 350, 353 (1904).
[2] Cf. *Trans. R. Scot. Arb. Soc.* xviii. 131 (1905).

of great age, about 800 years old, are said[1] to have existed there at the beginning of the 19th century.

In Corsica the silver fir occurs in the great forests of *Pinus Laricio*, but is not abundant, as it only grows, as a rule, in scattered groups in the gullies, where the soil is deeper and richer than elsewhere; and at Valdoniello I only saw a few trees, none of which were of large size. M. Rotges, of the Forest Service, informed me that it occurs in greatest quantity in the forest of Pietropiano, near Corte.

In Italy the silver fir is unquestionably wild on the Apennines, and considerable forests exist at Vallombrosa and Camaldoli, which are now owned by the government. That at Camaldoli is particularly fine, the total area covered by the silver fir being about 1600 acres. The trees are dense on the ground and very vigorous in growth; and this is easily explained by the heavy rainfall, which, as measured at St. Eremo, in the middle of the forest, at 3600 feet altitude, averages about 80 inches annually. I saw, when I visited Camaldoli, in December 1906, no trees of great size; but one was cut down in 1884, and a log of it shown at the National Exhibition at Turin in that year, which measured 140 feet in height and 17 feet in girth.

The silver fir also occurs in Sicily in small quantity, on the higher mountains, and specimens without cones, which I saw in the museum at Florence, are peculiar in the foliage, and form possibly a connecting link between *A. pectinata* and *A. numidica.*

In Germany, towards the northern part of its area of distribution, the silver fir is met with growing wild on the plains, as in Saxony, Silesia, and Thuringia. Towards the south it is entirely a tree of the mountains, occupying a definite zone of altitude, which, in the Bavarian forest, lies between 950 and 4000 feet. The largest forests, which are nearly pure, occur in the Black Forest and in Franconia; those in Bavaria, Bohemia, Thuringia, and Saxony being smaller in extent.

In Switzerland small forests occur at Zurich, Payerne, and on Mount Torat; the silver fir ascending in the Swiss Alps to 5300 feet altitude.        (A. H.)

As to the size[2] which the silver fir attains in its native forests, many particulars are given by French and German foresters, some of which have been quoted above. None exceed, however, what I have seen in the virgin forests of Bosnia, where I measured near Han Semec, at an elevation of about 3000 feet, a fallen tree over 180 feet long, whose decayed top must have been at least 15 to 20 feet more. Loudon states that he saw, in the museum at Strasburg, a section of a tree of the estimated age of 360 years, cut in 1816 at Barr, in the Hochwald, which was 8 feet in diameter at the base and 150 feet high.

The virgin forests of Silesia and Bohemia contain silver firs of immense size, of which very interesting particulars are given by Göppert,[3] who states that, in Prince Schwarzenberg's forest of Krummau, there existed many silver firs of from

---

[1] Willkomm, *Forstliche Flora*, 116, note (1887).

[2] Kerner, *Nat. Hist. Plants*, Eng. trans. i. 722, gives the "certified height" of *Abies pectinata* as 75 metres, or 250 feet; but this is not confirmed by other authorities.

[3] H. R. Göppert, *Skizzen zur Kenntniss der Urwälder Schlesiens und Böhmens*, 18 (1868).

120 to 200 feet high, free from branches up to 80 to 120 feet, and as much as 6 to 8 feet in diameter. He quotes Hochstetter,[1] who measured in the Greinerwald, near Unter-Waldau, at an elevation of 2563 feet, a silver fir blown down by a storm, which was 9½ feet in diameter at breast height and 200 feet long, and produced 30 klafter of firewood.

The silver fir is planted outside the area of its natural distribution in most parts of France, in Belgium, and in western and northern Germany, but not beyond lat. 51° in eastern Prussia. It is occasionally planted in Norway, and at Christiana has attained 68 feet in length by 3½ feet in girth. At Thlebjergene, near Trondhjem, where, on the side of a hill, sloping down to the sea, with an easterly exposure, a fine plantation,[2] mainly of spruce and Scots pine, was made in 1872 and subsequent years,—there are some splendid groups of silver fir, 30 to 40 feet in height, apparently exceeding in rapidity of growth the native spruce beside it. It is met with in gardens in the Baltic provinces of Russia, as in Lithuania where there is a small wood near Grodno, and in Courland and Livonia; here, however, it always remains a small tree, never bears cones, and is much injured by severe winters.

One of the most remarkable plantations in Europe is the one made by the Hanoverian Oberforster, J. G. von Langen, in the Royal Park of Jægersborg, near Copenhagen, about 1765. I visited this place in 1908, and measured some of the trees. I found that the largest now standing near the entrance at Klampenborg was 125 feet by 12 feet 10 inches. This tree is figured in a work[3] kindly sent me by Skovrider H. Mundt. There are, however, many taller trees on the south side of the main drive, two of which I found to be 140 feet by 9 feet, and 140 feet by 8 feet in girth, respectively. I measured the girth of twenty trees out of sixty-two which are growing on an area of 100 by 30 paces, and believe them to average over 130 feet high, with an average girth of 7½ feet. In Lütken's work full details are given of the measurements of these trees taken in 1893, and confirmed in 1898 by Oppermann, who found 432 trees, averaging 38·9 metres in height and containing 1400 cubic metres per hectare; which is equal to 20,000 cubic feet per acre in the round, or 15,700 feet English quarter-girth measure. My own hasty estimate on the spot was about 12,000 feet English quarter-girth measure per acre. These wonderful silver firs grow on a deep, sandy loam, on level ground near the sea, and seem to have passed their prime. Some of their timber has been used as rafters in the Secretariat hall of the new Raadhus at Copenhagen.

*A. pectinata*[4] was brought to the eastern United States early in the nineteenth century; but it is not hardy even in the middle states.

Witches' brooms and cankered swellings, due to the fungus *Æcidium elatinum*, De Bary, are common on the silver fir in the continental forests; and are often seen in Ireland and the south-west of Scotland,[5] though apparently rare in England, where they have been noticed in Norfolk[6] and at Haslemere.[7]

---

[1] Hochstetter, *Aus dem Böhmerwalde*, *Allg. Augsb. Zeit.* 1855, No. 182. Cf. Sendtner, *Die Vegetations-Verhältnisse des Bayerischen Waldes* (1860). [2] Seen by Henry in 1908.

[3] Lütken, *Den Langenske Forstordning*, p. 286, fig. 5 (Copenhagen, 1899).

[4] Sargent, *Silva N. Amer.* xii. 100, *adnot.* (1898).   [5] Somerville, in Hartig, *Diseases of Trees*, Eng. trans. 179 (1894).

[6] *Trans. Norfolk and Norwich Naturalists' Soc.* vii. p. 255.   [7] Specimens at Kew.

The swellings which affect the trunk or branches are due to the irritation of the fungus mycelium, which is perennial and stimulates the wood and bark to abnormal growth. These swellings become fissured and are entered by the spores of other fungi, which rot the wood; and the tree, if the stem is affected, is often broken off at the weakened spot by storms or falls of snow. The witches' brooms begin as young shoots, bearing small yellowish leaves, on the under surface of which two rows of *æcidia* are developed in August. These shed their spores at the end of that month and the leaves soon afterwards die and fall off. The affected shoots keep on growing, and develop into peculiar growths, set upright generally on the branches, and consisting of numerous twigs anastomosed together. The fungus passes one stage of its life on various species of *Stellaria*, *Cerastium*, and their allies, and Fischer[1] recommends the extirpation of these plants from nurseries in which the silver fir is raised.

The silver fir is very liable in its native forests to be attacked by the mistletoe. Modified roots, the so-called sinkers of the parasite, have been found in the wood enclosed in forty annual rings and as much as 4 inches long, showing that mistletoe may live on the tree for forty years. When the mistletoe dies the rootlets and sinkers survive for a time, but finally moulder and fall to pieces. The affected parts of the wood show numerous perforations, and exactly resemble the wood of a target that has been penetrated by shot or small bullets.[2]

The bark of the silver fir remains alive on the surface to an advanced age; and, on this account, when branches, stems, or roots of adjoining trees get into contact, they often become grafted together. This is the explanation of the curious phenomenon of the vitality of the stumps of certain trees in forests. After the stem is cut down, these stumps continue to increase in size and produce a callosity, which eventually covers the stump in the form of a hemispherical cap. Such a stump procures its nourishment from an adjoining tree, with which its roots have become grafted.[3]

## CULTIVATION

The silver fir[4] was introduced into England about the beginning of the seventeenth century; but the exact date is uncertain. The earliest trees recorded are two mentioned by Evelyn,[5] which were planted in 1603 by Serjeant Newdigate in Harefield Park in Middlesex. These had attained about 80 feet high in 1679, but from inquiries made by the late Dr. Masters, there is no doubt that they have long since been cut down.

Though in its own country the silver fir is a tree of the mountains, yet it attains its greatest perfection in the south and west of England, Scotland, and

---

[1] Abstract of Fischer's paper in *Journ. Roy. Hort. Soc.* xxvii. 272 (1902).

[2] See Kerner, *Nat. Hist. Plants*, Eng. trans. i. 210, fig. 48 (1898). We have never seen or heard of mistletoe on the silver fir in this country.       [3] See Mathieu, *Flore Forestière*, 529 (1897).

[4] Staves were found, in 1900, lining the ancient wells in the Roman city of Silchester, Hants; and the wood was identified by Marshall Ward with *A. pectinata*. The casks, from which the staves had been taken, were probably imported from the region of the Pyrenees, and had either contained wine or Samian ware. Cf. Clement Reid, in *Archaeologia*, lvii. 253, 256 (1901).       [5] *Sylva*, 106 (1679).

Ireland, under conditions of soil and climate very unlike those of its native forests. Though it will endure the severest winter frosts without injury, yet unless under the cover of other trees, or in very sheltered situations, it is often injured by spring frost, on account of its tendency to grow early. As regards soil it is somewhat critical, for though Boutcher[1] says that he has seen the largest and most flourishing silver firs on sour, heavy, obstinate clay, yet I have never myself seen fine trees on any but deep, moist, sandy soils, or on hillsides where the subsoil was deep and fertile. He also says it is vain to plant them in hot, dry, rocky situations, and this is my own experience on oolite formations, where I have never seen a large or well-developed silver fir. In the east and midland counties they usually become ragged at the top before attaining maturity, and in this country rarely attain a great age without suffering from drought and wind.

Though foresters of continental experience recommend this tree for underplanting, on account of its ability to grow under dense shade, yet from an economical point of view it cannot be recommended here; and I do not know of any place in England where the financial results of planting the silver fir are, or seem likely to be, such as would justify growing it on a large scale; partly because of its very slow growth when young, and partly because its timber is not valued as it is in France and Germany. Mr. Crozier's experience[2] is very noteworthy.

The silver fir seeds itself very freely in some parts of England, Scotland, and Ireland,[3] but the seedlings are so slow in growth and so delicate for the first few years, that few survive the risk of frost, rabbits, and smothering. Sir Charles Strickland tells me that in a wood of silver firs at Boynton, Yorkshire, which were mostly blown down in 1839, he remembers that a few years afterwards the growth of young seedlings was in places so dense that he could hardly force his way through them. Some of these self-sown trees are now 6 feet in girth and 60 to 70 feet high, but many are stunted from want of space. Their parents are rough and branchy, dying at the top, and 10 to 12 feet in girth.

## REMARKABLE TREES

Though the silver fir will probably be in time surpassed in height and girth by some of the conifers of the Pacific coast of America, yet at present it has no rival in size among coniferous trees in Great Britain. Perhaps the tallest which I have seen in England is the magnificent tree (Plate 208) which grows in Oates Wood, at the top of Cowdray Park, Sussex, at an elevation of 500 to 600 feet, and now owing to its being deprived of the shelter of the surrounding trees, likely to be blown down

[1] Boutcher, *A Treatise on Forest Trees*, 146 (1775).

[2] Formerly one of our most reliable trees, but now hopelessly unreliable as a timber crop, owing to its susceptibility to attack by Chermes. Like the larch, our old trees are practically immune to attack, but the difficulty in getting up young stock—experienced throughout the greater part of the country—is likely to lead to its extinction altogether as an economic species. Has been much recommended by continental trained foresters—even of late years—for the purpose of underplanting in our Scotch woods, and some of those experiments I saw lately. The result is a hopeless failure in all of them.—(J. D. CROZIER.)

[3] At Auchendrane, near Ayr, according to Mr. J. A. Campbell, there are several acres of self-sown seedlings; and in County Wexford I have also seen great numbers.

by the first severe gale.   I measured this tree in 1906 in company with Mr. Roberts, forester to the Earl of Egmont, as carefully as the nature of the ground would allow, and believe it to be still over 130 feet in height; when I first saw it in 1903 it was taller.   It is clear of branches to at least 90 feet and 10 feet 2 inches in girth.   In the background some spruce which are even taller may be seen in our illustration.

I am informed by Mr. F. H. Jervoise, of Herriard Park, Hants, that there was a silver fir there which probably exceeded this height before its top was broken off about sixteen years ago.   A photograph, taken in 1851, shows the height to have been then at least double what it now is, namely 70 feet, and another tree standing not far off measures approximately 140 feet.

In the Shrubbery at Knole Park, Kent, a very large silver fir is now about 110 feet high, with a clean bole about 80 feet by 12 feet; but its top is broken off, and it looks as if it might have been much taller.

At Longleat there are a great number of very fine silver firs near the Gardens, and also in the valley at Shearwater, the largest of which I measured in 1903, and found to be about 130 feet by 16 feet 5 inches in girth.[1]   Mr. A. C. Forbes estimated the contents of this tree at 550 feet, and in the *Trans. Eng. Arb. Soc.* v. 399, gives the measurements of a group of twenty-seven trees, 120 years old, growing on an area of ⅓ of an acre at the same place as follows:—Average height, 130 feet; average girth at 5 feet, 9 feet; average contents, 180 cubic feet.   Total, 5000 cubic feet.   I doubt whether any similar area of ground in England carries so much timber, except, perhaps, a group of chestnut and oak in Lord Clinton's park at Bicton.   Silver fir requires unusually good soil to attain these dimensions.   Plate 209 shows a part of this grove which stands at an elevation of about 500 feet on a greensand formation.

There is a row of very fine silver firs by the road on Breakneck hill in Windsor Park, one of which I measured as 130 feet by 11 feet, and no doubt many as large, or nearly so, can be found in other parts of the south and west of England; but, as a rule, when the tree attains about 100 to 110 feet its top ceases to grow and becomes ragged.

Near the great cedar at Stratton Strawless (see Plate 133) there are some tall silver firs, one of which in 1907 was 131 feet by 9 feet 7 inches; and Mr. Birkbeck informed me that another, believed to be the tallest tree in Norfolk, and measuring 135 feet, had been blown down in 1895 at the same place.

There are some very fine silver firs still standing at Eslington Park, Northumberland, which were planted about 1760, though Mr. Wightman, the gardener, informs me that the largest, which could be seen standing above all the other trees, was blown down in a gale in December 1894.   It measured 122 feet by 21 feet at five feet from the ground, and at fifty feet from the ground was still 9 feet in girth.

Almost equal to these are the trees in the Ladieswell Drive, near Alnwick Castle, Northumberland, which I saw in 1907; though not much exceeding 100 feet

[1] Loudon states that the tallest silver fir known in England in his time was believed to be at Longleat, and measured 138 feet high by 17 feet in girth; but this tree cannot now be identified.

in height, they measure from 14 feet to 16 feet in girth, the largest being estimated by Mr. A. T. Gillanders, forester to the Duke of Northumberland, to contain about 600 cubic feet each.

At Rydal Park, Cumberland, Mr. W. F. Rawnsley informs me that a silver fir was felled which contained 420 cubic feet, and doubtless there are others in the north-west of England as large.[1]

In Wales, however, I have seen none remarkable for size, though there are many places which seem as suitable as those I have mentioned.

In Scotland the silver fir attains its maximum of size in the south-west, and in a district where the climate is most unlike that of central Europe; being much warmer in winter, cooler in summer, and with a rainfall of 60 to 80 inches and even more in exceptional years.

On the Duke of Argyll's property at Roseneath are the champion silver firs of Great Britain, both as regards age and girth. Strutt figures them in *Silva Scotica* (plate 6), and states that the largest was then about 90 feet by 17 feet 5 inches. Loudon, twenty years later, gave the height as 124 feet, the age as 138 years, and the diameter of the trunk as 6 feet; but this height is almost certainly an error, as when I visited Roseneath in September 1906, a careful measurement made the largest about 110 feet by 22 feet 7 inches, and the other, which stands close by it, 105 feet by 22 feet 1 inch.[2] Plate 210, from a negative for which I have to thank Mr. Renwick, is the best I have been able to obtain of these noble trees, which grow close to sea-level in deep sandy soil. The Duke of Argyll believes them to have been planted about 1620 or 1630.

Near Inveraray Castle, on the lower slopes of Dun-y-Cuagh, Mr. D. Campbell, the Duke's forester, showed me some splendid silver firs, over 120 feet high and 15 feet in girth, and assured me that in his younger days he had helped to measure some which were much larger; one he believed to have been 24 feet in girth, containing over 800 feet of timber. On the Dalmally road, a little above the stables at Inveraray, are the tallest trees of the species that I have seen in Scotland; one measures 135 feet, or perhaps as much as 140 feet, by 16½ feet; another about 135 feet by 14 feet 3 inches; and there may be even taller ones here which I could not measure. These splendid trees were, as the Duke of Argyll informs me, probably planted by Duke Archibald in 1750, but their timber is so coarse that it is of little value, and is principally used by Glasgow shipbuilders for keel blocks.

Some of the most remarkable silver firs which I have seen in any country are at Ardkinglas, now the property of Sir Andrew Noble, near the head of Loch Fyne. They are described by J. Wilkie, and well illustrated in the *Trans. Scot. Arb. Soc.* ix. 174, and show a tendency, which I cannot explain, to throw out immense branches, which, after growing horizontally 10 to 15 feet from the main trunk, turn up and form an erect secondary stem. The largest of these (*op. cit.* plate 11), according to Wilkie's careful measurement in 1881, was 114 feet high by 18 feet in girth at

[1] Sir Richard Graham of Netherby Hall, Cumberland, showed me a very remarkable tree in a wood called Hog Knowe, which has large spreading branches, 80 paces in circumference, and measures 98 feet by 14½ feet. Mr. Watt of Carlisle has been good enough to send me a photograph of this tree, taken by his sister.

[2] See *Gard. Chron.* xxii. 8, fig. 1 (1884), and xxvii. 166, fig. 39 (1887), where good illustrations of these trees are given.

$2\frac{1}{2}$ feet. I made it in 1905 about 21 feet at the same height and 14 yards round the roots. Wilkie computed that the main stem contained 557 cubic feet and the branches 692 cubic feet, including bark, which exceeds the largest tree of the species recorded in this country. I certainly have never seen anything surpassing it in bulk, even in the virgin forests of Bosnia, though I have measured a fallen silver fir there which was at least 200 feet high. Another of these trees figured on plate 12 of the same volume, was estimated at 437 feet in the stem, and 449 feet in the ten principal limbs. At the same place is a very fine tree which Mrs. Henry Callender, who showed it to me, called "The Three Sisters," 115 feet high according to Wilkie,—I made it, twenty-four years later, 120 feet,—with a bole only 8 feet long, where it divides into three tall stems nearly equal in height and measuring just above where they separate, 8 feet 4 inches, 8 feet $5\frac{1}{2}$ inches, and 8 feet 7 inches respectively.

The Union trees,[1] in the avenue at Auchendrane, Ayrshire, planted in 1707, are six in number, the largest being, in 1902, 97 feet high and 16 feet 1 inch in girth. Another tree in the flower garden here, planted at the same time, was 110 feet by 16 feet in 1902.

In the island of Bute, James Kay describes, in *Trans. Scot. Arb. Soc.* ix. p. 75, some fine silver firs which grew in a clump north-east of the circle walk in the woods of Mountstuart, the seat of the Marquess of Bute. They were of immense height (120 feet), and could be seen for miles standing out like an island among this forest of sylvan beauty. There were nineteen silver firs, five spruce, one Scots pine, and two birches, all standing on a space of 60 yards square, where they were healthy and not overcrowded. They were very uniform in size, and ran from 10 to 12 feet in girth, ten being straight to the top and nine forked at 30 feet to 60 feet up.[2]

In other parts of Scotland the silver fir usually attains smaller dimensions, the largest that I have seen being on the banks of the Tay, near Dunkeld, and at Dupplin Castle, where I measured a tree over 100 feet high by $17\frac{1}{2}$ feet in girth. But Mr. W. J. Bean, in *Kew Bulletin*, 1906, p. 266, mentions an immense tree, which was blown down on November 17, 1893, near Drummond Castle, when 210 years old. The stump of this tree was $6\frac{1}{2}$ feet in diameter, and the cubic contents are said to have been 1010 cubic feet.

At Dawyck, near Peebles, in a cold situation at about 500 feet above the sea, Mr. F. R. S. Balfour showed me some large silver firs which far surpass the larches growing near them, which are believed to have been planted about 1730. The largest of the firs is 112 feet by $15\frac{1}{2}$ feet.

In most parts of Ireland the silver fir is a thriving tree wherever planted, and seems to be well suited to the climate. It was probably introduced early in the eighteenth century, as, according to Hayes, there were trees 100 feet high and 12 feet in girth in 1794 at Mount Usher, in Co. Wicklow. The largest silver fir in Ireland that we know of is at Tullymore Park, Co. Down, the seat of the Earl of Roden, growing in a sheltered valley below the house. Col. the Hon. R. Jocelyn, who showed me

---

[1] Cf. Renwick, in *Trans. Nat. Hist. Soc. Glasgow*, vii. 265 (1905).

[2] Mr. Kay informs me that many of the trees described by him thirty years ago have since been blown down, and I could not identify these silver firs when I visited Bute recently.

this tree in 1908, informed me that it was marked on a plan about 200 years old, and though still vigorous in appearance, it seems to be hollow for some way up. It measures from 115 to 120 feet high, with a girth of 18 feet 10 inches; and at about 20 feet from the ground throws out four large branches, which become erect, and form a tree of the candelabra type. (Plate 211.) At Carton, the seat of the Duke of Leinster, a tree was 16 feet 1 inch in girth in 1904, but the top had been blown off by the great gale of 1903. The finest silver firs in Ireland are probably those growing at Woodstock, Co. Kilkenny, where the biggest tree was in 1904 over 120 feet high by 15 feet 4 inches in girth. There are also here four trees standing so close together that they can be encircled by a tape of 30 feet; one of these is 133 feet high by 10 feet 10 inches in girth. At Avondale, Co. Wicklow, Mr. A. C. Forbes measured a tree in 1908, 125 feet in height and 15 feet 4 inches in girth. At Tykillen, Co. Wexford, the silver fir grows well and seeds itself freely, but does not attain anything like the dimensions above noted. There are fine trees at Castlemartyr, Co. Cork, one of which measures 114 feet by 14 feet 8 inches.

## Timber

Though on the Continent the wood of the silver fir is in some districts, and for purposes where strength combined with lightness is required,[1] valued more highly than that of the spruce or pine, yet in England it is little appreciated, because it seldom comes to market in any quantity, and the trees are rarely clean enough to make good boards. But I am assured by Dr. Watney that, when slowly and closely grown, it is distinctly superior in quality to that of the spruce, and that he uses it in preference on his own property for estate building; and Mr. H. E. Asprey, agent to the Earl of Portsmouth at Eggesford, Devonshire, where this tree grows very well, tells me that he finds the timber quite equal to that of spruce for all estate purposes. The Marquess of Bath informs me that a lot of 22 trees, averaging 140 feet each, were sold privately at 5½d. per foot, and used at Trowbridge for making tin-plate boxes; but most of his silver fir timber goes to the Radstock coal pits, where it is used underground.

Laslett says[2] that "the pinkish white and scarcely resinous wood works up well, with a bright silky lustre, and is of excellent quality for carpentry and ship-work. It is light and stiff, and like spruce takes glue well. Nevertheless it is as yet far less in request than the latter, though it is employed in the making of paper pulp, as well as for boards, rafters, etc."[3] So little is it known, however, to the English timber merchant that the author of *English Timber* does not even mention it, and I am not aware that it is imported to England as an article of commerce.

Strasburg turpentine, which was formerly extracted from the resinous glands found on its bark and largely used for the preparation of clear varnishes and at one time used as medicine, is now apparently superseded by other resins, though, according to Flückiger and Hanbury,[4] it was still collected to a small extent in the Vosges in 1873.

(H. J. E.)

[1] Cf. Mouillefert, *Essences Forestières*, 338 (1903).  [2] *Timber and Timber Trees*, 343 (1896).
[3] Christ, *Flore de la Suisse*, 255 (1907), says that its white wood is delicate and not so much in request as the more resinous wood of the spruce.  [4] *Pharmacographia*, 615 (1879).

## ABIES PINSAPO, Spanish Fir

*Abies Pinsapo*, Boissier, *Biblioth. Univ. Genève*, xiii. 167 (1838), and *Voyage Espagne*, ii. 584, tt. 167-169 (1845); Masters, *Gard. Chron.* xxiv. 468, f. 99 (1885), xxvi. 8, f. 1 (1886), and iii. 140, f. 22 (1888); Kent, Veitch's *Man. Coniferæ*, 534 (1900).
*Pinus Pinsapo*, Antoine, *Conif.* 65, t. 26, f. 2 (1842-1847).
*Picea Pinsapo*, Loudon, *Encycl. Trees*, 1041 (1842).

A tree attaining about 100 feet in height and 15 feet in girth. Bark smooth in young trees, becoming rugged and fissured on old trunks. Buds ovoid, obtuse at the apex, resinous. Young shoots glabrous, brownish, with slightly raised pulvini.

Leaves on lateral branchlets radially arranged, linear, flattened, but thick, rigid, short, $\frac{1}{2}$ to $\frac{3}{4}$ inch long by about $\frac{1}{10}$ inch wide, gradually narrowing in the upper third to the acute apex; upper surface convex without a median furrow and with eight to fourteen lines of stomata; lower surface with two bands of stomata, each of six or seven lines; resin-canals usually median.[1] In young plants the leaves are longer and end in sharp cartilaginous points. On cone-bearing branches the leaves are short and thick, lozenge-shaped in section, with twenty or more lines of stomata on the upper surface, and two bands of stomata of about ten lines each on the lower surface, which has a prominent keeled midrib.

Staminate flowers crimson, cylindrical, $\frac{1}{2}$ inch long, surrounded at the base by two series of broadly ovate obtuse scales.

Cones sessile or subsessile, brownish when mature, pubescent, cylindrical, tapering to an obtuse apex; 4 to 5 inches long by $1\frac{1}{4}$ to $1\frac{3}{4}$ inches in diameter. Scales: lamina three-sided, 1 inch wide by $\frac{7}{8}$ inch long, upper margin almost entire, lateral margins nearly straight, laciniate; claw short, obcuneate. Bract minute, situated at the base of the scale, ovate, orbicular or rectangular, denticulate, emarginate with a short mucro. Seed with wing $1\frac{1}{4}$ inch long; wing two to three times as long as the body of the seed. In cultivated specimens the cones and scales are usually considerably smaller than in wild trees.

Cotyledons[2] six, convex and stomatiferous on the upper surface, flattish and green on the lower surface.

### Hybrids

A series of hybrids have been obtained between *A. Pinsapo* and two other species, *A. cephalonica* and *A. Nordmanniana*, of which a full account is given by Dr. Masters in his valuable paper on hybrid conifers.[3]

1. *Abies Vilmorini*, Masters.[4] This is a tree growing at Verrières near Paris, which has the following history. In 1867, M. de Vilmorin placed some pollen of *A. cephalonica* on the female flowers of a tree of *A. Pinsapo*. A single fertile seed was produced, which was sown in the following year; germination ensued and the

---

[1] The resin-canals in this species are variable in position. Cf. Guinier and Maire, in *Bull. Soc. Bot. France*, lv. 190 (1908).
[2] Masters, *in litt.*  [3] *Journ. Roy. Hort. Soc.* xxvi. 99 *seq.* (1901).  [4] *Ibid.* 109.

seedling was planted out in 1868. M. Phillipe L. de Vilmorin[1] states that the tree was in 1905, 50 feet high by 5 feet in girth; and has three main stems, one of which, however, was broken by a storm two years ago. In its habit and foliage it resembles *A. Pinsapo* more than the other parent. The leaves, however, are longer and less rigid than in *A. Pinsapo*, and bear stomata only on their lower surface; moreover their radial arrangement on the branchlets is imperfect. The cones, which are produced in abundance and contain fertile seeds, resemble those of *A. cephalonica*, being fusiform in shape; they have longer bracts than in *A. Pinsapo*, in some years exserted, in other years shorter and concealed between the scales. Seedlings raised from this tree, now four years old, have acuminate sharp leaves like those of *A. cephalonica*.

2. *Abies insignis*, Carrière, *Rev. Hort.* 1890, p. 230. This hybrid was obtained in 1848 or 1849 in the nursery of M. Renault at Bulgnéville in the Vosges. A branch of *A. Pinsapo* was grafted on a stock of the common silver fir (*A. pectinata*); and after some years the grafted plant produced cones. Seeds from these were sown; and of the seedlings raised one-half were like *A. Pinsapo*, the remainder being intermediate in character, it was supposed, between *A. Pinsapo* and *A. pectinata*; and the variation was considered to be the result of graft hybridisation. However, at no great distance there was growing a tree of *A. Nordmanniana*; and it is more probable that the hybrid character of the seedlings was the result of a cross from *A. Pinsapo* fertilised by the pollen of *A. Nordmanniana*. A complete account of these seedlings is given by M. Bailly.[2]

3. *Abies Nordmanniana speciosa*, Hort.[2] This hybrid was raised in 1871-1872 by M. Croux in his nurseries near Sceaux, the cross being effected by placing pollen from *A. Pinsapo* on female flowers of *A. Nordmanniana*. A full account of this hybrid is given by M. Bailly.[2]

4. *Moser's hybrids.* Four different forms, all raised from *A. Pinsapo*, fertilised by the pollen of *A. Nordmanniana*, which were obtained in 1878 by M. Moser at Versailles. Full details are given in Dr. Master's paper, to which we refer our readers.

### DISTRIBUTION

*A. Pinsapo* has a restricted distribution, being confined to the Serrania de Ronda, a name given to the mountainous region around Ronda in the south of Spain. The late Lord Lilford informed Bunbury[3] in 1870 that he had seen it growing on the Sierra d'Estrella in Portugal; but we have not been able to confirm the statement.

There are three main forests of this species, none of considerable extent, occurring in localities at considerable distances apart. I visited these forests in December 1906, and explain the rare occurrence of the tree as due to the fact, that in the dry climate of the south of Spain, it can only exist on the northern slopes of mountain

---

[1] *Hortus Vilmorinianus,* 69, plate xii. (1906). See also *Gard. Chron.* 1878, p. 438; *Rev. Hort.* 1889, p. 115, and 1902, p. 162, fig. 66.

[2] *Rev. Hort.* 1890, pp. 230, 231.

[3] *Arboretum Notes,* 147 (1889).

chains running due east and west; and these are seldom met with.   In such situations the soil is never exposed to the direct rays of the noonday sun, and preserves in consequence a great deal of moisture.   The tree never grows even on north-west or north-east slopes, and is strictly limited to aspects looking due north.

The most important forest is in the Sierra de la Nieve, a few miles to the east of Ronda.   Here the tree extends for several miles in scattered groves on the north slope of the range, growing on dolomitic limestone soil, usually in gullies or under the shade of the cliffs.   It occurs mainly at elevations of 4000 to 5900 feet, though it occasionally descends to 3600 feet.   In shaded situations and where the soil is deep, there are dense groves of thriving trees, without any admixture of other species; but at the lower elevations, where there is more sun, the trees are scattered and mixed with oak and juniper.   In exposed situations, at high elevations, the trees are windswept, stunted, and more or less broken.   Seedlings are numerous in many places.   The largest trees, seen by me, were a group, on the road across the mountain from Ronda to Tolox, at a spot called *Puerta de las animas*.   One of these (Plate 212) was 106 feet in height and 13 feet 8 inches in girth; and another with a double stem, not so tall, girthed 16 feet 3 inches.   This group is overhung by a precipice, and is at 4700 feet altitude.   The stump of a tree, which had been cut down, showed 240 annual rings and was 32 inches in diameter.

The second forest, and by far the most picturesque, lies to the west of Ronda, on the northern slope of the precipitous peak, Cerro S. Cristoval or Sierra del Pinar, close to the mediæval town of Grazalema.   The fir grows here on a talus, composed of sharp angular white limestone stones; and the contrast between the dense mass of green foliage of the tree and the pure white ground from which it springs, is remarkably beautiful.   The stones and pebbles are loosely aggregated; and beneath the surface they are mixed with a mass of black mould, in which the roots of the tree freely spread.   The fir extends along the precipitous side of the mountain for about two miles, forming a band of continuous forest, which reaches nearly to the summit of the peak, attaining about 5800 feet altitude, and descending generally to 4000 feet, reaching in one gully to 3600 feet.   Seedlings are numerous. There is no undergrowth, except an occasional daphne; but climbers like ivy and clematis are common.   None of the trees are so tall as those in the Sierra de la Nieve; but many have gigantic short trunks, in one case girthing 25 feet, and are extremely old.   In this forest, trees with glaucous foliage, not seen elsewhere, are not at all uncommon.

The third wood of *A. Pinsapo* occurs on the Sierra de Bermeja, which overhangs the town of Estepona and the Mediterranean coast.   This wood, which covers only a small area, is most accessible from Gaucin, a station on the railway between Gibraltar and Ronda.   Here the soil is disintegrated serpentine rock, and the tree grows on the northern slope, between 4100 and 4900 feet, though stunted specimens occur up to 5400 feet.   The fir is pure on the precipitous upper part of the mountain; but lower down is mixed with *Pinus Pinaster*.   The largest tree, which I measured, was 90 feet high by 13 feet 5 inches in girth.

Isolated groups of a few trees, the remains of former forests, are reported to be growing on the Sierra de Alcaparain, near Carratraca, north-east of Ronda, and at Zahara and Ubrique, not far from Grazalema. Mr. Mosley of Gibraltar, who gave me valuable help and information, saw *A. Pinsapo* also growing on the Sierra Blanca de Ojen near Marbella.

## HISTORY AND CULTIVATION

This species was discovered by Edmond Boissier in 1837. He sent about half-a-dozen seeds to M. de Vilmorin in the same year, and from one of these was raised the very fine tree, which is now growing at Verrières[1] near Paris, and which is certainly the oldest cultivated specimen. This tree was in 1905, 70 feet high by 7 feet 3 inches in girth. *Abies Pinsapo* was introduced into England in 1839 by Captain Widdrington,[2] who was the first to obtain information about the existence of a new species of *Abies* in Spain, though he was anticipated in its discovery by Boissier.                                                                          (A. H.)

In cultivation this has proved to be, all over the southern, midland, and eastern counties, one of the most ornamental of its genus, and is perfectly hardy on dry soils throughout Britain, ripening seed at least as far north as Yorkshire. It is one of the few silver firs that seems to require lime to bring it to perfection, and though it will grow fairly well on sandy soils, it will not thrive without perfect drainage, or on heavy clay. It seems to have a great tendency to divide into several leaders and often forms a bushy rather than a clean trunk, unless carefully pruned. It is not often injured by spring frost, and, though not likely to have any economic value, is a tree that should be planted in all pleasure grounds on well-drained soil, and in a sunny situation.

The seedlings which I have raised grow at least as fast as those of *A. pectinata*, and are hardier when young, but require five or six years' nursery cultivation before they are fit to plant out.

The wood is soft and knotty like that of most of the silver firs when grown singly in cultivation.

## REMARKABLE TREES

Though specimens of this tree of from 50 to 60 feet high are found in many places all over England, we have not measured any which are specially remarkable. The largest recorded at the Conifer Conference in 1891 was a tree reported to be 62 feet high by 9 feet in girth, at Pampisford in Cambridgeshire; but these measurements were erroneous, as it now is only 56 feet high by 7 feet 3 inches in girth. Here there is a remarkable dwarf form[3] of this species, which is only a foot in height, with branches prostrate on the ground for 6 or 7 feet.

The largest tree we know of is growing in a sheltered position in moist soil, at Coed Coch, near Abergele in North Wales, the residence of the Hon. Mrs. Brodrick.

---

[1] *Hortus Vilmorinianus*, 69, pl. 7 (1906).                    [2] *Sketches in Spain*, ii. 239.

[3] This is var. *Hammondi*, Veitch, *Conifers*, ed. i. p. 105.

The gardener, Mr. Hunter, informs us that it is 82 feet high by 10 feet 2 inches in girth; but has never coned.    There is a fine tree at Oakly Park, Ludlow, measuring 70 feet by 5 feet 8 inches in 1908.

At Hardwicke, Bury St. Edmunds, Sir Hugh Beevor measured a tree in 1904, which was 63 feet high and 8 feet 11 inches in girth.    At Fornham Park, also in Suffolk, he found a tree, which was planted in 1866, 50 feet by 6 feet 7 inches; and says that its growth kept pace with that of an Atlantic Cedar close by. Col. Thynne has taken a photograph of a narrow, pyramidal, symmetrical tree at Longleat, which was 65 feet high by 7 feet 9 inches in 1906 (Plate 213).    At Dogmersfield Park, Hants, the seat of Sir H. Mildmay, I measured a well-shaped tree, 65 feet high by 6 feet 10 inches.

There are several good trees at Lilford Park, Oundle, growing on oolitic limestone; but Lord Lilford informed Henry that these were not raised from seed brought home by his father, and could give no confirmation of Bunbury's statement that the latter found the tree growing wild in Portugal.

At Essendon Place, Hertford, a slender tree was 68 feet high by 5 feet 1 inch in 1907.    At Merton Hall, near Thetford, Norfolk, there is a tall tree, 75 feet by 5 feet 10 inches, the stem being bare of branches for 30 feet.

At Highnam, Gloucestershire, Major Gambier Parry reports a fine specimen, growing in the pinetum, which measured 60 feet by 6 feet 8 inches in 1906.

At the Rookery, Down, Kent, the gardener, Mr. E. S. Wiles, reported in 1906 a fine specimen, 70 feet by 9 feet, which is growing on stiff yellow loam, intermingled with flint and clay, resting on chalk.

In Wales the best that I have seen is a tree at Bodorgan, Anglesey, the seat of Sir G. E. Meyrick, which in 1906 was about 70 feet high, and had some large witches' brooms growing on it.

In Scotland, we have seen none of more than average size, a tree at Scone being about the best, and, generally speaking, the climate seems too cold for this tree. Sir Archibald Buchan-Hepburn, however, reports one at Smeaton-Hepburn, East Lothian, which was 60 feet by 7 feet 9 inches in 1908.

In Ireland, there is a tree at Curraghmore, Co. Waterford, which the gardener, Mr. D. Crombie, reported in 1905 to be 65 feet high by 8 feet in girth.    At Carton, the seat of the Duke of Leinster, there is a good tree, 54 feet high by 8 feet in girth in 1903.    At Coollattin, Wicklow, another was 55 feet by 4 feet 10 inches in 1906.

Prof. Hansen states[1] that fine trees of 50 feet high or more may be seen in several Danish gardens, where it has produced cones; and that the tree exists in the south of Sweden and Norway.

In the eastern United States it[2] never really flourishes, although it is possible to keep it alive for many years in favourable situations, even as far north as eastern Massachusetts.                                                    (H. J. E.)

[1] *J. R. Hort. Soc.* xiv. 476 (1892).          [2] Sargent, *Silva N. Amer.* xii. 100, *adnot.* (1898).

## ABIES NUMIDICA, ALGERIAN FIR

*Abies numidica*, De Lannoy, ex Carrière, *Rev. Hort.* 1866, pp. 106, 203 ; Van Houtte, *Flore des Serres*, xvii. 9, t. 1717 (1867) ; Masters, *Gard. Chron.* iii. 140 (in part and excluding figures) (1888) ; Trabut, *Rev. Gén. Bot.* i. 405, ff. 17, 18 (1889) ; Kent, Veitch's *Man. Coniferæ*, 529 (1900).
*Abies Pinsapo*, Boissier, var. *baborensis*, Cosson, *Bull. Soc. Bot. France*, viii. 607 (1861).
*Abies baborensis*, Letourneux, *Cat. Arb. et Arbust. d'Algérie* (1888).
*Pinus Pinsapo*, Parlatore, DC. *Prod.* xvi. 2, p. 423 (in part) (1868).
*Picea numidica*, Gordon, *Pinet.* 220 (1875).

A tree attaining 70 feet in height and 8 feet in girth. Bark grey, smooth in young trees, becoming scaly and fissured on old trunks. Buds large, ovoid, acute at the apex, non-resinous ; scales ovate, acute, with white scarious margins, usually free at the apex. Young shoots brown, shining, glabrous, with slightly raised pulvini but without grooves.

Leaves on lateral branches pectinate below, the two lateral sets directed outwards in the horizontal plane ; those above shorter, crowded, directed upwards, and either, as on weak shoots, forming a narrow V-shaped pectinate arrangement, or, on strong shoots, with the median leaves directed backwards (not seen in any other species) and covering the upper side of the branchlet. Leaves short, $\frac{1}{2}$ to $\frac{3}{4}$ inch long, $\frac{1}{12}$ inch broad, linear, flattened, gradually tapering to the base, broadest above the middle or uniform in width in the upper three-fourths, rounded at the apex, which is sometimes entire but usually slightly bifid ; upper surface dark green, shining, with the median groove often faint and rarely continued to the apex, in many leaves with four to six broken lines of stomata in the middle line near the apex ; under surface with two white bands of stomata, each of about eight to nine lines ; resin-canals marginal. Leaves on cone-bearing branches all more or less upturned, those of the middle ranks also directed slightly backwards, short, rigid, rounded and entire at the apex.

Cones on short stout stalks, brownish, cylindrical with an obtuse apex and tapering base, about 5 inches long by $1\frac{1}{2}$ inch in diameter, with the bracts entirely concealed. Scales ; lamina fan-shaped, $1\frac{1}{4}$ inch wide, $\frac{3}{4}$ inch long, upper margin almost entire, lateral margins denticulate and either straight or with a wing on each side above ; claw short, obcuneate. Bracts, scarcely reaching half the height of the scales, with a broad oblong claw and an expanded ovate denticulate lamina, which is acuminate or cuspidate at the mucronate apex. Seed with wing about an inch long ; wing about $1\frac{1}{2}$ times as long as the body of seed. Cones of cultivated trees have smaller scales with more developed lateral wings ; and shorter bracts, scarcely reaching $\frac{1}{4}$ the height of the scale.

The seedlings of this species have been fully described by Fliche.[1]

---

[1] In *Bull. Soc. Forest. Franche-Comté et Belfort*, 1903, p. 168.

## IDENTIFICATION

The short broad leaves, which have usually four to six broken lines of stomata on their upper surface near the apex in the middle line, are a good mark of this species.  On strong shoots the backward direction of the median leaves, which densely cover the upper side of the branchlet, is also very characteristic.

## DISTRIBUTION

*Abies numidica* is very restricted in its distribution, being, so far as is known, confined to a small area towards the summit of the northern slope of Mount Babor, in the Kabylie range in Algeria.  It grows between 5000 and 6600 feet altitude in a climate where snow lies upon the ground from December to April.  In January, 1907, I visited Kerrata, at the head of the famous gorge of Chaba-el-Akra; and found that the ascent of the mountain, only a few miles distant, was impracticable. M. Bernard, Inspector of Forests at Bougie, who has charge of the forest of Mt. Babor, informed me that the northern slope contains an area of 4000 acres, and is clothed with a dense forest, composed mainly of cedar and *Quercus Mirbeckii* in the upper zone between 4700 and 6600 feet, and of *Q. Mirbeckii*, *Q. castaneæfolia*, and *Acer obtusatum*, in the lower zone below 4700 feet.  The total number of trees of *Abies numidica* scarcely exceeds 3000; and they only grow towards the summit, where they occur scattered amongst the cedars and oaks.  None of the trees are more than 70 feet high, and the largest is only 8 feet in girth.  The small size is due to their exposed position, and possibly to the destruction of larger trees by the natives in former times.  Seedlings are rare; and according to M. Bernard, this is accounted for by the poor germinating quality of the seed, as only 4 to 15 per cent of it produced plants with him.  The soil on which the tree grows is limestone, its surface being composed of stones and pebbles, underneath which there is a considerable mixture of mould.[1]

*Abies marocana*, Trabut,[2] discovered in January 1906 by M. Joly, in the mountains south of Tetuan, in Morocco, is intermediate in the characters of the foliage between *A. numidica* and *A. Pinsapo*.  M. Trabut showed me a branchlet, when I was in Algiers in 1907; but in the absence of cones, it is impossible to decide whether it deserves to rank as a new species.  Seeds of this should be readily procurable; and the attention of travellers is directed to the possibility of introducing a new silver fir.                                                    (A. H.)

## HISTORY AND CULTIVATION

The Algerian fir was discovered in 1861 by Captain de Guibert.  The first seeds were sent to France in 1862 by M. Davout, a forest officer; and another supply and six young plants were forwarded in 1864 by M. de Lannoy.

---

[1] M. Maurice de Vilmorin, in *Arbres Forestiers Etrangers*, 33 (1900) gives an account of *Abies numidica* on Mount Babor.  He noticed that many of the trees had short stout trunks, free of branches to 10 or 12 feet, occasionally more or less twisted, and often dividing into several stems.

[2] In *Bull. Soc. Bot. France*, liii. 154, t. 3 (1906).  In the plate, the name *Picea marocana*, Trabut, appears by mistake.

The tree is rare in cultivation in England. There are two or three young specimens at Kew; and Kent, writing in 1900, mentions small trees, about 20 feet high, growing at Bicton, and Streatham Hall in Devonshire.

At Pampisford, Cambridgeshire, there are two trees with fine healthy foliage, the larger of which, 37 feet high and 3 feet 2 inches in girth, bore cones in 1907. There is also a specimen at Highnam 35 feet by 3 feet 2 inches. Though we have not identified any specimens in Scotland Mr. Crozier speaks of it as a handsome and free-growing tree which bore cones in 1906 and seems quite at home at Durris.

In Ireland the finest we know of is at Fota, where a tree 39 feet by 6 feet was bearing cones in 1908. Lord Barrymore informs us that it was planted in 1878. There is a good specimen at Glasnevin, 38 feet by 3 feet 7 inches in 1906; and one at Castlewellan measured, in the same year, 25 feet by 3 feet.

At Verrières[1] near Paris, two trees, dating from the original introduction in 1862, were, in 1905, 46 feet in height by 4 feet 3 inches in girth. (H. J. E.)

## ABIES CEPHALONICA, Greek Fir

*Abies cephalonica*, Loudon, *Arb. et Frut. Brit.* iv. 2325 (1838); Masters, *Gard. Chron.* xxii. 592, f. 105 (1884); Kent, Veitch's *Man. Coniferæ*, 498 (1900); Halácsy, *Consp. Fl. Græcæ*, iii. 450 (1904).

*Abies Apollinis*, Link, *Linnæa*, xv. 528 (1841).

*Abies Reginæ Amaliæ*, Heldreich, *Gartenflora*, ix. 313 (1860).

*Abies Panachaica*, Heldreich, *Gartenflora*, x. 286 (1861).

*Picea cephalonica*, Loudon, *Gard. Mag.* xv. 238 (1839).

*Pinus cephalonica*, Endlicher, *Cat. Hort. Vindob.* i. 218 (1842).

A tree attaining about 100 feet in height. Bark greyish brown, smooth in young trees, in old trees fissuring into small oblong plates. Buds conical or ovoid, obtuse at the apex, composed of thick ovate acute keeled scales, with prominent tips, and covered with a layer of resin. Young shoots smooth, light brown, shining, glabrous.

Leaves on lateral branches radially arranged, but not so regularly as in *A. Pinsapo*, their apices pointing outwards and slightly forwards, those of the upper ranks shorter than those beneath. Leaves linear, flattened, curved, about 1 inch long, $\frac{1}{16}$ to $\frac{1}{12}$ inch broad, abruptly tapering at the base, narrowing gradually in the anterior two-thirds, and ending in a long cartilaginous point; upper surface dark green, shining, with the median furrow not continued to the apex, and usually with several broken lines of stomata; lower surface with two white bands of stomata, each of seven or eight lines; resin-canals marginal. Leaves on cone-bearing branches all upturned, curved, rigid, broad, with the apex simply acute and not prolonged into a fine cartilaginous point.

Cones, on short stout stalks, about 6 inches long by $1\frac{1}{2}$ inch in diameter, cylindrical, slightly tapering at both ends, brownish, with the bracts golden brown, exposed, and reflexed. Scales: lamina narrowly fan-shaped, almost triangular;

[1] *Hortus Vilmorinianus*, 69, pl. 1 (1906).

upper margin convex, undulate or entire ; lateral margins with two short denticulate wings ; base curving but not auricled on each side of the oblong claw.   Bracts : claw oblong, $\frac{1}{8}$ inch wide, extending $\frac{3}{4}$ the length of the scale ; lamina lozenge-shaped, $\frac{1}{4}$ inch wide, denticulate, ending in a triangular mucro, exserted and reflexed over the edge of the scale next below.   Seed-wing about twice as long as the seed ; seed with wing about 1 inch long.

Seedling ;[1] caulicle tapering upwards, reddish brown, erect, stout.   Cotyledons, five or six, acute not mucronate, about $1\frac{1}{4}$ inch long ; upper surface dotted irregularly with stomata and grooved in the middle line.   Primary leaves half the length of the cotyledons, not mucronate ; lower surface with stomata.

### Var. *Apollinis*, Beissner, *Nadelholzkunde*, 440 (1891).

*Abies Apollinis*, Link, *Linnaea*, xv. 528 (1841).

This variety differs from the type in the arrangement and shape of the leaves. On lateral branchlets the radial arrangement is imperfect, most of the leaves standing crowded on the upper side of the branchlet, with their apices directed upwards, those in the middle line straight and vertical, those on the sides curved and bending upwards ; on the lower side of the branchlet a few leaves are directed downwards and forwards.   Leaves thicker and broader than in the type, about $1\frac{1}{4}$ inch long by $\frac{1}{10}$ to $\frac{1}{12}$ inch broad, ending in a short acute point, bevelled off from behind ; upper surface with a continuous median groove and two to three short lines of stomata near the tip ; lower surface with two bands of stomata, each of ten lines.

The cones do not differ in any essential characters from those of the type ; and the differences noted by Murray[2] in the broader bract and expanded wing of the seed are trifling and inconstant.

Halácsy considers *Abies Reginæ Amaliæ* and *Abies panachaiaca* to be mere synonyms of *Abies cephalonica* ; and only allows the variety *Apollinis*, distinguished, according to him, by its acute leaves, those in the type ending in an acuminate or very sharp spine-like point.   According to other authorities, *A. Reginæ Amaliæ* is more akin to var. *Apollinis* than to the type.   In all probability there is a series of intermediate forms connecting the type and var. *Apollinis*.[3]

### DISTRIBUTION

According to Halácsy this species occurs in the sub-alpine region of almost all the higher mountains of Greece, between 2700 and 5700 feet elevation.   The type is met with in the island of Cephalonia on Mount Enos ; and on the mainland— in Doris on Mount Kiona, in Attica on Mount Parnes, and in Arcadia on

---

[1] Masters, in MS., who states that in var. *Apollinis* the cotyledons are seven in number, sub-acute at the apex, and about 1 inch long ; primary leaves shorter and more pointed than the cotyledons.

[2] *Proc. Roy. Hort. Soc.* iii. 141 (1863).

[3] Guinier and Maire, in *Bull. Soc. Bot. France*, lv. 187, figs. 2 and 3 (1908), describe a variety, with leaves like those of *A. cilicica*, which grows on Mount Pindus in Thessaly.

Mounts Mænalus, Madara, Thaumasion, and Rhudia. Var. *Apollinis* occurs, in Epirus on Mounts Tsumerka, Strungula, Peristeri; in Thessaly, on Pindus and Olympus; in Eubœa on Mount Dirphys. It has also been found in Hellas on Mounts Œta, Tymphrestus, Parnassus, Helicon, Cithæron, Pateras, and Parnes; and in Peloponnesus on Mounts Chelmos, Olenos, Malevo, Taygetos.

As Halácsy considers *Abies Reginæ Amaliæ* to be the same as the type, and not the var. *Apollinis*, his account of the distribution differs from that generally adopted, in which the view taken is that the type is confined to the island of Cephalonia, and that all the continental forms are referable to the var. *Apollinis*.[1]

In Cephalonia the forest of this species occurs on Mount Enos, along a ridge 4000 to 5000 feet above sea-level and about 12 to 15 miles in length. It was 36 miles in circumference in 1793; but its area was considerably reduced by disastrous fires in 1798. No recent account of this forest, of which full details were given by General Napier in 1833, has come under our notice.

The form which occurs in the mountains of Arcadia, distinguished as var. *Reginæ Amaliæ*[2] by some authors, is remarkable for its capacity of producing coppice shoots, when the trunk is felled; and the main stem, even when untouched, is said often to produce secondary stems and branches from the old wood.        (A. H.)

## CULTIVATION

Seeds[3] were first sent from Cephalonia to England by General Sir Charles Napier in 1824; and the first plants, few in number, were raised by Mr. C. Hoare of Luscombe Castle, who distributed them to various places.[4] Some time afterwards Mr. Charlwood[5] sold seeds to the public, having received a cask of cones from General Napier.

The form *Reginæ Amaliæ* was first noticed in 1856 by Schmidt of Athens, who found a forest of this tree near Tripolitza in Central Arcadia; its seeds have recently been introduced abundantly.

*A. cephalonica* seems to be quite hardy over the greater part of Great Britain, but it is rather more susceptible to spring frosts than *A. Pinsapo*, because it starts earlier into growth, and on this account should not be planted in low, damp, or exposed places. It seems to grow on limestone, but not to be so distinctly a lime-loving tree as *A. Pinsapo*. It ripens seeds in good years in the south of England, but the seedlings which I have raised do not grow so fast as those of *A. Pinsapo*.[6]

[1] With regard to the occurrence of this variety in Roumelia, Macedonia, and Thrace, see our remarks on p. 722 concerning the distribution of *A. pectinata* in the Balkan peninsula.

[2] See Regel, *Gartenflora*, ix. 299, fig. (1860); and Seemann, *Gard. Chron.* 1861, p. 755 fig.

[3] Loudon, *loc. cit.*

[4] A list of these places is given in Loudon, *Gard. Mag.* 1838, p. 31, and in *Pinetum Britannicum*, ii. p. 179.

[5] Loudon, *Gard. Mag.* 1839, p. 238.

[6] Owing to its susceptibility to late frosts and to attack by Chermes, it is now nearly impossible to grow this tree up to a planting size. Its timber, when closely grown and of some age, is, in my opinion, the best of the European silver firs, being hard, close in texture, and heavier in a dry state than any I have yet handled. Var. *Apollinis* is less subject to injury by frosts and attack by Chermes than the above, and seems well adapted for planting in the north of Scotland. In cultivation it maintains a more conical outline, and is easily distinguished from the type.—(J. D. CROZIER.)

## REMARKABLE TREES

Probably the finest tree in the British Isles is the one growing at Barton, Bury St. Edmunds, which in 1908 was 95 feet in height by 13 feet 3 inches in girth. This tree (Plate 214) is very symmetrical, branched to the ground, and in full vigour, though probably it has nearly attained its limit of height, as the top of the crown of foliage is flattened. This is one of the original plants raised at Luscombe, and was planted at Barton in 1838, being then about thirteen or fourteen years old. According to Bunbury,[1] it did not suffer in the slightest degree from the severe winter of 1860. In 1857, it was 35 feet in height; and in 1858 began to bear cones, which are confined to the topmost of the lateral branchlets. In 1867, the height had increased to 58 feet, and the trunk at three feet from the ground was 7 feet in girth. Seedlings have been frequently raised from its seed. One of these seedlings, which was sent many years ago to Lord Rayleigh, is now growing at Terling Place, Essex, and measures 53 feet high and 3 feet 3 inches in girth.

Another of the original trees is now growing at Luscombe Castle, near Dawlish, in a rather exposed place, about 200 feet above sea level; when I saw it, in April 1908, it was a healthy and well-shaped tree, 75 feet by 11 feet.

There is a very fine healthy tree at Blount's Court, Oxfordshire, which Henry measured in 1907, as 87 feet in height by 10 feet 8 inches in girth. Another planted at the Coppice, Henley, in 1860, measured in 1905, 62 feet high by 8½ feet in girth. At Pampisford, Cambridgeshire, there are two trees, the larger of which was, in 1908, 55 feet by 6 feet 1 inch. The Cephalonian fir has been largely planted on Lord Walsingham's estate at Merton, Norfolk, the largest specimen, 52 feet by 9 feet 7 inches, dating from 1852. On the Thetford road there is an avenue of these trees, growing in loose, shallow sand, which have attained at forty-eight years old an average girth of 8½ feet. The growth of the tap-root is stopped by the compact chalk sub-soil, wide-spreading horizontal roots being formed, which have no great hold in the shifting sand; and several trees have been uprooted by storms.

At Heron Court, near Christchurch, I measured in 1906 a very large tree with ragged top, 82 feet by 10 feet 8 inches. At Beauport, Sussex, there is a good tree, which in 1904 was about 80 feet high by 10 feet 3 inches in girth.

At Powderham there is a very large and spreading, but ill-shaped tree, which appears as though in the mild, soft climate of south Devon it would not be long lived. In 1892 it was recorded as the largest in Great Britain, being then 77 feet by 11 feet at 3 feet from the ground.

At Killerton there is a large tree which measured in 1903 80 feet by 11 feet 9 inches. It forks at about 25 feet. At Highclere another, in the same year, measured 75 feet by 11 feet. At High Canons, Herts, Mr. H. Clinton Baker measured a tree in 1908, which was 58 feet by 8 feet 3 inches. At Bayfordbury, a tree planted in 1847 was 70 feet by 6 feet 11 inches in 1905.

At Castle Kennedy there is a very wide-spreading tree, which in 1904 measured

[1] *Arboretum Notes*, 144.

59 feet by 9 feet 8 inches. Around it were several natural seedlings, from 1 foot to 5 feet in height. At Smeaton-Hepburn another measured, in 1905, 53 feet high by 10 feet in girth. A number of Cephalonian firs were planted at Blairadam, the seat of Sir Charles Adam, Bart., in Kinross-shire, by his ancestor Sir Frederick Adam, who was governor of the Ionian Islands in 1824, and who was censured by General Napier for not sufficiently protecting the forests in Cephalonia. Several of these trees still survive at Blairadam, the largest in the garden near the entrance gate being 49 feet high, and 8 feet 2 inches in girth at 4 feet. It divides into several stems at about 25 feet. Another measures 42 by 5½ feet, and there are several smaller ones, but the tops in most cases have been at various times injured by wind and frost. In other parts of Scotland the tree grows fairly well, but not so fast as in the south, the best I have heard of being at Abercairney, where Mr. Bean[1] records one 75 feet high in 1906. As this, however, was in 1892 only reported as 50 feet high there may be a mistake. Other good trees are growing at Whittingehame, East Lothian, at Haddo House, Aberdeenshire, and at Ochtertyre,[1] Perthshire.

In Ireland, the largest Cephalonian Fir known to us, is growing at Adare Manor, Co. Limerick, the seat of the Earl of Dunraven; and, in 1903, was 86 feet high by 9 feet 4 inches in girth.

At Powerscourt, Co. Wicklow, a tree measured, in 1903, 55 feet by 8 feet 9 inches; and at Hamwood, Co. Meath, there is a fair specimen which in 1904 was 50 feet by 9 feet 6 inches.

At Cahir Park, Co. Tipperary, there are four trees of nearly equal size, one measuring 46 feet by 6 feet 2 inches. Specimens sent in 1906 by Mr. Austin Mackenzie show that these trees belong to var. *Apollinis*.

In the Botanic Garden at Bergielund, near Stockholm, a tree, planted in 1890, was, when seen by Henry in August 1908, 30 feet in height and 1 foot in diameter, and exceeds in rapidity of growth all the other conifers in the garden. In the Botanic Garden, at Christiania, there is a tree, about 25 feet in height, which is, however, not quite hardy, being slightly browned by frost. Hansen[2] says that this species had attained in 1891 a height of 44 feet and a girth of 6 feet, at 40 years old, in the gardens at Carlsberg, near Copenhagen.

*A. cephalonica* has proved hardy[3] in eastern Massachusetts, where it has already borne cones.

Though General Napier stated that the wood of this tree in Cephalonia is very hard and durable, yet as grown in this country it is not likely to have any economic value, as it is too knotty and coarse for any but the commonest purposes.

(H. J. E.)

---

[1] *Kew Bulletin*, 1906, pp. 266, 267.    [2] *Journ. Roy. Hort. Soc.* xiv. 463 (1892).

[3] Sargent, *Silva N. Amer.* xii. 99, *adnot.* (1898). Sargent, however, states in his account of the *Pinetum at Wellesley in* 1905, p. 12, that the tree here, which is 51 feet by 6 feet, was considerably injured in the severe winter of 1903-4.

## ABIES CILICICA, Cilician Fir

*Abies cilicica*, Carrière, *Conif.* 229 (1855), and *Flore des Serres*, xi. 67, t. 1108 (1856); Tchihatcheff,
   *Asie Mineure*, ii. 494 (1860); Heuzé in *Rev. Hort.* 1856, p. 81, f. 14; Kent, Veitch's *Man.
   Coniferæ*, 500 (1900); Hickel, in *Bull. Soc. Dend. France*, 1908, p. 183.
*Abies selinusia*, Carrière, *Flore des Serres*, xi. 69 (1856).
*Pinus cilicica*, Kotschy, *Oestr. Bot. Wochenbl.* iii. 409 (1853).
*Picea cilicica*, Gordon, *Pin. Suppl.* 50 (1862).

A tree attaining in Asia Minor 100 feet in height and 7 feet in girth. Bark
ashy-grey in colour, smooth in young trees, deeply fissured and scaly in old trees.
Buds[1] small, non-resinous, ovoid, acute at the apex; scales few, keeled, with their
tips more or less free and not appressed. Young shoots smooth, greyish-brown, with
scattered short erect pubescence; bark fissuring slightly on the second year's
shoot.

Leaves on lateral branches usually pectinately arranged, the upper ranks pointing
outwards and upwards, thus forming a V-shaped depression above between the two
lateral sets; on vigorous shoots, the median leaves on the upper side are directed
forwards and upwards, and cover the branchlet, the V-shaped depression being
obliterated. Leaves thin and slender, 1 to $1\frac{1}{4}$ inch long, $\frac{1}{16}$ inch wide, linear,
flattened, uniform in width except at the tapering base, apex rounded or acute and
slightly bifid; upper surface light green with a continuous median groove and
usually without stomata, rarely with two to three short lines in the groove near the
apex; under surface with two narrow greyish bands of stomata, each of six to seven
lines; resin-canals marginal. Leaves on cone-bearing branches, upturned, curved,
more rigid and broader than those on barren branches, minutely bifid at the truncate
or obtuse apex.

Cones of wild trees subsessile or on short stout stalks, cylindrical,
tapering to an acute apex, 6 to 9 inches long by 2 to $2\frac{1}{2}$ inches in diameter,
brownish when ripe. Scales[2] larger than in any other species; lamina $1\frac{3}{4}$ inch
wide, $\frac{7}{8}$ inch long, fan-shaped, upper margin thin and entire, lateral margins convex,
denticulate, with a sinus on each side; claw short, obcuneate. Bract with an
oblong claw, expanding above into an ovate or quadrangular denticulate lamina,
tipped with a short mucro, extending to $\frac{1}{3}$ or $\frac{1}{2}$ the height of the scale. Seed-wing
about $1\frac{1}{2}$ times as long as the seed; seed with wing about $1\frac{1}{4}$ inch long. In
cultivated specimens, scales smaller, $1\frac{1}{2}$ inch wide by $\frac{3}{4}$ inch long; bracts with a
very short claw and a lamina not reaching more than $\frac{1}{4}$ the height of the scale; seed
with wing about 1 inch long.

### DISTRIBUTION

This species is confined to Asia Minor and northern Syria, occurring on the
Lebanon and the Antitaurus, and forming, in company with the cedar, great forests

---

[1] The buds are characteristic; and, as Hickel points out, distinguish this species from all the others.
[2] The peculiar hook-like processes of the scales which occur in some specimens are probably abnormal.

in the Cilician Taurus. It was first discovered by Kotschy[1] in the Cilician Taurus in August 1853, in the valley of Agatsch Kisse, at an elevation of 4000 to 5000 feet. It is known to the Turks as *Ak Illeden*, white fir, and grows in thick forests sometimes unmixed with other trees, sometimes in company with oaks, cedars, and junipers. Yew and *Pinus Laricio* also occur in these forests, which are protected from woodcutters by their inaccessibility. The climate of these mountains is extremely hot in summer, and cold in winter, with much snow in the upper region.

Post[2] says that it is found in alpine and subalpine Lebanon, and in the Amanus Mountains in the extreme north of Syria, but does not give any details of its size or the elevation at which it grows.

## CULTIVATION

The first seeds, received by the museum at Paris in 1854 from M. Blanch, French Consul at Saida, failed to germinate. Balansa sent a good supply of seed in 1855. From these or from Kotschy's seed the few trees which we have found in England were probably raised.

The Cilician fir is extremely rare in cultivation in this country. The best specimen we have found is a tree, growing at Welford Park, Newbury, which in 1908 was 51 feet by 4 feet 4 inches. Mr. Ross, the gardener, informs us that he found this tree as a small plant in a pot, when he came to Welford Park in May, 1860. The tree has been considered by many people to be *Abies homolepis*, and was figured in the *Garden*, for 1904, under that name. It is unquestionably, however, *Abies cilicica*, of which it has the foliage, and only differs slightly from wild specimens in the smaller size of the cones and scales. Mr. A. B. Jackson has identified two at Bicton, 48 feet by 4 feet and 47 feet by 3½ feet respectively; and another at the Heath, Leighton Buzzard, which is 48 feet by 3 feet 10 inches.

The finest in Scotland is a tree at Durris, which Mr. Crozier reports to be 55 feet high and 5 feet 8 inches in girth. It was incorrectly labelled *A. amabilis*. Another good specimen is growing at Castle Kennedy, which measured in 1904, 48 feet in height by 5 feet 1 inch in girth. A second tree here, not so tall, is very thriving. Kent mentions a tree at Rossdhu in Dumbartonshire.

A tree at Glasnevin was 34 feet by 3 feet 2 inches in 1907; and a specimen, at Powerscourt, 37 feet by 2 feet 8 inches in 1906, did not seem to be very thriving.

There is a good specimen at Verrières, near Paris, of which a figure is given by M. Philippe L. de Vilmorin in *Hortus Vilmorinianus* (plate 1). This tree is about 60 feet high. Another and slightly taller tree is growing in the Parc de Cheverny, in the department of Loir et Cher. Pardé says that at Harcourt (Eure) it reproduces itself naturally.

According to Sargent,[3] *Abies cilicica*, with the exception of *Abies concolor*, is the most beautiful of those silver firs, which are perfectly hardy and satisfactory in the north-eastern states of the U.S. Some trees are 40 feet in height, notably at

---

[1] *Reise in den Cilicischen Taurus* (1858).    [2] *Flora of Syria*, p. 751.    [3] *Silva N. Amer.* xii. 99 *adnot.* (1898).

Mr. Hunnewell's pinetum, Wellesley, Mass.;[1] Mr. Hall's garden, near Bristol, Rhode Island; and Mr. Hoope's pinetum, West Chester, Pennsylvania. Sargent states that the tree does not thrive in western Europe, as the young shoots, which appear early in the spring, are killed by late frosts; and in consequence it is not propagated by nurserymen. Seeds from wild trees are difficult to procure.    (A. H.)

## ABIES NORDMANNIANA, Caucasian Fir

*Abies Nordmanniana*, Spach, *Hist. Vég.* xi. 418 (1842); Regel, in *Gartenflora*, xx. 259, t. 699 (1871); J. D. Hooker, *Bot. Mag.* t. 6992 (1888); Masters, *Gard. Chron.* xxv. 142, f. 30 (1886); Kent, Veitch's *Man. Coniferæ*, 526 (1900).

*Pinus Nordmanniana*, Steven, *Bull. Soc. Nat. Moscow*, xi. 45, t. 2 (1838); Loudon, *Gard. Mag.* xv. 225 (1839).

*Picea Nordmanniana*, Loudon, *Encycl. Trees*, 1042 (1842).

A tree attaining in the Caucasus over 200 feet in height and 15 feet in girth. Bark in cultivated trees greyish brown, smooth when young, becoming slightly fissured in older trunks. Buds ovoid, acute at the apex, brown, non-resinous, with ovate, acute, slightly keeled scales. Young shoots grey, smooth, with very scattered short erect pubescence.

Leaves on lateral branchlets, pectinately arranged below, the two lateral sets spreading more or less in the horizontal plane; those above shorter, directed forwards and densely covering the branchlet in imbricated ranks. Leaves linear, flattened, about 1 to $1\frac{1}{4}$ inch long, $\frac{1}{10}$ to $\frac{1}{12}$ inch wide, uniform in width except at the gradually tapering base; apex rounded and bifid; upper surface dark green, shining, with a continuous median groove and without stomata; lower surface with two conspicuously white bands of stomata, each of eight or nine lines; resin-canals marginal. Leaves on cone-bearing branches all curved and upturned.

Staminate flowers ovoid-cylindric, $\frac{4}{10}$ inch long, each with three series of involucral bracts.

Cones sub-sessile, cylindrical, tapering at both ends, about 6 inches long by 2 inches in diameter, brown in colour, with the bracts exserted and reflexed. Scales: lamina, about $1\frac{1}{4}$ inch wide by $\frac{3}{4}$ inch long, either with a denticulate wing on each side or with straight lateral margins; claw obcuneate. Bract with oblong claw, expanding above into an almost orbicular lamina, which is denticulate and tipped with a long triangular mucro. Seed with wing about an inch long, the wing being twice the length of the body of the seed.

### Varieties and Hybrids

Several varieties are mentioned by Beissner, which are said to differ from the type in foliage, the leaves being shorter, glaucous, or yellow in colour. None of these appear to be in cultivation in England.

---

[1] Elwes saw this tree in May 1905, and remarked that it was very similar in growth to *A. Nordmanniana*, which has shorter darker leaves and denser habit. It had not suffered from the severe frost of the preceding winter which in some places had injured the Caucasian fir. According to Sargent, *The Pinetum at Wellesley in* 1905, p. 12, this tree is 49 feet high and 5 feet in girth.

Var. *equi-Trojani*, Guinier and Maire.[1]  A peculiar form, discovered by Sintenis on Mount Ida in north-west Anatolia.  It has reddish-brown glabrous shoots, leaves acute at the apex and only slightly emarginate, and cones with bracts much exserted and almost concealing the scales.

The hybrids, which have been obtained between *A. Nordmanniana* and *A. Pinsapo* are dealt with in our article on the latter species.

## DISTRIBUTION

This species is a native of the mountains in the southern and south-eastern shores of the Black Sea, including the western spurs of the Caucasus.  According to Radde,[2] it is entirely absent from the eastern parts of the Caucasus and Talysch, its easterly limit being longitude 42°.  It usually grows between 3000 and 6600 feet elevation, and either forms pure forests or is associated with *Picea orientalis*, being occasionally mixed both with that species and *Pinus sylvestris*.  It is said to prefer calcareous soil and to be dominant on the limestone formations, which are not so favourable to the growth of the oriental spruce and the common pine.  In pure forests, the trees stand very close together ; and in their deep shade underwood is absent and no light reaches the ground, which is very dry and covered with a thick layer of brown needles.  Such forests are the last hiding-place of the European bison in a truly wild condition.

The largest tree mentioned by Radde, the age of which is not given, grew in the valley of the Labba in the district of Kuban, and measured 213 feet in height and 15 feet in girth at breast height, and the stem alone had a volume of 1236 cubic feet. On an area of about $2\frac{1}{2}$ acres in this forest fifteen trees nearly as large were growing.  It thrives best and attains its largest size at high elevations, 5000 to 6000 feet ; where stems 150 to 170 feet in height, with a girth of 10 feet, are quite common.  The oldest tree, which is recorded by Radde, was 370 years old, and measured 170 feet high by 10 feet in girth.

*Abies Nordmanniana* was also found by Balansa[3] in Lazistan, and by Sintenis[3] at Kostambul in Paphlagonia.  Guinier and Maire[4] in 1904 found it growing on Mount Olympus in Bithynia, where, on the northern slope between 3700 and 6000 feet, it forms extensive forests, either pure or mixed with *Pinus Laricio*, beech, oak, and chestnut, and constituting the timber line at 6300 feet.  These botanists state that on Olympus, as well as in the Caucasus, it is a light-demanding tree, a least in the young stage, as the seeds everywhere germinate in open and unshaded places.  The discovery by these authors of *A. Nordmanniana* on Mount Olympus and of the var. *equi-Trojani* on Mount Ida extends the distribution of this species westwards through northern Asia Minor to the borders of the Ægean Sea.

[1] In *Bull. Soc. Bot. France*, lv. 186, fig. 1 (1908).  This variety was referred to *A. pectinata* by Boissier, in *Fl. Orientalis*, v. 701 (1881).

[2] *Pflanzenverb. Kaukasusländ*, 184, 222, 244, etc. (1899).  [3] Specimens in Kew herbarium.

[4] In *Bull. Soc. Bot. France*, lv. 185, fig. 1 (1908).  The silver fir on Mount Olympus was erroneously identified with *A. pectinata* by Boissier in *Flora Orientalis*, v. 701 (1881).

*Abies Nordmanniana* was first recognised as a distinct species by the Finnish botanist, Nordmann, Professor at Odessa, whose name it bears. He found it in 1836 in the Caucasian province of Imeretia. Pallas and other early botanists had referred the Caucasian silver fir to *Abies pectinata*. It was introduced[1] into Europe in 1848, when Alexander von Humboldt obtained seeds from the Caucasus, which were sown in the Berlin Botanic Garden. (A. H.)

## CULTIVATION

No other silver fir found in the Old World is more thoroughly at home in Great Britain, for it grows luxuriantly on soils where the common silver fir will not thrive; is absolutely uninjured by spring frost, even in a young state, and ripens seed as far north as Perthshire and County Down. It seems equally at home on rich loam in the south-east of England, on oolite gravel in the Cotswold Hills, and in the peaty soil and wet climate of Argyllshire. Out of 102 returns sent to the Conifer Conference from all parts of Great Britain, 78 mention this tree and nearly all speak well of it, though it is said to fail at about thirty years old on strong loam in Worcestershire, and to be liable to aphis at Durris in Kincardineshire.[2]

Sir Herbert Maxwell[3] states that the Crimean silver fir (a misleading name, as it does not occur wild in the Crimea), after it attains twenty to thirty years of age, frequently succumbs to the attacks of aphis, and gives as an instance in proof of this, that at Benmore, where large numbers were planted thirty to forty years ago, very few now remain. But I do not think that this is a fair example, as the climate of Benmore is very wet, and the soil in many places very shallow. In the warmer and drier parts of Scotland I have seen many flourishing specimens, though not so fine as in England.

Wilkie[4] says that at Tyninghame, in East Lothian, it is later in starting growth than the common silver fir, grows more freely when young, and either for use or ornament is certainly the more valuable of the two. Webster, also, whose experience was gained in Ireland, North Wales, and Kent, says,[5] "If *A. nobilis* be the best of the Californian silver firs, this is without doubt the finest of the European or Asiatic species." He expected that at no distant date it would supplant the silver fir for forest planting, the timber being of excellent quality, the tree more ornamental, and less exacting as regards soil. He says that it succeeds well on reclaimed peat bog, stiff loam, decomposed vegetable matter, and light gravelly soils.

---

[1] Hansen, in *Journ. Roy. Hort. Soc.* xiv. 471 (1892). In the catalogue of the Pinetum at Beernem, in western Flanders, Baron Serret says that he received his specimen in April 1847, from Lawson and Son, Edinburgh; and the earliest introduction would seem from this to have been prior to that stated by Hansen.

[2] *A. Nordmanniana*, the most susceptible of all silvers to attack by Chermes either in a seedling or older state. For general purposes this tree is doomed, and it is only by repeated spraying with insecticide that it will be possible to preserve even the largest specimens. In growth, it has proved itself much slower than *A. pectinata*, and being densely branched and of a shade-bearing nature, its timber when cut up has generally been coarse and knotty. In Scotland it has never been regarded by foresters as of economic importance.—(J. D. CROZIER.)

[3] Green's *Encyclopedia of Agriculture*, ii. 112 (1908). The erroneous statement that this fir occurs wild in the Crimea appears to have been first made in Veitch's *Man. Coniferæ*, 1st ed. 102 (1881), and has been repeated by Masters, Hansen, the *Kew Handlist of Conifers*, etc. No species of *Abies* grows wild in the Crimea. Cf. Démidoff, *Voyage Russie Meridionale et la Crimée*, ii. 231, 232, 375, 646 (1842).

[4] *Trans. Royal Scot. Arb. Soc.* xii. 211 (1889).        [5] *Ibid.* 257.

As the seed can now be procured in quantity and at a cheap rate, even when home-grown seed is not available, there seems to be no reason why this beautiful tree should not be raised at the same rate as the common silver fir and planted in preference to the latter, for though it has not yet had time to attain its full size in this country it grows quite as fast, and from what little we know of its timber is likely to be at least as valuable. Its average rate of growth is from 1 to 2 feet annually when once established; and though we have as yet no evidence that it will endure dense shade as well as the silver fir, yet the accounts of its growth in the Caucasus lead one to expect that it will do so.

## REMARKABLE TREES

Among the numerous specimens that we have measured in various places in England, I have seen none to surpass a very healthy and vigorous tree which grows in a wood facing east on the banks of the river at Eggesford, the property of the Earl of Portsmouth in Devonshire, which in April 1904 measured 84 feet by 5 feet 7 inches, and had produced cones. But a tree growing in a wood called Hook's Grove at Bayfordbury is perhaps taller; it was about 85 feet by 6 feet 10 inches in 1907.

At Strathfieldsaye, in the same year, I measured one as 78 feet by 6 feet 7 inches, and at Hemsted, in Kent, there is a tall but very slender specimen, not over forty years planted, which bids fair to become a very large tree. In 1905 it was 68 feet by only 3 feet 7 inches. At Lynhales, Herefordshire, the seat of S. Robinson, Esq., another is 70 feet by $5\frac{1}{2}$ feet and growing freely.

In Wales it is thriving at Penrhyn; where there are two trees, one with its top broken being about 75 feet by 10 feet; the other even taller measures 6 feet 10 inches in girth; and at Hafodunos, where it does well in plantations, Henry measured one 60 feet by 6 feet 7 inches in 1904.

In Scotland the largest recorded in 1891 was at Poltalloch, and then was said to measure 70 feet by 6 feet, but when measured by Mr. Melville in 1906 he made it only $73\frac{1}{2}$ feet by 7 feet 4 inches.

The finest I have seen myself is one at Moncreiffe which, in 1907, I made to be no less than 79 feet by $6\frac{1}{2}$ feet; a healthy tree from which many seedlings have been raised. This is stated by Hunter to have been planted about 1856, and in 1888 was only 30 feet by 2 feet 2 inches. It is said to have been hybridised by the silver fir, but I could not see anything in the seedlings to distinguish them.

In Ireland it also grows very well. A tree at Carton, the seat of the Duke of Leinster, was 74 feet by 5 feet 4 inches, and one at Fota 68 feet by about 6 feet in 1903. Another at Mount Shannon, Limerick, measured, in 1905, 75 feet by 8 feet 9 inches. A good specimen at Ballykilcavan, Queen's County, measured 68 feet by 5 feet 2 inches in 1907. There are many fine healthy specimens at Dereen in Co. Kerry.

In the University Botanic Garden at Upsala, in Sweden, a tree was seen by Henry in 1908 which was about 40 feet high and branched into three stems near the ground, the result evidently of injury to the leader by severe frost in early youth.

According to Hansen,[1] it is said to thrive at Trondhjem in Norway, but Henry saw no specimens at Trondhjem or Christiania. It is often planted in Danish gardens and forests, and is quite hardy in Denmark.

According to Sargent,[2] it is very hardy in the eastern United States, as far north, at least, as eastern Massachusetts, but although handsome when young, is apt to become thin and shabby here at an early stage. (H. J. E.)

## ABIES WEBBIANA, HIMALAYAN FIR

*Abies Webbiana*, Lindley, *Penny Cyclop.* i. 30 (1833); Griffith, *Icon. As. Pl.* t. 371 (1854); Masters, *Gard. Chron.* xxii. 467, f. 86 (1884), and x. 395, f. 47 (1891); Hooker, *Gard. Chron.* xxv. 788, ff. 174, 175 (1886), and *Flora Brit. India*, v. 654 (1888); Kent, Veitch's *Man. Coniferæ*, 543 (1900); Gamble, *Indian Timbers*, 718 (1902).

*Abies spectabilis*, Spach, *Hist. Vég.* xi. 422 (1842).

*Abies Mariesii*, Masters, *Bot. Mag.* t. 8098 (1906) (not Masters, *Gard. Chron.* xii. 788 (1879)).

*Pinus Webbiana*, Wallich, *ex* Lambert, *Genus Pinus*, 77, t. 44 (1828).

*Pinus spectabilis*, D. Don, *Prod. Fl. Nepal.* 55 (1825).

*Picea Webbiana*, Loudon, *Arb. et Frut. Brit.* iv. 2344 (1838).

A tree, attaining in the Himalayas 150 feet or more in height and 35 feet[3] in girth, with thick spreading horizontal branches; bearing a flattened crown of foliage. Bark speedily scaling on young stems; on old trunks, greyish brown, rough, irregularly fissured and very scaly. Buds large, globose, brownish, covered with resin, which conceals the keeled obtuse scales. Young shoots reddish brown with prominent pulvini, separated by deep grooves; pubescence short, erect, reddish, confined to the grooves and not spreading over the pulvini. In the second year's shoot the pulvini and grooves are more marked, the pubescence being retained.

Leaves on lateral branchlets pectinately arranged, in two lateral sets, each of several apparent ranks; the lower ranks on each side extending outwards in the horizontal plane; the upper ranks, with leaves becoming gradually shorter, directed outwards and upwards, and forming a V-shaped depression, in the bottom of which the upper side of the branchlet is plainly visible. Leaves 1 to $2\frac{1}{2}$ inches long, $\frac{1}{10}$ inch wide or more, linear, flattened, uniform in width except at the shortly tapering base, rounded and bifid at the apex; upper surface dark green, shining, with a continuous median groove and without stomata; lower surface with two broad conspicuously white bands of stomata; resin-canals marginal. Leaves on cone-bearing branchlets similar to those on barren branchlets.

Cones on short stout stalks, resembling in shape and colour those of *A. Pindrow*; in native specimens both from Sikkim and Kumaon, smaller than those on cultivated trees; scales fan-shaped, about $\frac{7}{8}$ inch wide and $\frac{5}{8}$ inch long (not including the short obcuneate claw); bracts extending to near the upper edge of the scale, with an oblong claw, expanding above into a suborbicular denticulate lamina, tipped with a

---

[1] In *Journ. Roy. Hort. Soc.* xiv. 472 (1892).

[2] *Silva N. Amer.* xii. 98, *adnot.* (1898). The trees, however, at Wellesley, one of which is 59 feet by 5 feet, were slightly injured during the severe winter of 1903-4. Cf. Sargent, *The Pinetum at Wellesley in* 1905, p. 12.

[3] Hooker, *Him. Journ.* ii. p. 108.

short triangular cusp and without any emargination. In cultivated specimens, cones very large, 6 to 8 inches long, 2 to 3 inches in diameter, bluish when growing, brownish when mature, cylindrical, slightly tapering to an obtuse apex; scales much broader than in wild specimens ($1\frac{1}{4}$ inch); bracts only extending to about half the height of the scale, with a broad rectangular claw, only slightly narrower than the broadly ovate denticulate lamina, which is tipped with a short triangular cusp: seed with wing about an inch long; wing broadly trapezoidal, shining brown, and about $1\frac{1}{3}$ times as long as the seed.

The cones of *Abies Pindrow* are very similar, the main difference being that in the latter the expanded portion of the bract is situated close to the lower edge of the scale, and is oval, less finely denticulate, and emarginate above with a minute mucro in the emargination.

## Varieties

The above description, which, as regards the leaves and branchlets, applies to ordinary cultivated specimens of *Abies Webbiana*, also fits exactly the form of that species which occurs in Sikkim, and does not differ from the original description which was founded on specimens from Nepal. The high-level silver fir, however, which occurs in the western Himalayas appears to be a much shorter-leaved tree than that which is common in Sikkim; and has been supposed by some to be a form of *Abies Pindrow*. This form, which is apparently the same as specimens collected on the Chor mountain near Simla by Sir George Watt, is met with occasionally in cultivation, and may be distinguished as follows :—

Var. *brevifolia*,[1] a tree with smooth bark on the stem and branches. Young branchlets grey, with only slightly prominent pulvini; pubescence short, erect, brown, confined to the indistinct fine grooves between the pulvini. Leaves much shorter than in the type, not exceeding $1\frac{1}{4}$ inch in length, greyish beneath with two inconspicuous stomatic bands.

This variety differs in appearance from the type, which has longer leaves, very silvery white beneath; but agrees with it in the arrangement, texture, and shape of the leaves. The grey colour and comparative smoothness of the branchlets, and the smooth bark on the stem and branches, suggest some affinity with *A. Pindrow*; but the long, slender, narrow leaves of the latter species, differently arranged on the glabrous branchlets, are entirely different.

I first received cultivated specimens of this variety from Glasnevin, Kilmacurragh, and Batsford Park, where there are young trees, which have not yet produced cones. The Glasnevin and Kilmacurragh trees were raised from seed, sent from the Himalayas in 1879, but without any record of the precise locality; and they resemble the type in habit. The origin of the Batsford tree is obscure.

[1] Brandis, in *Indian Trees*, 692 (1906), distinguishes two forms of *A. Webbiana*, viz. :—

(a) "*A. Webbiana*, Lindley. High Level Silver Fir of N.W. Himalaya." This is identical with our var. *brevifolia*, and is not the same as *A. Webbiana*, Lindley.

(b) "*A. densa*, Griffith. East Himalayan Silver Fir." This, from a comparison of type specimens in the Kew Herbarium, is identical with *A. Webbiana*, Lindley, which was founded on *Pinus Webbiana*, Wallich, described from Nepal specimens.

A tree of this variety, with smooth bark on the trunk and branches, is growing at Powerscourt, and measured in 1906, 58 feet high by 5 feet 5 inches in girth.   It bore numerous cones, similar to those of the type, but smaller in size and not so blue in colour.    Another tree, about 40 feet high, also with very smooth bark, is growing at Holker Hall, Cark-in-Cartmel, Lancashire.                              (A. H.)

## DISTRIBUTION

*Abies Webbiana* occurs in the inner Himalayas from Afghanistan to Bhutan at elevations of 10,000 to 14,000 feet, but rare[1] below 11,000 feet. In its western area, *i.e.* in the north-west Himalaya, it usually commences to grow at 1000 to 2000 feet above the line where *Abies Pindrow* disappears; and Gamble has never seen the two species growing together.   It is here usually stunted and gnarled, with very short leaves and short thick cones, and occurs commonly with *Betula utilis* and *Rhododendron campanulatum*.   Both it and the birch are the last trees to be seen before the treeless snowy wastes begin in the western Himalayas.   In Nepal, according to Don, it occurs on Gosainthan.

In its eastern area, Bhotan and Sikkim, it is apparently a larger and finer tree. Griffith[2] mentions it, under the temporary name of *Abies densa*, as constituting vast forests in Bhotan, remarkable for their sombre appearance, at 12,000 feet, being rare under 9500 feet.

It is slow in growth, the average rate in Sikkim being about 12 rings per inch of radius, and is of much less economic importance than *Abies Pindrow* is in the north-west.   Large quantities of planks, however, are exported from Lachoong to Tibet, and their preparation is an important native industry; but Hooker[3] says that the timber of Sikkim conifers is generally soft and inferior to that of European species.

In Sikkim this is the most abundant conifer in the interior; extending from a little above 8000 to 13,000 feet or more; at its lower limits scattered or isolated among other trees at 9000 to 11,000 feet, and forming forests which are sometimes almost pure, or in the Lachen and Lachoong valleys mixed with *Tsuga Brunoniana*. Higher up on drier slopes it occurs scattered among *Larix Griffithi* and many species of shrubs and rhododendrons.   On the Singalelah range which divides Sikkim from Nepal it begins to appear shortly before reaching Sandukpho, and on the boundary ridge north of that hill assumes a very wind-swept and often gnarled habit; the tops being often broken and covered with a dense mass of ferns, orchids, *Ribes*, begonias, and climbing plants of many species, and sometimes supporting shrubs and trees, which, favoured by the extremely moist summer climate, and from June to October almost constantly bathed in mist, become epiphytic. One tree which I specially noticed on this ridge at about 11,000 feet, bore on its decaying crown no less than four good-sized shrubs of different species, a Pyrus, an

---

[1] As in the Chor Hills, south of Simla.

[2] *Notulæ*, iv. 19 (1854), and *Itin. Notes*, ii. 141 (1848).   Griffith subsequently abandoned his name *A. densa* and adopted that of *A. Webbiana*.

[3] *Himalayan Journals*, ii. 45, *note*.

Aralia, a Rhododendron, and a birch, some of which had stems as thick as my leg. Plate 215 from photographs taken by my late friend, Mr. C. B. Clarke, at this spot, very well represents the trees I saw. The largest trees which I have found of this species were on the track from Lachoong to the Tunkra Pass, leading into the Chumbi valley, some of which must have been nearly 200 feet high, with stems clean up to 40 feet, and Sir Joseph Hooker measured a tree here no less than 35 feet in girth. Higher up, however, it assumes a stunted form and grows mixed with junipers.

## HISTORY AND CULTIVATION

*Abies Webbiana* was discovered early in the nineteenth century by Captain Webb. Seeds were repeatedly sent to England by Dr. Wallich, which probably came from Nepal, but none appear to have germinated till 1822, when some plants were raised in the Fulham Nursery. It is remarkable that most of the trees of *A. Webbiana* seen in this country resemble more nearly the Sikkim form, than the short-leaved Western form. It is probable that none of the original trees now exist, as they were planted in the vicinity of London, where the tree does not thrive, as it is very liable to be cut by spring frosts.

Though this tree is one of the most beautiful of its genus in the few parts of England where it really succeeds ; and will resist severe winter frosts without injury when on well-drained soil, yet its tendency to start into growth before the danger of spring frosts has passed, has caused its death in very many places. If seeds could be procured from the more alpine regions of Kashmir or the trans-Indus mountains, they might endure our climate better, but most of the trees now growing in England were probably raised from seed collected by Sir J. D. Hooker in Sikkim.

It ripens seed, however, in some parts of England and Scotland, and I have raised seedlings in 1901 from cones grown near Exeter, of which a few have survived though now not more than a foot high. A shady, elevated, and yet sheltered situation, is best for this species, and as regards soil a deep sandy loam.

## REMARKABLE TREES

The largest specimen of *A. Webbiana* recorded at the Conifer Conference in 1891 was at Howick Hall, Northumberland, the seat of Earl Grey, and was then said to be 51 feet by 8 feet. I am informed by Mr. Lambert that it has lost its leader several times since this date, and now measures about 50 feet by $8\frac{1}{2}$ feet.

The largest we have measured is a double-stemmed tree at Beauport, Sussex, 64 feet by $8\frac{1}{2}$ feet in 1904 ; but Mr. A. B. Jackson found a tree at Tregothnan in Cornwall which was 74 feet by $8\frac{1}{2}$ feet, and another tree at the same place 66 feet by 9 feet. Both of these bore cones in 1908.

At Menabilly, Cornwall, there is a healthy tree of no great size, which bore large cones in 1907, and these remained in perfect condition on the tree in April 1908, when I visited the place ; and at Pencarrow there is one 64 feet by $6\frac{1}{2}$ feet.

At Fulmodestone, Norfolk, in a wood where the Earl of Leicester has planted on a deep moist soil a large number of conifers, close enough to shelter each other, a tree was measured by Henry in 1905 as 69 feet by 6½ feet.   It has since been much damaged by snow, but has produced cones from which Capt. R. Coke has raised a few seedlings, and now measures only about 60 feet high.

At Enville Hall, Stourbridge, there is a tree, which now looks as if it were suffering from drought.   It measured, in 1904, 68 feet by 5 feet 4 inches.

A large tree, said to have been about 75 feet high, died and was cut down at Penrhyn in North Wales in 1902.   The stump, which I saw, was about 7½ feet round.   At Hafodunos, in North Wales, Colonel Sandbach states that this species is always nipped by the frost and forms new leaders when the old ones are killed, the growth being quite checked.

At Castle Kennedy there is a short avenue of trees of this species, averaging about 40 feet high by 6 feet in girth; but the tops had been cut off, as they had become bare and unsightly from exposure to wind.   Here *A. Webbiana* begins to produce cones at an early age; and there is a seedling 20 feet high, with many smaller ones near it.

At Poltalloch, Argyllshire, I measured a fine healthy tree 61 feet by 5 feet 3 inches which in 1906 bore no cones.

At Keir, Perthshire, there are two trees, one with a broken top, the other 57 feet by 4 feet 10 inches, and more narrowly pyramidal than is usually the case. Seedlings were raised from the seed of this tree about ten years ago; but as a rule it bears very small cones (only 3 inches in length) with unfertile seed.   A tree at Dunphail, Morayshire, has also produced small cones, which slightly resemble those of *Abies Mariesii*.[1]   This tree was probably planted in 1856, and is now only 33 feet in height; but has a double leader.[2]

In Ireland, *A. Webbiana* thrives well, and there is a good number of fair-sized trees.   It is said, however, to be slightly touched by frost at Fota, in the south of Ireland, where the temperature fell to 14° Fahr. during four nights in the winter of 1901-1902.   A tree at Fota was in 1903, 47 feet high by 3 feet 7 inches in girth.

At Churchhill, Armagh, the seat of Mr. Harry Verner, there was growing in 1904 a tree laden with cones, even on the lowermost branches; it measured 53 feet by 4 feet 10 inches.

At Courtown, Co. Wexford, a tree was recorded at the Conifer Conference of 1891, as being 52 feet by 6 feet 3 inches.                                        (H. J. E.)

[1] The Dunphail tree has been described and figured by Masters as *Abies Mariesii*, in *Bot. Mag.* t. 8098; but there is no doubt that this is erroneous.   Specimens which I have seen show ordinary foliage and branches of *A. Webbiana*; and some of the cones are as large as those usually produced by this species.—(A. H.)

[2] Thrives at Durris only in partial shade, when exposed it suffers much from late spring frosts, both top and branch shoots become clubbed and unsightly.   Cones at a comparatively early age—about 25 years.   Is of no economic value.

(J. D. Crozier.)

## ABIES PINDROW, PINDROW FIR

*Abies Pindrow*, Spach, *Hist. Vég.* xi. 423 (1842); Masters, *Gard. Chron.* xxv. 691, f. 154 (1886); Kent, Veitch's *Man. Coniferæ*, 533 (1900); Gamble, *Indian Timbers*, 719 (1902); Brandis, *Indian Trees*, 692, 720 (1906).
*Abies Webbiana*, Lindley, var. *Pindrow*, Brandis, *Forest Flora Brit. India*, 528 (1874); Hooker, *Flora Brit. India*, v. 655 (1888).
*Pinus Pindrow*, Royle, *Illust. Bot. Himalaya*, 354, t. 86 (1839).
*Picea Pindrow*, Loudon, *Arb. et Frut. Brit.* iv. 2346 (1838).

A tree attaining in the Himalayas over 200 feet in height, with a girth of 25 feet. Narrowly pyramidal in habit, with the branches small and short. Bark smooth and silvery grey when young; greyish brown, deeply and longitudinally fissured on old trunks.

Buds large, globose, covered with white resin. Young shoots quite smooth, grey, glabrous, the bark fissuring slightly in the second year. Leaves on lateral branches mostly pectinate below, pointing forwards and outwards in the horizontal plane, some of the median leaves, however, being directed downwards and forwards; above, covering the shoot, those in the middle line much shorter and directed forwards and slightly upwards. Leaves, soft in texture, up to $2\frac{1}{2}$ inches long, very narrow ($\frac{1}{16}$ inch wide), linear, flattened, shortly tapering at the base and narrowing gradually in the anterior third to the acute apex, which is bifid with sharp unequal cartilaginous points; upper surface dark green, shining, with a continuous median groove and without stomata; lower surface paler with two greyish bands of stomata, each of about eight lines; resin-canals marginal. Leaves on cone-bearing shoots all upturned and more or less directed forwards, covering the shoot in the middle line above, shorter than on barren branches and only slightly bifid at the apex.

Cones on short stout stalks, bluish when growing, brown when mature, cylindrical, about 6 inches long by 3 inches in diameter. Scales; lamina about $1\frac{1}{4}$ inch wide by $\frac{3}{4}$ inch long, fan-shaped, variable in form, with two slight wings in cultivated specimens, not winged and with the lateral edges straight or curved in wild specimens, base auricled. Bracts with the expanded portion situated on the scale just above the claw, oval, denticulate, emarginate above with a minute mucro. Seed with wing about 1 inch long, the wing narrowly trapezoidal and about $1\frac{1}{2}$ times as long as the body of the seed.

### IDENTIFICATION

*Abies Pindrow* is remarkably different in most characters from *Abies Webbiana*, with which it has been united by many authors. The trees are very distinct in habit, *A. Pindrow* forming, both in the Himalayas and in cultivation in England, a narrow pyramid with short branches; while *A. Webbiana* is a broader tree with wide-spreading branches. The bark of the former is smooth, that of the latter scaly. The former has smooth, glabrous, grey shoots; the latter has shoots with prominent pulvini, separated by pubescent furrows. The cones are similar in size and colour ; but differ in the shape and position of the bracts. The arrangement and character of the foliage are entirely different.

Var. *intermedia.*

> *Pinus (Abies) sp. nova* (?), M'Nab, in *Proc. Roy. Irish Acad.* ii. 692, f. 19 (1876).

A tree at Eastnor Castle, planted thirty-seven years ago, and about 60 feet in height with a girth of 3 feet 4 inches, is apparently identical with the form described by M'Nab, who mentions two specimens, one collected in the western Himalayas by Hooker and Thomson, and another from a tree, which formerly grew at Castle Kennedy. Mr. Mullins, the gardener at Eastnor Castle, informs me that the tree is narrowly pyramidal in habit, with dark green foliage, and smooth bark on the stem and branches.

Specimens, which I have received, show the following characters :—Branchlets, buds, bark, and habit, as in *A. Pindrow.* Leaves more pectinate than in that species, and arranged on the branchlets as in *A. Webbiana* ; about $2\frac{1}{8}$ inch in maximum length, dark shining green above ; gradually tapering in the upper third, as in *A. Pindrow* ; thicker than in this species ; lower surface convex ; resin-canals median, in which respect this variety differs from both *A. Pindrow* and *A. Webbiana.* Cones about 4 inches long and 2 to 3 inches in diameter, resembling those of *A. Webbiana* in the position and shape of the bracts.

This variety is intermediate in many respects between *A. Pindrow* and *A. Webbiana*, and is possibly a hybrid.      (A. H.)

## DISTRIBUTION

*Abies Pindrow* is more restricted in distribution than *A. Webbiana* and occurs at a lower elevation. It is met with in the outer Himalayas from Chitral to Nepal, at elevations of 7000 to 9000 feet, occasionally ascending to 10,000 feet ; and commonly grows in ravines with a northerly or westerly aspect. It is often associated with *Picea Morinda, Quercus dilatata,* the deodar, and *Pinus excelsa* ; but more often is accompanied by broad-leaved trees, such as the walnut, maples, bird cherry, and Indian horse-chestnut. Madden says that it forms dense forests on all the great spurs of Kumaon and occurs in Kashmir. According to Gamble, it has the same narrowly pyramidal habit with short branches which it assumes in cultivation in England. It grows very tall, but does not attain so great a height as the deodar. The largest trees correctly noted were measured in the Mundali forest in Jaunsar, and had heights varying from 188 to 206 feet with girths of 19 to 25 feet. The rate of growth averages 13 rings per inch of radius. It bears intense shade and its natural reproduction is excellent. The timber is employed indiscriminately with that of *Picea Morinda,* though not quite so good. It is used for planking, tea-boxes, packing-cases, and makes excellent shingles, and would be suitable for railway sleepers if creosoted.

### Remarkable Trees

The finest trees in Britain are probably two, which I saw in 1906, growing in the grounds of Mr. Victor Marshall at Monk Coniston, the largest of which is now 69 feet high and 4 feet 9 inches in girth. The climate here is mild and damp, but the soil dry and slaty. Mr. S. A. Marshall wrote to Kew that one of these trees coned for the first time in 1902, and in 1904 produced many cones.[1]

There is a well-shaped tree at The Coppice, Henley, the seat of Sir Walter Phillimore, Bart. It came from Dropmore as a young plant in 1858, and grew very slowly at first, sustaining some damage from the frost of May 1867. It is now healthy and thriving, and measured, in 1907, 62 feet high by 3 feet 8 inches in girth. At Bury Hill, Dorking, the seat of R. Barclay, Esq., there is a well-grown specimen, which he informed me had been raised from a tree at Denbies in the same neighbourhood, and which in 1908 measured 58 feet by 3 feet 2 inches. It grows, like so many of the best conifers in this country, on greensand. At Dropmore, a tree 48 feet by 3 feet 10 inches in 1905, coned for the first time in 1907. At High Canons, Herts, Mr. H. Clinton Baker measured a tree, which was bearing cones in February 1908, as 48 feet high by 2 feet 8 inches in girth; and in a wood, on his own property at Bayfordbury, there is a thriving young tree, 32 feet high and 1 foot 5 inches in girth. At Cuffnells, near Lyndhurst, a good-sized but badly-grown tree was bearing cones in 1907. At Leighton Hall, near Welshpool, the seat of Mrs. Naylor, there are three, at a considerable elevation, the best of which in 1908 was about 50 feet high. In the pinetum at Lyndon Hall, Oakham, Rutland, Henry measured in 1908 a fine specimen, 58 feet by 3 feet 3 inches, which was planted, according to Mr. E. L. P. Conant, by his father in 1864.

In Scotland I measured a tree at Conon House in Ross-shire, in 1907, which was about 50 feet by only 3 feet 4 inches. At Smeaton-Hepburn, East Lothian, a tree planted in 1844, according to Sir Archibald Buchan-Hepburn, Bart., was, in 1908, 56 feet high by 4 feet 1 inch in girth. Seedlings have been raised from it. At Durris, Mr. Crozier says that it suffers from the same causes as *A. Webbiana*. In Wigtownshire, this species is tender, and a tree at Galloway House, 48 feet by 3 feet 10 inches, as measured by Henry in 1908, has been much injured by frost. Sir Herbert Maxwell reports a good specimen at Stonefield, Loch Fyne; and another, in the Quarry garden at Gordon Castle, which measures 69 feet by 4 feet 9 inches.

In Ireland, the tallest tree is at Charleville, Co. Wicklow, the seat of Viscount Monck. In 1904 it was 60 feet by 5 feet 1 inch. It produced cones in 1903, and since then has been in an unhealthy state. This tree has sent forth from the stem epicormic branches. At Kilmacurragh, Co. Wicklow, the seat of Mr. Thomas Acton, there is a healthy tree, which, in 1904, was 51 feet high by 5 feet 5 inches. At Powerscourt, there are two good trees, one 55 feet by 5 feet in 1904, which bore cones in 1902, 1903, and 1904; the other is 42 feet by 6 feet 3 inches in girth. At Brockley Park, Queen's County, a tree, planted thirty-five years ago, measured in 1907, 51 feet by 5 feet 6 inches. (H. J. E.)

[1] A tree at Kenfield Hall, near Canterbury, produced cones in 1886. Cf. *Gard. Chron.* xxvi. 85 (1886).

## ABIES SIBIRICA, SIBERIAN FIR

*Abies sibirica*, Ledebour, *Fl. Alt.* iv. 202 (1833); Masters, *Journ. Linn. Soc.* (*Bot.*) xviii. 519
(1881); Kent, Veitch's *Man. Coniferæ*, 539 (1900).
*Abies Pichta*, Forbes, *Pin. Woburn.* 113, t. 39 (1840).
*Abies Semenovii*, Fedtschenko, *Bot. Centralblatt*, lxiii. 210 (1898), and *Bull. Herb. Boissier*, vii. 191
(1899).
*Pinus sibirica*, Turczaninow, *Bull. Soc. Nat. Mosc.* xl. 101 (1838).
*Pinus Pichta*, Endlicher, *Syn. Conif.* 108 (1847).
*Picea Pichta*, Loudon, *Arb. et Frut. Brit.* iv. 2338 (1838).

A tree attaining about 100 feet in height. Bark smooth, greyish, and covered with resin-blisters, even in old trees. Buds small, globose, brownish, smooth, and covered with resin. Young shoots ashy-grey, with a scattered minute erect pubescence, quite smooth, the pulvini not being at all prominent; in the second year, the bark fissures slightly, and the pubescence is retained.

Leaves on lateral branches resembling in arrangement those of *A. Veitchii*, but more irregular; the lower ones pectinate, and directed outwards and forwards, a few, however, in the middle line with their apices directed forwards and downwards; on the upper side the leaves cover the branchlet and are directed forwards and upwards in the middle line, being about three-fourths the length of the lower leaves. Leaves linear, flattened, slender, up to $1\frac{1}{2}$ inch long, $\frac{1}{20}$ inch wide, uniform in width except at the slightly narrowed base; apex rounded, slightly bifid or entire; upper surface light green, shining, with a continuous median groove and rarely two to three short lines of stomata near the apex in the middle line; lower surface greyish in colour, with two narrow bands of stomata, each of four to five lines; resin-canals median. Leaves on cone-bearing branches all upturned, curved, thick, short ($\frac{3}{4}$ inch long), acute at the apex.

Cones sessile, cylindrical, obtuse at the apex, 2 to 3 inches long, $1\frac{1}{4}$ inch in diameter, bluish when growing, brown when mature, with the bracts concealed. Scales; lamina fan-shaped, thin, $\frac{5}{8}$ to $\frac{3}{4}$ inch wide, $\frac{1}{2}$ inch long; upper and lateral margins denticulate; base with a sinus on each side of the obcuneate claw. Bract, at the base of the scale, rectangular or reniform, coarsely denticulate, $\frac{3}{16}$ inch broad, with a short triangular mucro. Seed with wing about $\frac{5}{8}$ inch long; wing broad, purplish, about twice as long as the body of the seed.

The form[1] described by Fedtschenko as a new species (*A. Semenovii*) occurs in Turkestan. Specimens show longer leaves, more pubescent branchlets, and slightly different cone-scales and bracts. Korshinsky, however, in a note in the Kew herbarium, states that the Turkestan tree is identical with *A. sibirica* from the Ural and Altai; and the differences noted would probably disappear if there were more material to examine.

A weeping variety of this species was seen by Conwentz[2] in 1881 in Regel and Kesselring's nursery at St. Petersburg.

This species, with long slender leaves covering the branchlet above, is best

---

[1] Cf. Guinier and Maire's remarks on this form in *Bull. Soc. Bot. France*, lv. 184 (1908).
[2] *Seltene Waldbäume in Westpreussen*, 161 (1895).

distinguished by its ashy-grey smooth shoots which are minutely pubescent, and its small globose resinous buds. It can only be confused with *A. sachalinensis*, which has shoots with prominent pulvini, and leaves with broader and whiter bands of stomata below. *A. lasiocarpa*, which has, like this species, median resin-canals, is distinguished by the peculiar arrangement of the leaves, which have conspicuous lines of stomata on their upper surface.

## DISTRIBUTION AND CULTIVATION

This species is the most widely distributed of all the silver firs, occupying large areas of both the plains and mountains of north-eastern Russia and Siberia. In European Russia it forms forests in company with spruce and larch, or rarely with birch and aspen; and occurs through the governments of Archangel, Vologda, Kostroma, Perm, Ufa, Kazan, and Orenburg. On the mountains it does not go as high as the timber line, and does not extend so far south in European Russia as the spruce. It is common in the Ural range, and attains perhaps its maximum development in the Altai,[1] where it forms vast forests between 2000 and 4500 feet elevation. In Turkestan it is found in the Thianshan mountains, and is reported by Korshinsky to form forests at low altitudes in the province of Ferghana, where it grows in mixture with *Picea Schrenkiana*. Its distribution in Siberia is not clearly known, but it appears to be widely spread from west to east, its northern limit on the Yenisei being 66° lat. and on the Lena 60° lat. It occurs on the high lands of Dahuria, and, according to Komarov, reaches its most easterly point in the Yablonoi mountains, being replaced in Kamtschatka, by *A. gracilis*, Komarov; and on the borders of the sea of Okhotsk and in Manchuria by *A. nephrolepis*, Maximowicz.

According to Loudon this species was introduced from the Altai into England in 1820. It is very rare in cultivation, and does not grow for any length of time in the south of England, where the climate is unsuitable to it. Even at Durris, Mr. Crozier describes it as "a slow-growing, many-headed, and evidently short-lived tree." There is an unhealthy specimen, about 15 feet high, at Ochtertyre in Perthshire. At Pampisford, Cambridgeshire, there is a small tree, 30 feet high by 13 inches in girth in 1907, narrowly pyramidal in habit, with the lower branches layering and producing five independent stems about six feet in height. This tree has been badly damaged by the snowstorm of April 1908. Another at Bicton measured 28 feet by 1 foot 8 inches in 1908.

In August 1908 I saw a fine specimen in the University Botanic Garden, Upsala, Sweden, which was 70 feet high and 1 foot in diameter, forming a very narrow pyramid, closely resembling the habit of *A. Pindrow*. Hansen[2] says that specimens, forty years old, have attained a height of about 40 feet in Denmark; and that there are beautiful examples in the Botanic Gardens at Helsingfors, Finland (lat. 60°), where many seedlings have sprung up around the old trees.     (A. H.)

---

[1] I saw this tree in the forests on the north slopes of the Altai, where the climate was damp, but it did not strike me as a fine or large tree, and was not seen in the drier valleys towards Mongolia.—(H. J. E.)

[2] *Journ. Roy. Hort. Soc.* xiv. 477 (1892).

## ABIES SACHALINENSIS, Saghalien Fir

*Abies sachalinensis*, Masters, *Gard. Chron.* xii. 588, f. 97 (1879), and *Journ. Linn. Soc. (Bot.)* xviii.
517 (1881); Mayr, *Abiet. jap. Reiches*, 42, t. 3, f. 6 (1890); Sargent, *Forest Flora of Japan*,
83 (1894); Kent, Veitch's *Man. Coniferæ*, 537 (1900).
*Abies Veitchii*, Lindley, var. *sachalinensis*, F. Schmidt, *Mém. Acad. St. Pétersbourg*, sér. 7, xii. 175,
t. 4, ff. 13-17 (1868).

A tree attaining in Yezo 130 feet in height. Bark smooth, grey in colour. Buds small, ovoid-globose, rounded at the apex, covered with white resin. Young shoots grey, with prominent pulvini and grooves; pubescence, short, dense, and confined to the grooves. In the second year the pulvini, grooves, and pubescence are well-marked.

Leaves on lateral shoots arranged similarly to those of *A. Nordmanniana*; those below longest, pectinate in two sets in the horizontal plane, directed outwards and forwards; those above covering the branchlet in imbricated ranks, the median leaves shortest, directed forwards and appressed to the shoot. Leaves linear, flattened, slender, very thin, about $1\frac{1}{2}$ inch long, $\frac{1}{20}$ inch wide, uniform in width except at the shortly tapering base; apex rounded and shortly bifid; upper surface grass-green, shining, with a continuous median groove and without stomata; lower surface with two narrow bands of stomata, each of seven to eight lines; resin-canals median. Leaves on cone-bearing shoots upturned, acute or rounded at the minutely bifid apex.

Cones, $3\frac{1}{2}$ inches long, $1\frac{1}{4}$ inch in diameter, cylindrical, tapering to an obtuse or slightly acute apex, conspicuously marked externally by the reflexed greenish bracts, which leave little of the surface of the scales visible. Scales crescentic, small; lamina, $\frac{1}{2}$ to $\frac{5}{8}$ inch wide, nearly $\frac{1}{4}$ inch long, deeply auricled by two basal sinuses, the denticulate wings ending in a sharp point on each side of the sinus; upper margin entire; outer surface densely tomentose. Bract with a broad cuneate claw, expanding above into an almost orbicular lamina, emarginate and mucronate on its upper margin. Seed with wing $\frac{3}{8}$ inch long; wing broader than long and shorter than the body of the seed.

### Varieties

Var. *nemorensis*, Mayr, *loc. cit.* This variety is met with in north-eastern Yezo and the Kurile Isles, and is distinguished by its smaller cones, about $2\frac{1}{2}$ inches in length, with minute concealed bracts. In the cones this variety resembles *A. Veitchii.*

Sargent mentions a curious variety found by Miyabe in central Yezo, in which the bark, wood, and bracts of the cone are red in colour.

### Identification

*Abies sachalinensis* agrees in many technical characters with *A. Veitchii*, but owing to its longer and more slender leaves looks different from that species, and would not in practice be confused with it. The prominent pulvini of the young branchlets, which are only pubescent in the grooves, will distinguish it at once from *Abies sibirica*, which it resembles in general appearance. The latter species has quite smooth branchlets, provided with a scattered minute pubescence. (A. H.)

## DISTRIBUTION AND CULTIVATION

This species was discovered in the Island of Saghalien by Schmidt in 1866, and was subsequently found, in 1878, in Yezo by Maries, who sent home seeds in the following year.[1] The tree is known in Japan as *Todo-matsu*.

It is a native of the Kurile Islands, Saghalien, and the northern island of Japan.[2] In Saghalien it either forms pure woods or is mixed with one or both of the spruces (*Picea ajanensis* and *Picea Glehnii*) which occur in that island.

This is the common, and perhaps the only, silver fir of Hokkaido, where it extends from nearly sea-level up to 4000 or 5000 feet altitude, and all over the island in suitable places; in the south usually as a scattered tree in mixed forests of deciduous trees; in the north and some parts of the west central districts in dense pure forests, or with a mixture of birch and poplar. The finest areas of this species are in the Imperial domains at Tarunai, Uryu, Kushiro, and in the State forests at Shari, and Kunajiri. I endeavoured to visit some of these under the guidance of Mr. Shirasawa, but owing to the torrential rains which flooded the country in the middle of July and broke the railway in many places, I was unable to do so. The country where these forests occur is much like parts of eastern Siberia, having a hot, moist summer, a warm autumn, and a very heavy snowfall which lies for four to five months; the climatic conditions, therefore, are such that the tree is not likely to be a success in Great Britain, and, so far as I could see, it has no special beauty to recommend it. The largest that I saw were about 100 feet by 9 feet, but it grows taller in some places.

The timber is of fair quality, and is used in house- and ship-building, also for furniture and paper-making; and is worth at Tokyo about 10d. per cubic foot.

The Saghalien fir is rare in cultivation, the largest specimen we have seen being one at Fota, in the south of Ireland, which was about 25 feet high in 1907. It looks healthy, but begins to grow early in the season, and is said to be frequently hurt by spring frost. We have measured no specimens in Scotland, but one at Murthly Castle, about 16 feet high, is reported by Mr. Bean[3] as not looking healthy. According to Kent, the tree is, like most of the conifers coming from similar climates, unable to thrive in England. In New England, however, it grows much better, and I saw healthy young trees at Mr. Hunnewell's place at Wellesley, Massachusetts. (H. J. E.)

---

[1] *Hortus Veitchii*, 337 (1906).

[2] It was reported by Matsumura (*Tokyo Bot. Mag.* xv. (1901), p. 141), to occur in Formosa, on Mt. Morrison; but this was a mistaken identification, as the silver fir in this locality is *A. Mariesii*, according to Hayata, in *Tokyo Bot. Mag.* xix. (1905), p. 45.      [3] *Gard. Chron.* xli. 117 (1907).

## ABIES FIRMA, Japanese Fir

*Abies firma*, Siebold et Zuccarini, *Fl. Jap.* ii. 15, t. 107 (1844); Masters, *Gard. Chron.* xii. 198, 199
(1879), and *Journ. Linn. Soc. (Bot.)* xviii. 514 (1881); Mayr, *Abiet. Jap. Reiches*, 31, t. 1, f. 1
(1890); Shirasawa, *Icon. Ess. Forest. Japon*, text 17, t. 6, ff. 1–21 (1900); Kent, Veitch's
*Man. Coniferæ*, 506 (1900).
*Abies bifida*, Siebold et Zuccarini, *Fl. Jap.* ii. 18, t. 109 (1844).
*Abies Momi*,[1] Siebold, *Verhand. Batav. Gen.* xii. 101 (1830) (*nomen nudum*).
*Pinus firma* and *Pinus bifida*, Antoine, *Conif.* 70, 79 (1846).
*Picea firma*, Gordon, *Pinet.* 147 (1858).

A tree, attaining 150 feet in height and 16 feet in girth. Bark of branches
and trunk early becoming scaly, in old trees fissuring into small plates. Buds small,
ovoid, obtuse at the apex, brown, glabrous, slightly resinous. Young shoots
brownish grey, with the pulvini slightly raised and separated by grooves; pubescence
short, erect, scattered, confined to the grooves. Older shoots retaining the pubescence
and fissuring between the pulvini, which are not very prominent.

Leaves on lateral branches pectinately arranged; those below extending
laterally outwards in the horizontal plane; those on the upper side gradually
shortening to nearly one-third of the length of the lower leaves, and directed in two
sets laterally outwards and slightly upwards, forming a shallow V-shaped arrangement.[2] Leaves up to $1\frac{1}{2}$ inch long, linear, flattened, very coriaceous, shortly tapering
to the base, broadest about the middle ($\frac{1}{8}$ inch or more), gradually narrowing to the
acute apex, which ends in two sharp cartilaginous points, unequal in size; upper
surface dark green, shining, with a continuous median groove and without stomata;
lower surface with broad greyish bands of stomata, each of about ten to twelve lines;
resin-canals close to the epidermis of the lower surface. Leaves on cone-bearing
branches upturned, rounded and entire or only minutely bifid at the apex.

Staminate flowers, $\frac{1}{2}$ inch long, ovoid-conic, surrounded at the base by two to
three series of broadly ovate scales.

Cones on stout short stalks, cylindrical, tapering shortly at the base, and obtuse
or flattened at the slightly narrowed apex, yellowish-green before ripening, brown
when mature, 4 to 5 inches long by $1\frac{1}{2}$ to $1\frac{3}{4}$ inches in diameter, with the tips of the
bracts exserted between the scales but not reflexed. Scales: lamina $1\frac{1}{16}$ inches
wide by $\frac{5}{8}$ inch long, broadly trapezoidal; upper margin thin, minutely denticulate;
lateral margins convex, denticulate; base broad with a sinus on each side of the
obcuneate claw. Bract extending either nearly up to the edge of the scale or
beyond it, always visible externally between the scales, oblong in the lower half,
expanding above into an oval lamina, which ends in a triangular cusp. Seed-wing
broadly trapezoidal, about twice the length of the body of the seed. Seed with
wing nearly $\frac{7}{8}$ inch long. **Cotyledons four.**

---

[1] This name, which has been adopted by Sargent, *Silva N. Amer.* xii. 101, *adnot.*, was published without any description, and cannot be maintained. Cf. Masters, *Gard. Chron. loc. cit.*

[2] On vigorous shoots, the leaves are directed more upwards so that the V-shaped depression is very acute.

The broad coriaceous leaves, ending in two cartilaginous points of unequal size, and pectinately arranged, are characteristic of this species, which when once seen, can scarcely be confused with any other.

*Abies holophylla*, Maximowicz,[1] which has been identified by Dr. Masters with *A. firma*, is considered by Komarov,[2] the latest observer, to be a distinct species. It differs in the leaves not being bifid, and also in the bracts of the cone, which are short, scarcely extending more than one-third the length of the scale. This species, according to Komarov, attains 150 feet high, and grows in mountain woods at elevations not exceeding 1800 feet above sea-level, in the Manchurian provinces of Ussuri, Kirin, and Mukden, and also in northern Korea. It was introduced into cultivation in Russia by Komarov, who sent seeds in 1898.

Other specimens of Abies from the Chinese provinces of Hupeh, Shensi, and Yunnan[3] have also been considered by Dr. Masters to be *A. firma*; but this identification is doubtful. (A. H.)

## DISTRIBUTION

This, the best known fir in Japan, is widely distributed in the south, and, according to Mayr, does not extend north of lat. 40°, and attains perfection in the warm sub-tropical provinces of Kii, Shikoku, and Kiusiu. It is very commonly planted in temple grounds and parks, but few of these specimens looked as if the isolated situation agreed with them; and wherever I saw the tree growing naturally, it was scattered among deciduous trees and other conifers in more or less shady places in the forest. It grows to a great size in the sheltered valleys and moist, deep soils of the central and southern provinces. I measured one at Myanohara, on the Nakasendo road near Wada, which was 135 feet by 16 feet, but this tree was dying at the top, and may have been planted or have been a natural seedling in a temple grove. Another in the forest near the entrance to Koyasan was about 120 feet high by 15 feet 9 inches in girth, but the average size of the mature trees that I saw was not over 100 feet by 9 feet. A third, growing close to a temple at Narai (Plate 216), measured 125 feet by $11\frac{1}{2}$ feet. As the timber is of little value except for packing-cases, tea-boxes, and pulp-wood, the tree is not much planted at the present time except for ornament. It reproduces itself freely from seed whenever the conditions are suitable, and its large greenish-yellow cones are fully formed in August.

According to Rein[4] its natural habitat is from 1000 to 1500 metres, but though this may be the case in the southern island, I should say it was too high for the central provinces, as in Kisogawa I saw it much lower, and I do not think it there reaches 4000 feet.

This species was introduced into Europe in 1861 by J. Gould Veitch,[5] but has never become common in cultivation, though it seems to be hardy even in some parts of Scotland. It undoubtedly requires a warm, moist climate to bring it to

---

[1] *Mél. Biol.* vi. 22 (1866).  [2] *Flora Manshuriæ*, 204 (1901).
[3] *Journ. Linn. Soc.* (*Bot.*) xxvi. 557 (1902); and *Journ. Bot.* 1903, p. 270.
[4] *Industries of Japan*, 235 (1889). Mayr says that it ascends to 700 feet in the north and to nearly 7000 feet in the south.
[5] *Hortus Veitchii*, 335 (1906), where it is stated that it was also sent in 1878 by Maries.

perfection, and seems to be in most places a slow grower. The seedlings which I have raised from Japanese seed will not grow on my soil, from which I infer that lime is distasteful to them.

REMARKABLE TREES

The largest tree that we know of is at Carclew in Cornwall. This was reported in 1891 to be 45 feet by 2 feet 8 inches, and when I measured it in 1902 had increased to about 60 feet by 4 feet. Another at Pencarrow, in the same county, was about 59 feet by 6 feet 5 inches in 1908. At High Canons in Herts, Mr. Clinton Baker showed me a specimen which bore cones in 1907 and measured 47 feet by $3\frac{1}{2}$ feet.

There is a good-sized tree at Grayswood, with longer and less sharp-pointed leaves than usual, and another at Tortworth which in 1905 was 30 feet by 3 feet 9 inches. A tree planted at Bagshot Park by the late Emperor of Germany on July 10, 1880, was, when I saw it in 1907, 36 feet by 3 feet 11 inches.

In Scotland[1] the best that we have seen is at Castle Kennedy, which, in 1904, Henry found to be 44 feet by 5 feet 5 inches. Another in a wood at Munches, Dalbeattie, was 30 feet by $2\frac{1}{2}$ feet. Trees were reported to be growing in 1891[2] at Balmoral, and at Haddo House in Aberdeenshire, but we have not identified them.

In Ireland there were thriving trees at Fota 25 feet high, and bearing cones in 1907 ; at Hamwood, Co. Meath, 36 feet by 2 feet 10 inches in 1904 ; and at Powerscourt, which in 1906 was bearing cones, and measured 39 feet by 3 feet 11 inches.

(H. J. E.)

ABIES HOMOLEPIS

*Abies homolepis*, Siebold et Zuccarini, *Fl. Jap.* ii. 17, t. 108 (1844); Masters, *Gard. Chron.* 1879, p. 823, and *Journ. Linn. Soc. (Bot.)* xviii. 518 (1881).
*Abies Tschonoskiana*, Regel, in *Index Sem. Hort. Petrop.* 32 (1865).
*Pinus Harryana*, M'Nab, *Proc. R. Irish Acad.* ii. 689, Pl. 47, f. 16 (1876).

This species, imperfectly described by Siebold and Zuccarini, is considered by Mayr to be a form of *A. brachyphylla*. It is different in the pulvini of the branchlets, in the shape and arrangement of the leaves, and in the position of the resin-canals in the latter. Specimens in cultivation, described below, agree with the type of Siebold and Zuccarini's species in the Leyden Museum. The cones are unknown ; and it is possible that it may be a juvenile form or variety of *A. brachyphylla* ; but in the present state of our knowledge, it is best kept distinct.

---

[1] Has been tried at Durris repeatedly, but does not live beyond a year or two. Quite unfitted for our climate.—(J. D. CROZIER.)
[2] *Gard. Chron.* x. 458 (1891).

As seen in cultivation at Kew, it is a small tree, resembling in bark and habit *A. brachyphylla*. The foliage, however, is rather like that of *A. firma*, and the tree is occasionally cultivated under that name.[1]

Buds ovoid, obtuse at the apex, whitened with resin, much larger than those of *A. firma*. Young shoots grey, glabrous, with prominent pulvini and grooves, which become less marked in the second year.

Leaves on lateral branches arranged as in *A. firma*, those of the upper rank about half the length of those of the lower rank, linear, flattened, rigid and slightly coriaceous, up to about $1\frac{1}{4}$ inch long, $\frac{1}{12}$ inch wide (much narrower than in *A. firma*), tapering gradually to the base, and narrowing near the rounded or acute apex, which is bifid with two short unequal cartilaginous points; upper surface dark green, shining, with a continuous median groove and without stomata; lower surface with two raised narrow white bands of stomata, each of about eight lines; resin-canals marginal.

Though this species has been in cultivation[2] since 1876 or earlier, we have seen no large specimens; and are ignorant as to whether it changes in character as it grows older or is short-lived. Its distribution in Japan is not known.

(A. H.)

## ABIES BRACHYPHYLLA, Nikko Fir

*Abies brachyphylla*,[3] Maximowicz, *Mél. Biol.* vi. 23 (1866); Masters, *Gard. Chron.* xii. 556 (1879), and *Journ. Linn. Soc.* (*Bot.*) xviii. 515 (1881); Hooker, *Bot. Mag.* t. 7114 (1890).

*Abies homolepis*,[1] Mayr, *Abiet. Jap. Reiches*, 35, t. 2 f. 3 (1890); Shirasawa, *Icon. Ess. Forest. Japon*, text 14, t. 3, ff. 1-12 (1900); and Kent, Veitch's *Man. Coniferæ*, 513 (1900) (not Siebold et Zuccarini, *Fl. Jap.* ii. 17, t. 108 (1844)).

*Pinus brachyphylla*, Parlatore, in D.C. *Prod.* xvi. 2, p. 424 (1868).

*Picea brachyphylla*, Gordon, *Pinetum*, 201 (1875).

A tree attaining in Japan over 100 feet in height and 16 feet in girth. Bark fissuring and scaly on young branches and on the stems of young trees, becoming like that of a spruce on old trees. Buds ellipsoid or broadly conical, obtuse at the apex, smooth, brownish, resinous. Young shoots greyish, glabrous, with prominent pulvini, separated by deep grooves, the pulvini and grooves becoming more marked in older shoots.

Leaves on lateral branches pectinate; those below extending laterally outwards in the horizontal plane, with a few in the middle line directed forwards and downwards; those on the upper side of the branchlet directed upwards and outwards, in

---

[1] It is readily distinguishable from *A. firma* by its glabrous shoots, larger buds, and much less coriaceous and narrower leaves.

[2] M'Nab, *loc. cit.* mentions plants of this species, which were growing under the name of *A. Veitchii* in several nurseries.

[3] The following description applies to the tree, described by Maximowicz, which is, in my opinion and that of Dr. Masters, very different from *Abies homolepis*, S. et Z., which is treated by us as a distinct species.

two lateral sets, separated by an acute V-shaped depression.   Lower ranks with the longest leaves (about $\frac{7}{8}$ inch), those in the other ranks gradually diminishing in size as they approach nearer the middle line above.   Leaves linear, flattened, uniform in width except at the gradually tapering base, about $\frac{1}{15}$ inch, rounded and slightly bifid at the apex; upper surface dark green, shining, with a continuous median groove and without stomata; lower surface with two broad conspicuously white bands of stomata, each of ten to twelve lines; resin-canals median.

Leaves on cone-bearing branches shorter than on barren branches, those on the upper side of the shoot crowded and directed upwards, so that the V-shaped depression between the lateral sets is scarcely visible.

Cones on short stalks, cylindrical, slightly narrowed at the base and apex, 4 inches long, $1\frac{1}{2}$ inch in diameter, purple when growing, brown when mature. Scales very thin and flat; lamina fan-shaped, $1\frac{1}{8}$ inch long by $\frac{3}{4}$ inch wide, upper margin entire, lateral margins with denticulate short wings; claw short and obcuneate. Bract short, not extending half-way up the scale; with a sub-orbicular finely denticulate lamina, tipped by a minute mucro, and a short obcuneate claw.   Seed with wing about $\frac{3}{4}$ inch long, the wing about $1\frac{1}{2}$ times as long as the body of the seed.   In cultivated specimens, the scales of the cone and the seeds are smaller than in wild specimens.   The very thin flat scales with the short minutely denticulate bract distinguish well this species.

## IDENTIFICATION

This species has short leaves very white underneath, with an acute V-shaped depression between the lateral sets on the upper side of the branchlet, and is best distinguished by the very prominent pulvini and grooves on the branchlets. The bark of the branches and young stems begins to scale very early, an unusual character in the silver firs, and conspicuous in this species, in *A. homolepis*, and in *A. Webbiana*.   The resinous buds, glabrous shoots, and leaves with median resin-canals are additional points in the discrimination of *A. brachyphylla*.

(A. H.)

## DISTRIBUTION

According to Mayr this tree occurs on the main island of Japan, between lat. 36° and 38°, in the interior of the  mountainous provinces; where it attains its maximum development in the zone of the beech forests, some trees attaining as much as 130 feet in height.   The Japanese informed Mayr that it was also present on the highest peaks of Shikoku; but Shirasawa limits its distribution to the central chain of Honshu above 3000 feet elevation, and says that it grows in mixture with broad-leaved trees.

To most Europeans it is the best known of the Japanese silver firs, as it grows abundantly at Chuzenji, a favourite tourist resort.   Here at 4000 to 5000 feet it is scattered through the forest of deciduous trees and attains a height of 100 feet or more, the largest that I measured being 105 feet high by 16 feet in girth, and 95 feet by 11 feet. Higher up the mountains it becomes mixed with *Abies Veitchii* and diminishes in size. Its range of distribution is not accurately known, for though Japanese botanists distinguish it from the other species, as *Dake-momi*, the foresters and woodmen, who call all silver firs *momi*, do not seem, so far as I could learn, to distinguish it from *A. Veitchii* and *A. Mariesii*.

## CULTIVATION

The date of introduction of this tree is not certainly known. Kent gives it as about 1870, and Mr. H. J. Veitch tells me that he believes that the first seeds were sent by Dr. Regel from St. Petersburg, but it was at first grown under other names.

It seems to thrive in most parts of England as well as or better than the other Japanese firs, but neither the trees I have planted, nor the seedlings I have raised from Japanese seed, will live long on the calcareous soil at Colesborne; a moist climate in summer, and a deep sandy soil free from lime being apparently the most favourable conditions for its existence. At Kew it seems to grow faster than other firs.

At Pampisford, Cambridge, a narrow conical tree measured in 1907, 44 feet by 3 feet 6 inches. At Grayswood, Haslemere, a tree, obtained from Messrs. Veitch in 1882, measured in 1906, 41 feet by 3 feet 3 inches. Both these trees bear cones freely. A specimen at Bicton is about 47 feet high, and bore cones in 1902. There is a very thriving one on the lawn at Eridge Park, Kent, planted by the Duke of Manchester in 1885, which now measures 30 feet high by 3 feet. At Dropmore there is one which in 1908 was 32 by 2 feet. At Kew, where there are several thriving trees, this species first produced[1] cones in this country in 1887. At Pencarrow,[2] a tree measured 40 feet by 3 feet 10 inches in 1907.

In Scotland we have seen no specimen of any size.[3]

Kent figures a handsome tree at Castlewellan, Co. Down, which was about 35 feet high and coning freely in 1907. Henry measured one at Fota, which was, in 1903, 40 feet in height and 2 feet 8 inches in girth. Another at Glasnevin was, in 1906, 38 feet by 2 feet 6 inches.

This species[4] is very hardy in eastern Massachusetts, U.S., where it has already produced cones. A tree in Mr. Hunnewell's pinetum at Wellesley was 35 feet high in 1905.

(H. J. E.)

---

[1] *Gard. Chron.* ii. 248 (1887).

[2] This tree is figured under the erroneous name of *A. Veitchii*, in *Hortus Veitchii*, plate opposite p. 83 (1906).

[3] *A. brachyphylla* seems in every way adapted for cultivation in the north of Scotland, but too little is yet known of rate of growth to enable an opinion to be formed of its economic value.—(J. D. CROZIER.)

[4] Sargent, in *Silva, N. Amer.* xii. 102 *adnot.* (1898), and *The Pinetum at Wellesley in* 1905, p. 12.

## ABIES UMBELLATA

*Abies umbellata*, Mayr, *Abiet. Jap. Reiches*, 34, t. 1, f. 2 (1890).
*Abies umbilicata*, Beissner, *ex* Mayr, *Fremdländ. Wald- u. Parkbäume*, 258 (1906).

This species is very imperfectly known ; and there are no specimens in the Kew herbarium.    The plants distributed in England some years ago under the name by Messrs. Veitch are identical with *Abies homolepis*, and, when seen by Mayr, were pronounced by him to be a form of *A. brachyphylla*.

According to Mayr this species is distinguishable with difficulty, when in the young state, from *A. brachyphylla*, with which it agrees in the disposition and form of the leaves, and in the characters of the buds and shoots.    Mayr states, however, that the leaves are not so white underneath as in *A. brachyphylla*.    I have received from Herr Späth of Berlin, a specimen of reputed *A. umbellata*, which agrees generally in buds, shoots, and foliage with *A. brachyphylla* ; but has leaves slightly longer than, and not so conspicuously white beneath, as is usual in that species.

The cones, according to Mayr, resemble those of *A. firma*, and are very different in size, colour, scales, and bracts, from those of *A. brachyphylla* ; and are described by him as greenish-yellow when growing, brown when mature, about 4 inches long by $1\frac{1}{2}$ inch in diameter, cylindrical, the flattened apex having in its centre a raised umbo ; scales about $1\frac{1}{4}$ inch broad by $1\frac{3}{8}$ inch long ; bracts narrowed in the middle, slightly shorter than the scales, only exserted at the base of the cone.

According to Mayr, this species is only found in a few localities in Japan, but grows in considerable quantity on Mount Mutzumine in the province of Musashi, where it occurs with *A. brachyphylla* in the beech forests.    It grows also on the Iumonji-toge, leading from Musashi to Shinano, and is also probably not uncommon on the neighbouring mountains of Hida and Kai.

This species has been united by Sargent, Kent, and others with *A. firma* ; but it is very different from that species in foliage and shoots.    It is possibly a hybrid between *A. brachyphylla*, of which it has the foliage, and *A. firma*, which it resembles in its cones.                                                                                  (A. H.)

## ABIES VEITCHII, Veitch's Fir

*Abies Veitchii*, Lindley, *Gard. Chron.* 1861, p. 23 ; Masters, *Gard. Chron.* xiii. 275 (1880), and
    *Journ. Linn. Soc. (Bot.)* xviii. 515, t. 20 (1881); Mayr, *Monog. Abiet. Jap. Reiches*, 38, t. 2,
    f. 4 (1890) ; Kent, Veitch's *Man. Coniferæ*, 541 (1900) ; Shirasawa, *Icon. Ess. Forest. Japon*,
    text 16, t. 5, ff. 23-42 (1900).
*Abies Eichleri*, Lauche, *Berlin. Gartenzeit.* 1882, p. 63.
*Pinus selenolepis*, Parlatore, DC. *Prod.* xvi. 2, p. 427 (1868).
*Picea Veitchii*, Murray, *Proc. Roy. Hort. Soc.* ii. 347, ff. 52-62 (1862).

A tree, attaining 60 to 70 feet in height.    Bark of trunk greyish and remaining smooth even in old trees.    Buds small, subglobose, purplish, resinous.    Young branchlets smooth, brown, covered with moderately dense, short, erect pubescence, retained on the older branchlets, the bark of which becomes slightly fissured.

Leaves on lateral branches arranged almost as in *A. Nordmanniana*; those on the under side of the branchlet pectinate; those on the upper side shorter and covering the branchlet, the median ones pointing upwards and forwards, and not appressed so much as in *A. Nordmanniana*. Leaves, about $\frac{1}{2}$ to 1 inch long, $\frac{1}{16}$ inch wide, linear, flattened, gradually tapering to the base, uniform in width in the anterior half, with a truncate bifid apex; upper surface dark green, shining, with a continuous median groove and without stomata; lower surface with two conspicuously white, broad bands of stomata, in nine to ten lines; resin-canals median. On cone-bearing branches, the leaves are more crowded, and less plainly pectinate below, than is the case in barren branches.

Staminate flowers[1] $\frac{1}{4}$ inch long on a stalk of the same length; anthers stalked, connective developed into a saddle-shaped flap, from the back of which projects a horizontal or deflexed spur-like process.

Cones sessile or sub-sessile, cylindrical, flattened at the apex, 2 to $2\frac{1}{2}$ inches long, $\frac{3}{4}$ to 1 inch in diameter, bluish before ripening, brown when mature. Scales small; lamina $\frac{5}{8}$ inch wide, $\frac{3}{8}$ inch long, crescentic, with two lateral denticulate wings, which are separated from the narrow obcuneate base by rounded deep sinuses. Bract as long as the scale, obcuneate below, dilated above into a two-winged denticulate lamina, ending in a short mucro, slightly exserted and reflexed. Seed-wing very broad and short, scarcely the length of the body of the seed; seed with wing about $\frac{5}{8}$ inch long.

## VARIETIES

Mayr distinguishes two forms of cones:—

1. Var. *typica*. Cones large, about $2\frac{1}{2}$ inches long; bracts exserted and reflexed.

2. Var. *Nikkoensis*. Cones small, 2 inches long; bracts scarcely visible, their fine points projecting only slightly between the scales.

*Abies nephrolepis*, Maximowicz,[2] has been united with *Abies Veitchii* by Masters, and is perhaps a geographical form of the latter species, occurring in Amurland. According to Maximowicz it differs in the leaves of cone-bearing branchlets being sometimes acute and not bifid, and in the smaller ovoid-cylindrical cones, the scales of which are longer than the bracts and less in size than those of the Japanese tree. This Manchurian tree has not apparently been introduced into cultivation and is still imperfectly known.

*Abies Eichleri*, Lauche, was supposed to have been raised from seeds sent from Tiflis to Potsdam; and was considered to be a new species from the Caucasus. Some error, however, had arisen, as the plants turned out to be identical with *Abies Veitchii*.

*Abies Veitchii* has been collected according to Beissner[3] by Père Giraldi at

---

[1] Masters, *loc. cit.*     [2] *Mél. Biol.* vi. 22 (1866).     [3] See *Journ. Linn. Soc.* (*Bot.*) xxvi. 557 (1902).

9800 feet elevation in the Peling Mountains in the province of Kansu in China. The identification of herbarium specimens of Abies is difficult, and the Kansu plant will probably turn out to be a new and distinct species.

## IDENTIFICATION

This species, with leaves covering the branchlet on the upper side, which are very white beneath, truncate and bifid at the apex, and less appressed than is the case in *A. Nordmanniana*, is further characterised by its small resinous buds, median resin-canals in the leaves, and smooth branchlets with short erect pubescence. The distinctions between it and *A. Mariesii* are given under the latter species.

(A. H.)

## HISTORY AND DISTRIBUTION

*Abies Veitchii* was discovered on Fuji-yama by J. Gould Veitch in 1860. According to Sargent[1] it was introduced by Mr. T. Hogg into Parson's nurseries in Flushing, New York, in 1876, and a plant raised there was 16 feet high in 1889. It was cultivated in the United States for a time under the name of *Abies japonica*. It was not known in England or on the continent until 1879, when seeds were sent home[2] to Messrs. Veitch by their collector Maries.

The best account of the distribution is given by Mayr, who considers the tree to be the typical silver fir of the cold region of Japan, a zone which does not occur in Kiushu, where there are no mountains high enough. In Shikoku, *A. Veitchii* is very rare, only about 200 trees being known, which grow on the summit (6600 feet elevation) of Ishitzuchi-yama. It extends in the main island of Japan over the central mountain chain, from Fuji-yama to lat. 39°, growing at elevations of 6600 feet and upwards. Mayr denies its occurrence beyond lat. 39°; and states that north of this line it is replaced by *Abies Mariesii*,[3] which thus intervenes over three degrees of latitude between the southern region, occupied by *A. Veitchii*, and the northern region, occupied by *A. sachalinensis*, these two species not meeting at any point, and having no transitional forms. *A. Veitchii* either forms pure woods or is associated with *Picea hondoensis* and *Picea Alcockiana*, but never with *Picea polita*. Sometimes it is mixed with *Tsuga diversifolia* or with *A. Mariesii*.

Shirasawa gives its lower limit of altitude in the main island as 5000 feet, and states that it attains about 70 feet in height by 7 feet in girth.

According to Mayr and Matsumura, the Japanese name, which is exclusively applied to this species, is *Shirabiso*. *Shiramomi* is also another name for the tree.

So far as I could learn the tree is of no special economic value in Japan.

---

[1] *Garden and Forest*, ii. 589 (1889). In this journal, x. 511 (1897) the statement is made that Mr. Hogg introduced it some forty years earlier, evidently a mistake for twenty years. *A. Veitchii* is very hardy in the United States, where it has produced cones.                                                    [2] *Hortus Veitchii*, 337 (1906).

[3] Prof. Miyabe showed me, in his herbarium, a barren specimen with small leaves, from Samani, near Cape Erimo, in eastern Hokkaido, which he believed to be *A. Veitchii*. This was *A. sachalinensis*, Masters, var. *nemorensis*, Mayr.

## CULTIVATION

It seems to grow fairly well though rather slowly on soils which contain no lime; but it will not live on the calcareous soil at Colesborne.

We have seen no trees of considerable size. One at Tregrehan near St. Austell, measured 27 feet by 2 feet in 1908; and another at Ochtertyre, Perthshire, measured 30 feet by 2 feet in the same year. Mr. Bean[1] noticed in 1906 a specimen 31 feet high at Murthly Castle, and another 20 feet high at Dalkeith Palace. Small specimens will be found in most collections of conifers; and the young trees at Kew at present appear to thrive better than most species of Abies.

(H. J. E.)

## ABIES MARIESII, MARIES' FIR

*Abies Mariesii*, Masters,[2] *Gard. Chron.* xii. 788, f. 129 (1879), and *Journ. Linn. Soc. (Bot.)* xviii. 519 (1881); Mayr, *Abiet. Japan. Reiches*, 40, t. 2, f. 5 (1890); Shirasawa, *Icon. Essences Forest. Japon*, text 15, t. 4. ff. 15-28 (1900); Kent, Veitch's *Man. Coniferæ*, 520 (1900).

A tree, attaining in Japan about 80 feet in height and 6 feet in girth. Buds small, globose, resinous; terminal buds on strong shoots are girt at the base by a ring of ovate, acuminate, rusty-red pubescent scales. Young shoots densely covered with a rusty-red tomentum, retained more or less in older shoots, the bark slightly fissuring in the third year.

Leaves on lateral branches arranged as in *Abies Nordmanniana*, the median leaves on the upper side almost appressed to the stem in imbricating ranks, and about $\frac{1}{2}$ to $\frac{2}{3}$ the length of the lower leaves, which spread pectinately outwards and slightly forwards in the horizontal plane. Leaves linear, flattened, tapering at the base and gradually widening beyond the middle, so that their broadest part is in the upper third; about $\frac{3}{4}$ inch in maximum length, $\frac{1}{10}$ to $\frac{1}{12}$ inch wide; apex rounded and bifid; upper surface yellowish green, shining, with a continuous median groove and without stomata; lower surface with two white bands of stomata, each of eight or nine lines; resin-canals marginal. Leaves on cone-bearing branches all appressed more or less to the shoot, upturned, and shorter than on barren branches.

Cones sessile, deep blue with a velvety lustre before ripening, dark brown when mature, ellipsoid, with an obtuse apex, about 4 inches long by 2 inches in diameter. Scales fan-shaped; lamina 1 inch wide, $\frac{7}{8}$ inch long, upper margin undulate, lateral margins with two denticulate wings; claw broadly obcuneate. Bract with a broad obcuneate claw, expanding just above the base of the scale, into a broadly oval lamina, which is emarginate at the apex with a short mucro. Seed-wing nearly twice the length of the body of the seed; seed with wing about $\frac{7}{8}$ inch long.

The cones show that the tree is nearly related to *Abies Webbiana*; but it differs entirely from that species in the characters of the branchlets and foliage.

---

[1] Cf. *Kew Bulletin*, 1906, pp. 260, 268.

[2] *Abies Mariesii*, Masters, *Bot. Mag.* t. 8098 (1906), described from a tree at Dunphail, Morayshire, is referable to *A. Webbiana*, as mentioned in our account of the latter species.

## IDENTIFICATION

This species is similar in the arrangement and size of the leaves to *A. Veitchii*; but is distinguishable from that and from all other species of *Abies*, by the rusty-red or chocolate colour of the densely tomentose branchlets. The leaves are shorter and broader in proportion than those of *A. Veitchii*, being widest in their upper third, with their apex rounded and not truncate as in that species. The two species differ also in the position of the resin-canals. (A. H.)

## HISTORY AND DISTRIBUTION

This species was discovered[1] in 1878, by Charles Maries, when collecting for Messrs. Veitch, on Mount Hakkoda near Aomori in northern Hondo; and for some years it was supposed to occur only in the main island of Japan, where Mayr gives its distribution as from lat. 36° to the extreme northerly point of the island. It has since been found, according to Sargent,[2] by Tokubuchi in 1892 in one place on the shores of southern Yezo;[3] and Dr. Honda lately discovered it in Formosa on Mount Morrison at 10,000 feet elevation.

Sargent saw it on Mount Hakkoda, and says that it is common at about 5000 feet, scattered amongst deciduous trees, and is the only species of *Abies* in this locality, where it forms a compact pyramid, 40 to 50 feet high, with crowded branches and many large dark purple cones. Maries also found it on Nantai above Nikko, which I had not time to ascend; Mr. Tome Shirasawa, who was my companion in North Japan, says that it grows here in company with *Abies Veitchii* on the upper zone of the mountain at 7000 to 8000 feet. The tree according to Mayr is the smallest of all the Japanese silver firs, its maximum height being 80 feet, with a girth of about 6 feet. It is known in Japan as *Aomori-todo-matsu*, and, so far as I could learn, has no economic value.

## CULTIVATION

Seeds were sent home by Maries in 1879, but gave poor results; and we have not found anywhere in this country a single tree of any size; but Mr. Bean[4] has seen a small but healthy tree at Scone Palace in Perthshire. As seen in the nursery at Kew and Coombe Wood, it is very slow and feeble in growth, and apparently is not suited to the English climate, young plants usually having very small leaves and short shoots. There are, however, three flourishing young trees at Bayfordbury, which were obtained from Hesse's nursery at Weener, in Hanover. It seems to do very much better in America, where I saw a vigorous tree[5] growing at Mr. Hunnewell's pinetum at Wellesley, Massachusetts. Reputed trees of *Abies Mariesii* usually turn out on examination to be *Abies Veitchii*. (H. J. E.)

[1] *Hortus Veitchii*, 336 (1906).      [2] *Forest Flora of Japan*, 82 (1894).
[3] But Prof. Miyabe told me in 1904 that he had seen no specimens from this place, and doubted its occurrence in Hokkaido. He had specimens in his herbarium from Nambu near Morioka.      [4] *Gard. Chron.* xli. 117 (1907).
[5] Reported to be 9 feet high, by Sargent, in *The Pinetum at Wellesley in 1905*, p. 13.

## ABIES GRANDIS, GIANT FIR

*Abies grandis*, Lindley, *Penny Cycl.* i. 30 (1833); Masters, *Gard. Chron.* xv. 179, ff. 33-36 (1881) xvii. 400 (1882), and xxiv. 563, ff. 128-131 (1885), and *Journ. Linn. Soc. (Bot.)*, xxii. 174 (1886); Sargent, *Silva N. Amer.* xii. 117, t. 612 (1898), and *Trees N. Amer.* 60 (1905); Kent, Veitch's *Man. Coniferæ*, 510 (1900).
*Abies Gordoniana*, Carrière, *Conif.* 298 (1867).
*Abies amabilis*, Murray, *Proc. Roy. Hort. Soc.* iii. 310 (1863) (not Forbes).
*Pinus grandis*, Hooker, *Fl. Bor. Amer.* ii. 163 (1839).
*Picea grandis*, Loudon, *Arb. et Frut. Brit.* 2341 (in part) (1838).

A tree attaining in America in the coast regions 300 feet in height and 16 feet in girth; but on the mountains of the interior rarely more than 100 feet high by 6 feet in girth; often smaller and stunted at high elevations. Bark of young trees smooth, thin, and pale; of older trees in America, brownish, divided by shallow fissures into low flat ridges roughened by thick appressed scales; in cultivated trees fissuring into thin irregular plates, exposing the reddish brown cortex. Buds small, conical, obtuse at the apex, resinous, roughened by the raised tips of the scales. Young shoots olive-green, smooth, with a minute, erect, not dense pubescence.

Leaves on lateral branchlets pectinate, in two lateral sets in the horizontal plane, each set of apparently two ranks, the upper rank with leaves about half the length of those below. Leaves linear, flattened, up to about $1\frac{1}{2}$ to 2 inches long, $\frac{1}{10}$ to $\frac{1}{12}$ inch in width, narrowed at the base, uniform in breadth elsewhere, with a rounded and bifid apex; upper surface dark green, shining, with a continuous median groove and without stomata; lower surface with two white bands of stomata, each of about eight lines; resin-canals marginal. Leaves on cone-bearing branches crowded, less spreading or nearly erect, blunt or bifid at the apex, shorter than on sterile branches.

Cones 2 to 4 inches long by 1 to $1\frac{1}{4}$ inch in diameter, cylindrical, slightly narrowed towards the rounded or retuse apex, bright green in colour, with the bracts concealed. Scales resembling those of *Abies Lowiana*, but smaller. Bract situated a little above the base of the scale, quadrangular; upper margin broad, denticulate, deeply emarginate, and with a minute mucro. Seeds $\frac{3}{8}$ inch long, light brown, with pale shining wings about $\frac{5}{8}$ inch long.

### IDENTIFICATION

*Abies grandis* is readily distinguished by the very flat pectinate arrangement of the leaves; those of the upper rank being about half the length of those in the lower rank. *Abies Lowiana*, when growing feebly, resembles it somewhat in arrangement; but in this species the upper surface of the leaves has stomatic lines, absent in *A. grandis*, and the leaves in the upper rank are only slightly shorter than those in the lower rank. (A. H.)

## DISTRIBUTION

On the north-west coast of America this magnificent tree has a wide range, from Vancouver Island, where it grows at low levels and is not, so far as I saw, a conspicuous feature in the forest; through Washington and Oregon as far south as Mendocino County in California, where it does not extend far from the coast, and grows in company with *Sequoia sempervirens* and *Picea sitchensis*.

Inland it is less abundant on the eastern slopes of the Cascade Mountains, but extends to the Cœur d'Alene and Bitter-root Mountains of Idaho and Montana. In the Flathead Lake Country it is a comparatively small tree, attaining only 12 to 15 inches in diameter, and ascending to about 3500 feet.

It reaches its maximum development in a damp climate and in sheltered valleys, where I have measured trees much over 200 feet in height, and where, according to Sargent and Sheldon, it sometimes reaches as much as 300 feet. So far as I saw, and Sargent confirms this observation, it never grows gregariously, but scattered among other species; and rarely forms an important element in the timber.

It is easy to recognise when young by the flat arrangement of the leaves, but when its branches are far above one's head I could not distinguish it from *A. amabilis* in the Cascade mountains, or from *A. Lowiana* which seems to take its place in southern Oregon and northern California.

It grows very fast in its own country, a specimen measured at 2500 feet altitude on the Cascade Mountains being 140 feet by 16 feet on the stump, at only 106 years old. Though the timber is not much valued by lumbermen, it is used for various purposes locally, and, according to Sheldon, makes the most durable shakes—a name used for large shingles cleft with the axe—used in Oregon.

The tree figured (Plate 218) was growing in 1904 on Swallowfield farm, about fifty miles north of Victoria, in Vancouver Island, and when I measured it, was 215 feet by 19 feet.

*Abies grandis* was discovered on the Columbia river by Douglas in 1825, though he does not seem to have sent seeds to the Horticultural Society until 1831 or 1832. Very few of these germinated, and it is doubtful if any of the original seedlings are still living.[1] The next consignment[2] of seed was sent by William Lobb in 1851 to Messrs. Veitch at Exeter; and about the same time seeds were received by the Scottish Oregon Association from their collector Jeffrey.

## REMARKABLE TREES

There are many fine trees of this species in the warmer and moister parts of England, Scotland, and Ireland; and, with *A. nobilis* and *A. Lowiana*, it seems best suited of all the American firs to our climate.[3]

---

[1] Murray, *Proc. Roy. Hort. Soc.* 311 (1863), states that there were then living no authentic seedling specimens of *A. grandis* raised from the seeds sent by Douglas, but a multitude of young plants existed which had been raised from cuttings.

[2] *Hortus Veitchii*, 336 (1906).

[3] The most vigorous of all the genus. Thrives admirably on gneiss, free from all trace of disease, is not susceptible to frosts or Chermes, and as a shade bearer has no equal amongst silver firs. Produces timber, which is white and

The tallest that I have seen in England grows in Oakly Park near Ludlow, the property of Lord Plymouth, on the rich flat by the river Teme, and measured 102 feet by 8½ feet in 1908. Other fine trees are at Fonthill Abbey, which was 98 feet by 8 feet in 1906; and at Madresfield Court[1] and Eastnor Castle, both of which are over 95 feet high and 7½ feet in girth. The latter is figured (Plate 217). I have seen several others over 90 feet, of which perhaps the one at Heanton Satchville is the largest, though it is too spreading to be a typical specimen. In 1903 it was about 94 feet by 9 feet 7 inches, and 56 yards in circumference of the branches. At Castlehill there are some fine trees, one of which measured 92 feet by 7 feet 10 inches in 1904. At Petworth there is a very tall but not a well-grown tree, 94 feet by 6 feet 6 inches. At Eridge Park a very handsome tree, planted by Mr. Disraeli in 1868, measures 76 feet by 6½ feet. At Youngsbury, Ware, in Herts, Mr. H. Clinton Baker measured a tree in 1907, which was 91 feet in height and 9 feet 8 inches in girth; and at his own place, Bayfordbury, there is another, 73 feet by 5 feet 9 inches in 1905. He also reports two good trees at The Heath, Leighton Buzzard, 98 feet by 8 feet 6 inches, and 88 feet by 7 feet 10 inches respectively. There is also a very large tree in a belt by the road at Flitwick Manor, near Ampthill, Bedfordshire, the seat of Miss Brooks, which is 95 feet by 10 feet. At Welford Park there are two trees which though only planted in 1878, are now about 90 feet high by 7 feet in girth.

At Barton a thriving tree, 68 feet by 4 feet 3 inches, is the best we know in the eastern counties, and this is sheltered and drawn up by tall trees around it.

At Golden Grove, Carmarthenshire, there is a fine tree which in 1892 was 60 feet by 7 feet 8 inches, and when I saw it in 1905 had increased to 80 feet.

In Scotland there are many fine specimens, of which one at Riccarton, in Midlothian, was reported at the Conifer Conference in 1891 to have been 83 feet 3 inches high and only 3 feet 8½ inches in girth, as carefully measured by the owner, Sir James Gibson Craig; and stated by him to have grown 53 feet in twelve years. Soon after this it was attacked by Chermes and was cut down.

At Glenlee, near New Galloway, Mr. T. R. Bruce informs me that there is a tree, planted by Mrs. Melville in 1864, which in 1905 measured no less than 95 feet by 10 feet, though, having lost its leader four years previously, it has now three leads. At Castle Kennedy this species grows much faster than any of the other numerous firs planted there. In 1904, one of two trees, nearly equal in size, was 78 feet high by 6 feet in girth. This tree[2] was only twelve years old in 1891, when it measured 30 feet by 1 foot 7 inches.

At Benmore, in Argyllshire, one of the wettest places in Scotland, a tree said to be only thirty-five years planted was, in 1907, 80 feet by 7 feet 4 inches; but the trunk was infested with scale and did not seem to be healthy when I saw it. At Poltalloch, there is a fine specimen over 80 feet high, and at Inveraray and

somewhat soft, in great volume, and which is found useful in connection with box-making and other industries in Aberdeen. Specially adapted for cultivation for profit where a large volume of timber is a desideratum.—(J. D. Crozier.)

[1] A note signed J. N. in the *Trans. Scot. Arb. Soc.* xx. 126 (1907) states that this tree, in Sept. 1906, was 114 feet by 8 feet 4 inches. When I measured it in 1904, I made it 96 feet by 7½ feet; and though owing to the ground I could not get a level base line, I can hardly believe that it is now so tall as stated.

[2] *Journ. Roy. Hort. Soc.* xiv. 547 (1892).

Ardkinglas, in the same county, are trees over 70 feet, which in that wet climate flourish exceedingly.

In the Keillour Pinetum, near Balgowan, in Perthshire, a tree growing in boggy soil was, in 1904, 90 feet high by 7 feet 3 inches. It is not well furnished above, and is perhaps beginning to suffer from the nature of the soil. At Keir, Dunblane, there is a tree which in 1904 measured 82 feet high by 9 feet 3 inches in girth. This tree[1] was twenty-eight years old in 1891, and then measured 55 feet by 4 feet 2 inches. At Abercairney, Perthshire, there is a fine tree, which in 1904 was 91 feet by 8 feet 4 inches. This tree[2] was about thirty years old in 1891, and then measured 58 feet by 4 feet 6 inches.

At Durris, Aberdeenshire, there is a good tree, which Mr. Crozier measured in 1904 as 82 feet high by 9 feet 6 inches in girth. When I saw it in 1907 it had increased to nearly 90 feet.

The largest tree in Ireland was formerly at Carton, which was reported in 1891 to be 80 feet high by 6 feet in girth. The top was blown off by the gale of February 1903, and when seen by Henry in the autumn of that year, the tree measured 67 feet by 9 feet 6 inches. At Kilmacurragh, Co. Wicklow, a fine specimen was, in 1906, 86 feet by 7 feet 2 inches ; and at Coollattin, in the same county, another measured 63 feet by 6 feet 4 inches. At Powerscourt I measured one in 1903 which was about 87 feet by 7½ feet.

*Abies grandis* thrives very well in north-western Germany, and according to Count Von Wilamitz-Möllendorf[3] grows at Gadow faster than any other silver fir, a specimen figured being 25 metres by 1.40 metre when only twenty-five years old. It also succeeds in some parts of Denmark, where Hansen[4] states that a specimen planted in 1864 had attained, in 1891, 53 feet by 6 feet.    (H. J. E.)

[1] *Journ. Roy. Hort. Soc.* xiv. 531 (1892).
[2] *Ibid.* 527.
[3] *Mitt. D. Dendr. Ges.*, 1907, p. 138.
[4] *Journ. Roy. Hort. Soc.* xiv. 469 (1892).

## ABIES CONCOLOR, Colorado Fir

*Abies concolor*,[1] Lindley and Gordon, *Journ. Hort. Soc.* v. 210 (1850); Masters, *Journ. Linn. Soc.* (*Bot.*) xxii. 177, ff. 8-11 (1886), and *Gard. Chron.* viii. 748, ff. 147, 148 (1890); Sargent, *Silva N. Amer.* xii. 121, t. 613 (1898) (in part), and *Trees N. Amer.* 62 (1905) (in part); Kent, Veitch's *Man. Coniferæ*, 501 (1900).

*Picea concolor*, Gordon, *Pinetum*, 155 (1858).

*Picea concolor*, var. *violacea*, Roezl, *ex* Murray, *Gard. Chron.* iii. 464 (1875).

*Pinus concolor*, Engelmann, *ex* Parlatore, in DC. *Prod.* xvi. 2, p. 427 (1868).

A tree attaining in America 100 to 125 feet in height, with a girth of 9 feet. Bark of old trees fissuring into small irregular plates. Buds, much larger than those of *A. Lowiana*, broadly conical, rounded at the apex, brownish, resinous, and slightly roughened by the raised tips of the scales. Young shoots smooth, yellowish-green, with a minute scattered pubescence, variable in quantity and often absent from the greater part of the branchlet. Second year's shoot greyish and irregularly fissuring.

Leaves on lateral branchlets irregularly arranged and not truly pectinate; most of the leaves extending laterally outwards and curving upwards, a few on the lower side directed downwards and forwards, some on the upper side directed upwards and forwards; those above shorter than those below. Leaves up to 2 to 3 inches long, $\frac{1}{12}$ inch broad, glaucous on both surfaces, linear, flattened, slightly tapering at the base, uniform in width elsewhere; apex acute or rounded and not bifid, though occasionally a slight emargination is discernible with a lens; upper surface slightly convex, not grooved, with fifteen to sixteen regular lines of stomata; lower surface convex with two bands of stomata, each of about eight irregular lines, not conspicuously white; resin-canals marginal. Leaves on cone-bearing branches shorter, thicker, falcate, all curving upwards.

Cones, 3 to 5 inches long, $1\frac{1}{4}$ inch in diameter, cylindrical, narrowed at both ends, rounded or obtuse at the apex; greenish or purple before ripening, brown when mature. Scales of native Colorado specimens much broader than long; lamina about 1 inch wide by $\frac{1}{2}$ inch long, upper margin entire, lateral margins rounded and denticulate, gradually passing into the obcuneate claw or with a slightly auricled truncate base. Bract, at the base of the scale, rectangular, denticulate, with truncate upper margin and a minute mucro; in some specimens deeply bifid above. Seeds $\frac{1}{3}$ inch long, with broad shining pinkish wings, about $\frac{1}{2}$ inch long. In cultivated specimens, both brown and purple cones occur.

The following varieties have arisen in continental nurseries :—

1. Var. *falcata*, Beissner,[2] leaves sickle-shaped, curving upwards.

2. Var. *glabosa*, Beissner,[2] globose in habit, with symmetrical short branches.

---

[1] According to the view taken here, *Abies concolor* includes only the tree found in Colorado, Utah, and Southern California. Sargent and other American botanists combine with this species the tree found in the Californian Sierras, which is considered by us to be a distinct species, *A. Lowiana*. The two forms differ remarkably in buds and foliage; and it is most convenient to regard them as distinct species.      [2] *Mitt. Deut. Dendr. Ges.* 1905, p. 112.

3. Var. *aurea*, Beissner,[1] young foliage golden yellow, gradually changing to a silvery grey colour.

4. Var. *brevifolia*, Beissner,[1] leaves short, thick, obtuse, twice as broad as in the typical form.

## DISTRIBUTION

*Abies concolor* occurs in the Rocky Mountains of southern Colorado and extends southwards over the mountains of New Mexico and Arizona into northern Mexico, being the only silver fir in the arid regions of the Great Basin and of southern New Mexico and Arizona. It occurs also in Utah in the Wasatch Mountains, and in southern California, in the San Bernardino and San Jacinto Mountains. It is accordingly confined to dry regions, while *Abies Lowiana*, which is in all probability only a geographical form of it, occurs in the more rainy regions of the Sierra Nevada of California and the southern mountains of Oregon. According to some opinions, the three species, *Abies grandis*, *Abies Lowiana*, and *Abies concolor* are only geographical forms of one large species.

Sargent says, of *Abies concolor*, that it endures heat and dryness best of all the silver firs of North America, and its distribution is accordingly more southerly than that of the other species, which occur in the United States.

## HISTORY AND CULTIVATION

This species was discovered by Fendler, near Sante Fé, in 1847, and was first clearly described by Parlatore, who adopted for it Engelmann's MS. name, *Pinus concolor*. It does not appear to have been introduced[2] into cultivation until about 1872. Syme mentions[3] two-year-old seedlings of it as a new species in 1875. Roezl, apparently in 1874, sent specimens and seeds, which were labelled *Picea concolor violacea*,[4] from New Mexico to Messrs. Sanders and Co., St. Albans. This species has been much confused with *A. Lowiana*, which was introduced considerably earlier. It is probable that there are no trees of true *A. concolor* in cultivation, older than 1873 or 1874.

*Abies concolor*, according to Sargent, is the only American silver fir, which is really successful in cultivation in the eastern part of the United States, where it grows better than *A. Lowiana*.

We have seen few trees of large size, though one at Highnam Court, Gloucestershire, of no great age, was 44 feet by 2 feet 9 inches in 1908.

It is less common in cultivation than *A. Lowiana*, which it much excels in beauty of foliage. Mr. Crozier says that young trees growing at Durris are quite healthy.                                                                  (A. H.)

---

[1] *Mitt. Deut. Dendr. Ges.* 1906, p. 144.

[2] Roezl sent a few seeds in 1872. Cf. Lavallée, *Nouveaux Conifères du Colorado et de la Californie*, in *Journ. Soc. Cent. Hort. France*, viii. (1875).          [3] *Gard. Chron.* iii. 563 (1875).          [4] *Ibid.* 464.

## ABIES LOWIANA, Californian Fir

Abies Lowiana, A. Murray, *Proc. Roy. Hort. Soc.* iii. 317 (1863).
Abies *lasiocarpa*, Masters (not Nuttall or Murray), *Gard. Chron.* xiii. 8, f. 1 (1880).
Abies *grandis*, Lindley, var. *Lowiana*, Masters, *Journ. Linn. Soc. (Bot.)*, xxii. 175, ff. 6, 7 (1886).
Abies *concolor*, Sargent, *Silva N. Amer.* xii. 121 (1898), and *Trees N. Amer.* 62 (1905) (in part).
Abies *concolor*, Lindley and Gordon, var. *lasiocarpa*, Beissner, *Handb. Conif.* 71 (1887).
Abies *concolor*, Lindley and Gordon, var. *Lowiana*, Lemmon, *W. Amer. Cone-Bearers*, 64 (1895);
    Kent, Veitch's *Man. Conif.* 502 (1900).
Picea *Lowiana*, Gordon, *Pinet. Suppl.* 53 (1862).
Picea *Parsonsiana*, Barron, *Catalogue*, 1859, and *Gard. Chron.* v. 77 (1876).
Pinus *Lowiana*, M'Nab, *Proc. R. Irish Acad.* ii. 680 (1877).

A tree, attaining on the Californian Sierras 200 to 250 feet in height, with a trunk often 18 feet in girth. Bark in cultivated specimens as in *A. concolor*; in wild trees becoming, near the ground, on old trunks, very thick and deeply divided into broad, rounded, scaly ridges. Buds ovoid, blunt at the apex, brownish, resinous, roughened by the raised tips of the scales. Young shoots yellowish green, smooth, covered with a minute scattered pubescence.

Leaves on lateral branchlets pectinately arranged, each lateral set of about two ranks, directed almost horizontally outwards, or curving upwards and outwards, so as to assume above a V-shaped arrangement. None of the leaves are directed irregularly in the middle line; and those of the upper rank are only slightly shorter than those of the lower rank. Leaves, up to $2\frac{1}{2}$ inches long, about $\frac{1}{12}$ inch broad, linear, flattened, slightly tapering at the base, uniform in width elsewhere, rounded and bifid at the apex; upper surface with a wide median furrow, usually not continued to the apex, and with eight lines of stomata in the furrow; lower surface with two white bands of stomata, each of eight to nine lines; resin-canals, marginal. Leaves on cone-bearing branches, upturned.

Cones, according to Sargent, not distinguishable from those of *Abies concolor*. Wild specimens, however, from California slightly differ, in having larger scales and broader bracts. Cultivated specimens in England bear cones which are chestnut-brown, and apparently never purple, as is often the case in *Abies concolor*.

### Identification

*Abies Lowiana* is regarded by Sargent and other American botanists as a form of *A. concolor*. As seen in cultivation it is very distinct from that species: moreover, it has a different distribution in the wild state. We have kept it separate, as being more convenient to cultivators.

In practice it can only be confused with *A. grandis*, and true *A. concolor*. The characters distinguishing it from *A. grandis* are given under this species on p. 773.

In *A. concolor* the arrangement of the leaves is irregular, not being truly pectinate. Many of the leaves in the middle line, both above and below, are not directed outwards, but point forwards parallel to the axis of the branchlet. In *A. concolor* the leaves are entire at the apex, and their convex upper surface shows sixteen lines of stomata, and is without a groove; whereas, in *A. Lowiana*, the apex of the leaves is bifid, and their upper surface is grooved, showing eight lines of stomata. The buds are smaller in the latter species. (A. H.)

## DISTRIBUTION

*Abies Lowiana* is found on the Siskiyou Mountains in southern Oregon, and on Mt. Shasta and the Sierra Nevada ranges in California. Its northern limit is the dry interior of southern Oregon, near the divide between the headwaters of the Umpqua and Rogue rivers, which, according to Sargent, is the real northern boundary of the Californian flora.[1] With *Abies magnifica* it forms in great part one of the principal forest belts on the west slope of the Sierra Nevada Mountains for 450 miles, and extends from 4000 to 9000 feet above sea-level. Here I saw it on my way into the Yosemite Valley in 1888, but did not then measure any trees. I found it in September 1904 in company with *A. magnifica, Pinus ponderosa*, and *Pinus Lambertiana* abundant on Mount Shasta, from about 3000 to 6000 feet; and here it was of moderate size, the largest that I measured being 140 feet by 11 feet 8 inches. It attains, however, 200 to 250 feet on the Sierra Nevada, and as much as 200 feet in Oregon.

## HISTORY AND CULTIVATION

*Abies Lowiana* was introduced from the Sierra Nevada of California by William Lobb in 1851; and about the same time seeds were sent from southern Oregon by John Jeffrey, who collected for the Scottish Oregon Association. The plants raised from Lobb's seeds were distributed by Messrs. Veitch of Exeter as *Picea lasiocarpa*, while those raised in Scotland from Jeffrey's seeds were distributed as *Picea grandis*.[2]

Messrs. Parsons of Flushing, United States, received seeds from California in 1853; and plants raised from these were imported to England in 1855 by Messrs. Low of Clapton. These passed into commerce as *Picea Parsonsiana*, a name which first appeared in Barron's Catalogue in 1859, and as *Picea Lowiana*, the name given by Gordon in 1862.

Of all the western silver firs this seems to be the most accommodating to the varied conditions of England, growing well on soils where *A. nobilis* will not thrive, and in a drier climate than *A. grandis* prefers. It is usually grown under the name of *A. lasiocarpa*, in the pineta which I have visited, and generally seen in good health and with a symmetrical top; as it is not so liable to become stunted by the production of cones as *A. nobilis*.

According to Sargent, the Californian form of *A. concolor* grows in the eastern

---

[1] The fir named *A. concolor* by Plummer in his Report on the Mt. Rainier Forest Reserve, p. 101 (Washington, 1900), is evidently *A. grandis*, which he does not mention, and all his references to white fir no doubt relate to that species.

[2] Cf. *Hortus Veitchii*, 39, 335 (1906).

States with less vigour and rapidity than the Colorado form; but is equally hardy, and has attained 40 to 50 feet in height in New England.

## REMARKABLE TREES

Among the numerous trees that I have measured I find it difficult to say which is the finest specimen. The one at Linton Park was the largest recorded at the time of the Conifer Conference, when it was 64 feet by 8 feet 7 inches. In 1902 I found it to be 85 feet by 10 feet 6 inches, a great increase in ten years (Plate 219).

There is, however, a tree at Fonthill Abbey which I believe to be *A. Lowiana*, though I could not reach the branches in order to identify it, which, in 1906, measured 90 feet by 6½ feet, and resembled, by its short branches, the typical habit of *A. magnifica*.

At Highnam Court, Gloucester, there is a fine specimen which was figured by Kent; according to Major Gambier Parry, it measured 77 feet by 9 feet 2 inches in 1906. I made it 80 feet by 9½ feet in 1908. Another at Eastnor Castle is about 88 feet by 7 feet 4 inches.

In Herts there are several good trees, one at Essendon Place being 82 feet high by 5 feet 9 inches in 1907; another at Youngsbury, Ware, which was planted in 1866, being 68 feet by 5 feet 6 inches in the same year; and a third at Bayford-bury, which was 69 feet by 6 feet 9 inches in 1905.

A very remarkable specimen, narrow and almost columnar in habit, which was planted twenty-six years ago, was seen by Henry at Crowsley Park, Oxfordshire, the seat of Colonel Baskerville, and measured, in 1907, 71 feet by 6 feet.

In Wales there is a very fine tree at Hafodunos, which Henry measured, in 1904, as 87 feet by 7 feet 9 inches; and I saw one at Glanusk Park in Breconshire, which was over 80 feet high in 1906.

In Scotland this species is not usually so large as in the south, though it grows well even in the west, where I have seen good trees at Inveraray and Poltalloch; and in the reports of the Conifer Conference it is generally described as thriving, and several trees of 40 to 50 feet high are mentioned. The largest we have heard of is at Abercairney, mentioned by Mr. Bean[1] as 65 feet by 5 feet.

In Ireland the tree does not appear to have been often planted, and the largest reported at the Conifer Conference in 1891 was growing at Abbeyleix in Queen's County, and measured 45 feet by 6 feet 10 inches. At Coollattin, Co. Wicklow, another was, in 1906, 52 feet by 4 feet 9 inches; and, in the same year, a fine specimen at Castlewellan, Co. Down, measured 67 feet in height and 9 feet in girth. (H. J. E.)

---

[1] *Kew Bulletin*, 1906, p. 258, and *Gard. Chron.* xli. 168 (1907), where it is named through inadvertence *A. concolor*.

## ABIES AMABILIS, Lovely Fir

*Abies amabilis*, Forbes, *Pinet. Woburn.* 125, t. 44 (1840); Masters, *Journ. Linn. Soc. (Bot.)* xxii.
171, t. 2 (1886), and *Gard. Chron.* iii. 754, f. 102 (1888); Sargent, *Silva N. Amer.* xii. 125,
t. 614 (1898), and *Trees N. Amer.* 59 (1905); Kent, Veitch's *Man. Conif.* 489 (1900).
*Abies grandis*, Murray, *Proc. Roy. Hort. Soc.* iii. 308 (1863) (not Lindley).
*Pinus amabilis*, Douglas, *Comp. Bot. Mag.* ii. 93 (name only) (1836); Antoine, *Conif.* 63
(1846).
*Pinus grandis*, Don, in Lambert, *Pinus*, iii. t. (1837).
*Picea amabilis*, Loudon, *Arb. et Frut. Brit.* iv. 2342 (in part) (1838).

A tree sometimes attaining in America 250 feet in height and 18 feet in girth,
but at high altitudes and in the north usually not more than 80 feet.   Bark thin,
smooth, pale or silvery white; becoming, on very old trunks, thick near the ground
and irregularly divided into small scaly plates.   Buds small, globose, resinous,
smooth, with purple scales all immersed in the resin, except occasionally two or
three, small and keeled, at the base of the bud.   Young shoots grey, smooth,
densely covered with short, loose, wavy pubescence.

Leaves on lateral branches arranged as in *A. Nordmanniana*, up to $1\frac{1}{4}$ to $1\frac{1}{2}$ inch
long by $\frac{1}{14}$ inch broad, fragrant, linear, flattened, gradually tapering from the
middle to the base, slightly broader in the anterior half, with a truncate and bifid
apex; upper surface very dark green and lustrous, with a continuous median groove
and without stomata; lower surface with two broad white bands of stomata, each of
eight to ten lines; resin-canals marginal.   Leaves on vigorous leading shoots acute
with long rigid points, closely appressed or recurved near the middle.   Leaves on
cone-bearing branches upturned, acute or acuminate.

Cones ovoid-cylindric, slightly narrowing to the rounded apex, dark purple
when growing, brown when mature; $3\frac{1}{2}$ to 6 inches long by 2 to $2\frac{1}{2}$ inches in
diameter.   Scales, 1 to $1\frac{1}{8}$ inch wide, nearly as long as broad, inflexed at the upper
rounded margin, gradually narrowing towards the base.   Bracts rhombic or obovate-
oblong; lamina situated just above the base of the scale and ending in a long
acuminate tip, which reaches half the height of the scale.   Seeds light yellowish
brown, $\frac{1}{2}$ inch long, with oblique pale brown shining wings about $\frac{3}{4}$ inch long.

*Abies amabilis* resembles *A. Nordmanniana* in the arrangement and size of the
leaves; but is readily distinguished from it by the small globose resinous buds.
The leaves are also much darker, shining above, more truncate at the apex; and
emit, especially when bruised, a strong fragrant odour which resembles that of
mandarin orange peel.                                                    (A. H.)

### DISTRIBUTION AND HISTORY

*Abies amabilis* occurs on mountain slopes and terraces from British Columbia
southward along the Cascade Mountains to northern Oregon, and on the
coast ranges of Oregon and Washington.   According to Sargent, it attains its
largest size on the Olympic Mountains, where it is the most common silver fir,

extending from 1200 feet up to timber line at about 4500 feet, and forming, with the Western Hemlock, a large. part of the forest between 3000 and 4000 feet. In the Cascade Mountains it extends south to about 20 miles north of Crater Lake where Mr. Coville found it on the east side of Diamond Mountain. It occurs[1] in the extreme south-eastern end of Alaska, at the Boca de Quatre inlet, ranging from sea-level to 1000 feet altitude; but has not yet been found between this point and the northern end of Vancouver Island. It is the common fir[2] in south-western Vancouver Island, where it grows abundantly from sea-level up to the summits of the highest mountains. Near the sea it often forms groves of almost pure growth, the trees standing close together and having very tall slender trunks, about 3 feet in diameter at the base, and often unbranched to a height of 100 feet or more. At an altitude of 3000 feet it is a comparatively small tree, often clothed with branches to the base. Plate 220, taken from a photograph, for which I am indebted to Mr. J. M. Macoun of the Geological Survey of Canada, shows the tree as growing near Kamloops, in British Columbia.

Sargent says, "unsurpassed among fir trees in the beauty of its snowy bark, dark green lustrous foliage, and great purple cones, *Abies amabilis* can never be forgotten by those who have seen it in the alpine meadows covered with lilies, dog's-tooth violets, heaths, and other flowers which make the valleys of the northern Cascade Mountains the most charming natural gardens of the continent."

Engelmann in a letter, dated "Portland, Or., August 6," 1880, and quoted in *Gardeners' Chronicle* of December 4, 1880, says of it :—"*A. amabilis*, on the same mountain where Douglas discovered it, just south of the Cascades of the Columbia, is a magnificent tree, at about 4000 feet, attaining 150 to 200 feet high with a trunk 4 feet in diameter, branching to the ground and forming a perfect cone. The bark of old trees is 1½ to 2 inches thick, furrowed and reddish grey, that of younger trees, less than 100 years, is quite thin and smooth, light grey or almost white. It is certainly very closely allied to *A. grandis*, but readily distinguished by its very crowded dark green foliage and its large dark purple cones. It has the purple cones and sharp-pointed leaves (on fertile branches) of *A. subalpina*, but this latter has much smaller cones, and not such crowded leaves."

Though I saw this tree in abundance on Mount Rainier I cannot say that I know how to distinguish it in the forest from *A. nobilis* without the leaves and cones. It has, according to Plummer, a wider range of elevation than that species, and grows from 800 up to 5500 feet. The cone is as large as that of *A. nobilis*, but without the projecting bracts. From *A. lasiocarpa*,[3] with which it was mixed in the upper part of its range, it is distinguished by its habit, which is much less slender and spiry, by its greater size, and by its cones, which are nearly twice as large. Plummer says that it attains 200 feet in height by 15 feet in girth, but I saw none so large as this that I could identify. It is a slow-growing tree, one 20 inches in diameter having 288 rings.

[1] *U.S. Forest Service, Sylvical Leaflet*, 22, p. 1 (1908). Its most southerly point in the coast range is Saddle Mountain, 25 miles south of the mouth of the Columbia River.

[2] Cf. Butters, *Conifers of Vancouver Island*, in *Postelsia*, 187 (St. Paul, Minn., 1906).

[3] Sargent, *Silva*, xii. 126, *adnot.*, mentions the occurrence in a wild state of a hybrid between these two species.

As there were no cones on any of the firs on this side of Mount Rainier in 1904 I was unable to procure seed of either of these species, though Prof. Allen sent me both of them in 1905.

The wood is yellowish and can, according to Plummer, be distinguished from that of *A. lasiocarpa*, by its darker colour. It is soft and perishable, and of no commercial importance at the present time.

*Abies amabilis* was discovered in 1825 by Douglas on a high mountain south of the Grand Rapids of the Columbia River; but it was not until 1830 that he succeeded in sending to England seed, from which a few plants were raised in the garden of the Horticultural Society at Chiswick; and of these original trees hardly any now survive. For many years afterwards the tree was not seen by any traveller or collector; and seeds of reputed *A. amabilis* sent to Europe invariably turned out to be some other species; and much confusion resulted in the nomenclature of the western American silver firs. In 1880 the tree was re-discovered by Sargent in company with Engelmann and Parry, who found it on Silver Mountain near Fort Hope on the Fraser River; and a few days later Sargent himself observed it on the mountain where it had first been seen by Douglas. Large supplies of seed were sent from Oregon in 1882, and young trees are not now uncommon.

## REMARKABLE TREES

Of the original trees, those raised from seed sent home by Douglas, Kent knew only of two surviving in 1900, one at Dropmore and another at Orton Longueville. The latter, as I was told by Mr. Harding, was cut down in 1905, when it measured 5 feet 9 inches in girth.

The tree at Dropmore, which was received from the Royal Horticultural Society, and planted in 1835, was cut down four years ago. Mr. Page informs us that the trunk in the timber yard measured 36 feet long by 8 feet 4 inches in girth at 5 feet from the lower end. A cutting from the tree was raised in 1847 by the late Mr. Frost, and is now growing at Dropmore, and measures 50 feet high by 7 feet 3 inches. For a time, up to 1873, it promised to be a better tree than its parent; but it is now a miserable object, being badly affected by "knotty" disease.[1] This disease has attacked also all the young trees of this species at Dropmore, some fourteen or fifteen in number, which were planted a few years ago.

---

[1] Dr. Masters, in *Gard. Chron.* xvii. 812, xviii. 109, figs. 19, 20 (1882) states that Mr. Barron had proved the gouty swellings on branchlets of *A. amabilis* and *A. nobilis* to be due to a woolly aphis, and had succeeded in killing the pest, in his nursery at Borrowash, by applications of fir-tree oil. A petroleum emulsion is recommended in *Gard. Chron.* xxvii. 190 (1900). I am indebted to Prof. Borthwick of Edinburgh for a paper on the subject (in *Nat. Zeitschr. Forst. u. Landwirth-schaft*, 1908, p. 151, figs. 1-4) by Dr. E. Wolz, who states that these swellings are caused by a Chermes which Cholodkovsky has named *C. piceæ*, var. *Bouvieri*. The life-history of this insect does not seem to have been fully worked out; and it may not be identical with the *Chermes piceæ*, which attacks the bark of silver firs, and is said by Gillanders (*Forest Entomology*, 333) to be common in the nursery on young plants of *A. pectinata* and of *A. Nordmanniana*. The figures given by Wolz, however, of *Abies nobilis*, attacked by the disease, represent exactly the swellings which I have seen on that species at Carlisle, and which is present on most of the trees of *A. amabilis* in England. E. R. Burdon, in *Journal of Economic Biology*, 1908, ii. 132, states that Cholodkovsky's drawing looks more like the effect attributed to *Æcidium elatinum*. Cf. Hartig, *Diseases of Trees*, 180, fig. 109 (1894), who states that no formation of spores ever takes place on these swellings.

We have, however, found several other old trees, none of which are fine specimens, and may have been planted later.

A tree at Bayfordbury, with a broken top, is about 20 feet high. At Brickendon Grange, Herts, there is a remarkable specimen, only a foot in height, with long branches spreading over the ground for about 12 feet. This curiosity is probably very old; and its peculiar form is possibly due to the leader having been repeatedly bitten by animals.

At Pencarrow, Cornwall, a tree is growing, which I made in 1905 47 feet high by 7 feet 10 inches in girth. Mr. Bartlett, in a letter dated February, 1906, gives the following interesting particulars concerning this tree:—"According to Sir W. Molesworth's catalogue of the trees at Pencarrow, the *Abies amabilis* was planted in 1843. The soil is well-drained loam, and the tree stands in a sheltered position. For many years it was a strikingly beautiful specimen, quite symmetrical and feathered to the ground. A few years ago it was attacked by Chermes, and is now in a poor state and likely to be completely ruined by the disease in a few years. The tree bore a few cones near the top, four years ago; but these contained no good seeds. The cones were resinous, dark blue in colour when growing, fading to a dull brown towards autumn. The bark of the trunk and branches is covered with resin-blisters, which exude a liquid resembling golden syrup in colour and consistency. The buds are late in unfolding." Mr. Bartlett states that there is, at Lamellan, in north Cornwall, a perfectly healthy but stunted example of *Abies amabilis*, growing on very poor soil on the edge of a quarry. This tree was probably raised from a cutting of the Pencarrow tree. At Menabilly, in the same county, there is another tree, the flowers of which have been figured.[1] In 1908 it measured 37 feet by 3 feet 7 inches.

At Brocklesby Park, Lincolnshire, Henry measured in 1908 a tree, 50 feet by 4 feet, the date of planting of which is unknown. Though very healthy in general appearance, some of the lower branches are beginning to suffer from knotty disease. The bark is very smooth and covered with numerous resin blisters, differing markedly from the rough bark of an *A. Nordmanniana*, of the same size, growing beside it.

At Smeaton-Hepburn, in East Lothian, there is a tree,[2] which was planted in 1843; but its top was blown off in 1859, and it is now only 31 feet high, but has a girth of 8 feet 10 inches. It produced staminate flowers in 1886.

At Castle Kennedy, *Abies amabilis* takes on a low creeping bushy habit, possibly due to the plants being raised from cuttings, and I saw a similar dwarf stunted plant at Moncreiffe, which I believe to be *A. amabilis*.

On the whole this species appears to be a failure in cultivation, in Europe; and does not succeed any better in New England, where, according to Sargent,[3] it has proved rather tender and grows very slowly. (H. J. E.)

---

[1] *Gard. Chron.* iii. 755, f. 102 (1888).

[2] Cf. Sir Archibald Buchan-Hepburn's account in *Proc. Berwickshire Naturalists' Club*, xviii. 207, 210 (1904). It was 8 inches high at the time of planting, when it was supposed to be *A. grandis*.

[3] Sargent, in *The Pinetum at Wellesley in 1905*, p. 12, mentions a small healthy specimen, which was raised in the Veitchian nurseries near London, from seeds collected in Oregon by C. S. Pringle in 1882.

## ABIES NOBILIS, Noble Fir

*Abies nobilis*, Lindley, *Penny Cycl.* i. 30 (1833); Masters, *Gard. Chron.* xxiv. 652, f. 146 (1885),
and *Journ. Linn. Soc.* (*Bot.*) xxii. 188 (excl. habitat Mt. Shasta, and var. *magnifica*) (1886);
Sargent, *Silva N. Amer.* xii. 133, t. 617 (1898), and *Trees N. Amer.* 65 (1905); Kent,
Veitch's *Man. Coniferæ*, 521 (1900).
*Pinus nobilis*, Douglas, in *Comp. Bot. Mag.* ii. 147 (1836).
*Picea nobilis*, Loudon, *Arb. et Frut. Brit.* iv. 2342 (1838).

A tree, attaining in America occasionally 250 feet in height with a girth up to 24
feet, but more usually 150 to 200 feet high. Bark smooth on young trees, becoming on
old trunks reddish-brown and deeply divided by broad flat ridges, irregularly broken
by cross fissures and covered with thick closely appressed scales.

Buds concealed by the leaves at the tips of the branchlets, ovoid-globose;
terminal bud resinous above and surrounded at the base by a ring of lanceolate
acuminate or subulately pointed pubescent brown scales; lateral buds with ovate
basal scales. Young shoots smooth, densely covered with minute rusty brown
tomentum, which is retained in the second year.

Leaves on lateral branches pectinate below, extending outwards in the horizontal
plane in two lateral sets; above, the leaves in the middle line, much shorter, completely cover the shoot, from which they arise curving upwards, after being appressed
to the branchlet for a short distance near their bases, their tips usually having a
slight inclination forwards. Leaves up to about $1\frac{1}{4}$ inch long, $\frac{1}{16}$ inch wide, linear,
flattened, narrowed at the base, uniform in width elsewhere, rounded and entire at
the apex; upper surface with a continuous median groove and variable as regards the
stomata, which are sometimes in two definite bands each of six to eight lines or sometimes present as a few irregular lines, or rarely absent; lower surface with two
narrow bands of stomata, each of five to six lines; resin-canals marginal.

Leaves on cone-bearing branches all upturned, thickened, and with sharp
cartilaginous points.

Staminate flowers reddish. Pistillate flowers with broad rounded scales, much
shorter than the nearly orbicular bracts, which are erose in margin and contracted
above into slender elongated reflexed tips.

Cones cylindrical, but narrowing towards the full and rounded apex; 4 to 5
inches long by 2 inches in diameter on wild trees, 6 to 10 inches long by 3 to 4 inches
in diameter on cultivated trees; pubescent, purplish-brown with green bracts when
growing, the bracts becoming bright chestnut brown in the mature fruit. Scales:
lamina, $1\frac{1}{4}$ to $1\frac{1}{2}$ inch broad, 1 inch long, variable in shape; gradually narrowing to
the base with straight lateral margins, or rounded and denticulate on the sides above
the middle and contracted below; claw short, clavate. Bracts exserted and strongly
reflexed, covering the greater part of the scale next below; lamina, broad, full and
rounded above, fimbriate in margin, and with a conspicuous midrib prolonged into a
mucro about $\frac{1}{2}$ inch long; claw long and cuneate. Seeds pale brown, about $\frac{1}{2}$ inch

long, with similarly coloured obovate-cuneate wings, which in cultivated specimens are considerably longer than the body of the seed.

This tree can only be confused with *A. magnifica*, which has a different habit. The difference between these two trees in the shape and disposition of the leaves is given in the Key, p. 718. (A. H.)

## DISTRIBUTION

According to Sargent, this species forms extensive forests on the Cascade Mountains in Washington, extending southwards to the valley of the Mackenzie River, Oregon. It also occurs on the coast ranges of Washington, and the Siskiyou Mountains of California. It is most abundant on the western slopes of the Cascade Mountains, and ranges from 2500 to 5000 feet above sea level, attaining its largest size at 3000 to 4000 feet. It is less abundant and of smaller size on the northern and eastern slopes of these mountains. It commonly attains 200 feet in height; and often grows to 250 feet; Sheldon says, even to 300 feet.

In the Cascade Range Forest Reserve[1] the noble fir forms about 6 per cent of the total, and is an important element in the mixed forests of the middle zone on the western slope, where it often comprises 15 or 20 per cent of the forest. It crosses the summit in lat. 45° where a moist climate prevails, but cannot compete with pine and larch in the drier areas. It is closely associated with the lovely fir, and among lumbermen both species are called larch. Some individuals attain as much as 8 feet in diameter, but the average size is about 150 feet high by 12 feet in girth at the base. Langille states[2] that this tree cannot hold its own against the lovely fir (*A. amabilis*) and hemlock, which are superseding it, and that a sapling is seldom seen. A tree growing at 6000 feet elevation was 163 years old and 125 feet in height, with a diameter of 4 feet 5 inches at the base.

In the forests of Mt. Rainier in Washington, Plummer says that the noble fir is the finest timber tree and is found from 1800 to 5200 feet. The largest that he measured was 225 feet by 18 feet. But when I ascended this mountain from Longmire's Springs I did not see it, or perhaps I did not distinguish it in the absence of cones from *Abies amabilis*. In the watershed of the Washougal and Rock Creek rivers, however, which are very heavily timbered, it forms, according to Plummer, 25 per cent of the timber. The cones here measure about 4½ inches long by 2¼ inches wide, not so large as some I have seen in England.

I saw this tree at its best in the Cascade Mountains above Bridal Veil in northern Oregon in June 1904. In this district the tree is known to the lumbermen as larch, and grows in thick forest, more or less mixed with Douglas fir and hemlock; with *Acer circinatum* and other shrubs as underwood, where there is light enough for any to exist. The largest trees I saw here were above 200 feet in height, and were clear of branches for at least two-thirds of their height, as in the illustration, which was taken from a tree at this place which measured 210 feet by 13

[1] *Forest Conditions of the Cascade Range Forest Reserve, U.S. Geological Survey*, Washington, 1903.
[2] *Ibid.* p. 35.

feet. (Plate 221.) A stump close by showed 360 rings on a diameter of 4 feet, the first fifty being twice as wide as any of the later ones. I could find no seedlings of the noble fir in this part of the forest, and my guide said that he had seen none except at higher elevations.

The wood of this tree, though not of equal value to that of Douglas fir, is beginning to be more appreciated, and I saw it being cut up at the mill at Bridal Veil where the owner, Mr. Bradley, told me it was worth twenty to twenty-five dollars per 1000 feet, and was sent east to be used for the same purposes as white pine.

## HISTORY

This tree was discovered by David Douglas on the south side of the Columbia river in September 1825, and introduced by him five years later on his second journey. Ravenscroft,[1] after quoting Douglas's account of the collection of the seeds, which was published in the *Companion to the Botanical Magazine*, vol. ii. p. 130, says that the seeds arrived in good condition, and were successfully grown and distributed among the Fellows of the Royal Horticultural Society for whom at that time Douglas was working. "Extravagant prices were paid for the plants, fifteen and twenty guineas being then no unusual price." As it usually does, the demand called forth a supply, but for a long time this supply was in a great measure obtained by making grafts and cuttings from the older plants. Plants grown from this source, however, seldom have the same beauty as seedling trees.

The next importation was a small package of seed sent by Mr. Peter Banks, who was drowned soon after. After him Jeffrey sent a quantity to the Oregon Association, but not a plant came up, as the seeds had been destroyed in the cone by the larva of a hymenopterous insect, *Megastigmus pini*, and the same thing happened to the greater part of the seeds sent by William Murray and Beardsley. Afterwards Lobb and Bridges sent more consignments.

Ravenscroft says that plants raised from home-grown seeds are not so strong and healthy as those from imported seed, and have often died from a fungoid attack.

## CULTIVATION

Among the silver firs of North America none has had a greater success as an ornamental tree than this, but it is only after many years of cultivation that we are able to say with confidence, what are the conditions of soil under which it will preserve its beauty.

When first introduced it became so popular that seedlings could not be procured in sufficient quantity to supply the demand, and grafting was resorted to by nurserymen ; the silver fir being usually the stock selected. These trees grew well for a good many years, and some grafted trees are still thriving ; but the majority of

---

[1] In Lawson, *Pinet. Brit.* ii. 184.

them have shown a tendency to produce cones in such quantity and so prematurely, that the trees have ceased to produce a straight leader, and have often become unsightly and ragged. This applies specially to those which were planted on lawns or on pleasure-grounds, without much shelter.

An avenue of this tree was planted in 1868 at Madresfield Court, Worcestershire, with grafted trees of the glaucous variety from the Worcester nurseries. It was figured in Veitch's *Manual of Coniferæ*, ed. 2, p. 524. Though every care has been taken by top-dressing, and removing the cones to keep these healthy, they do not seem likely to remain so, as the lateral branches are, in many cases, covered with the knotty swellings described under *Abies amabilis*, p. 784, note 1.

Mr. W. E. Gumbleton of Belgrove, near Queenstown, tells me that many years ago when *Abies nobilis* was still scarce, the Duke of Leinster, whose tree was one of the first to produce cones, sold the seed of it for £40. The cones were artificially fertilised by shaking out the pollen from the male catkins at the foot of the tree, and dusting it from a ladder on the female flowers at the top.

It is often stated that this is one of the few silver firs which grows well on limestone, but my own experience disproves this, and I have never seen a really fine tree where there was much lime in the soil. A deep sand resting on rock or a hillside, where good drainage is combined with plenty of humus, seem to be the best conditions for the noble fir ; and if the glaucous variety, of which seedlings are difficult to obtain, is desired, I would graft it on *A. Nordmanniana*, which is usually a most vigorous grower, and endures spring frosts better than the common silver fir.

In woods the noble fir is often healthier than in the open, and in some cases has reproduced itself, though not abundantly. I have raised numbers of seedlings from grafted trees, but they were always sickly and died young on my soil, and in any case their growth is slow at first, six to ten years being required to produce trees fit to plant out. But in Scotland seedlings raised from home-grown seed are healthy and vigorous.

The tree is quite hardy in all parts of the country, even in the severe climate of upper Deeside, where at Balmoral it thrives well, and has endured several degrees below zero without injury.[1] It enjoys a fairly wet climate, but will also grow well in the drier parts of England if the soil is deep and cool.

### Remarkable Trees

The largest noble fir that I know of in England is at Tortworth, where, on a deep bed of sand sloping down to the lake, it had attained in 1901 a height of about 100 feet and 9 feet 6 inches in girth in forty-seven years from the date of planting.

[1] *A. nobilis*, one of the hardiest and best wind-resisting conifers in cultivation, thrives well on gneiss or granite, and may be planted on the most exposed sites. It is the most prolific of all silvers in seed bearing, and readily reproduces itself. Commercially it may be placed next to *A. grandis* amongst exotic firs. The timber, like all the west North American trees of the genus, is white, soft, and light, but closer in texture than *A. grandis*. Root formation ruined by frequent transplanting.— (J. D. Crozier.)

This tree has suffered to some extent from an attack of Chermes, with which the trunk was covered in 1903, but when I last saw it this had mostly disappeared. Lord Ducie had the tree accurately measured by a man climbing it in May 1908, and informs me that it was then 103 feet 9 inches high, by 9 feet 11 inches in girth. It was planted in 1854, and was 7 feet high in 1855 and 23 feet in 1864.

At Highnam Court, Gloucestershire, there is also a fine specimen in the pinetum, measuring 75 feet by 8 feet, but the trees here seem, as they do in many other places, to have nearly exhausted the soil they grow in, and are beginning to go off. At Miserden Park, the seat of A. Leatham, Esq., in the same county, there is an avenue of grafted trees on dry oolite soil, which were so laden with cones in the year 1900 that they have suffered much in consequence, though hitherto they have borne the exposed situation well.

At Chatsworth Mr. Robertson has measured a tree 85 feet by 8 feet 5 inches with a fine clean stem containing 195 cubic feet. At Walcot, the seat of the Earl of Powis, in Shropshire, I measured in 1906 a very fine glaucous specimen which, though grafted, was 86 feet by 10 feet 9 inches. At Beauport, Sussex, there is a tree, also grafted, 86 feet by 8 feet 1 inch in 1905. At Linton, Kent, there is a tree 90 feet by 8 feet 6 inches in 1902. At Barton there is a tree 80 feet by 7 feet, sheltered in a high wood, and growing fast.

In Fulmodestone Wood, on the Earl of Leicester's property, there is a tree 74 feet by 9 feet 6 inches, from which a self-sown seedling has sprung up, which at eleven to twelve years old was, in 1903, 3 feet 6 inches high ; another self-sown seedling in the same place was 20 feet high at about 23 to 25 years old.

At Sandringham there are two fine trees in a shrubbery near York House, the largest of which, in October 1907, measured 85 to 90 feet by 8 feet 10 inches.

At Twizell, Northumberland, once the property of Selby, the author of *British Forest Trees*, I saw in 1906 a tree 80 feet by 8 feet, the top of which, however, was damaged by wind.

In Wales it seems to thrive well both at Penrhyn and Hafodunos, in the north ; and at Dinas Mawddwy in Merionethshire, where in 1906 I measured a very flourishing tree 75 feet by 5 feet 8 inches.

In Scotland[1] it generally succeeds better than in England, and where it has sufficient shelter seems likely to attain a great size and age. By far the finest that I have seen, are some trees growing at the foot of a sheltered bank on deep sandy soil, in the Dolphin walk at Murthly, four of which certainly exceed 100 feet in height, and the tallest was, as nearly as I could measure it, from 105 feet to 110 feet by 7 feet 11 inches in September 1906.

A tree growing at Ballindalloch Castle, Banffshire, the seat of Sir J. Macpherson-Grant, is said to be the finest in the north of Scotland, and is stated to have measured in August 1907, 94 feet by 9 feet 11½ inches, and to be only forty-seven years planted.[1]

The next largest we have seen is at Keir,[2] Perthshire, which was, in 1905, 99

---

[1] *Trans. Roy. Scot. Arb. Soc.* xxi. 98 (1908).

[2] This tree was reported to be forty years old in 1891, and then measured 82 feet by 5 feet 8 inches (*Journ. Roy. Hort. Soc.* xiv. 531 (1892)).

feet high by 7 feet 5 inches in girth, remarkable for its clean stem and short branches occurring only on the upper half of the tree. Another tree at Keillour, in the same county, was 91 feet by 7 feet 1 inch in 1904; and at Castle Kennedy, Wigtonshire, another measured in the same year 80 feet by 7 feet 10 inches. Sir Archibald Buchan-Hepburn reports one at Smeaton-Hepburn, East Lothian, which measured, in 1908, 84 feet by 8 feet 10 inches. It was planted in 1843.

At Blair Castle, a tree planted by the Duke of Atholl about forty-two years ago was, in 1904, 70 feet by 5 feet; and at Balmoral, though of no great size, it seems to be the best of the silver firs, and has endured a temperature of $-15°$ without injury.

In Ireland, *Abies nobilis* thrives well. At Churchill, Armagh, there is a magnificent specimen, which in 1904 was covered with cones, and measured 73 feet by 8 feet 4 inches. At Curraghmore, Co. Waterford, a tree measured, in 1907, 75 feet in height by 10 feet in girth. At Powerscourt, a tree in 1903 was 59 feet by 6½ feet. At Carton, in the same year, a tree measured 61 feet by 6 feet. At Birr Castle, King's County, there is a very tall tree, which was reported[1] in 1891 to be 83 feet high and 6 feet in girth. There is an avenue of this species at Woodstock, Kilkenny; and good specimens are growing at Castlewellan in Down. In a plantation behind the old deer park at Castle Martyr, Co. Cork, there is a very large tree of the glaucous variety, which, though I could not measure the height accurately, seems to be about 75 feet high and is 10 feet 4 inches in girth. (H. J. E.)

[1] *Journ. Roy. Hort. Soc.* xiv. 557 (1892).

ABIES MAGNIFICA, Red Fir, Shasta Fir

*Abies magnifica*, A. Murray, *Proc. R. Hort. Soc.* iii. 318, ff. 25-33 (1863); Masters, *Gard. Chron.*
    xxiv. 652, f. 148 (1885); Sargent, *Silva N. Amer.* xii. 137, tt. 618, 619 (1898), and *Trees
    N. Amer.* 66 (1905); Kent, Veitch's *Man. Coniferæ*, 516 (1900).
*Abies nobilis*, Lindley, var. *magnifica*, Kellogg, *Trees of California*, 28 (1882); Masters, *Journ. Linn.
    Soc. (Bot.)* xxii. 189, t. 5, ff. 19-21 (1886).
*Abies shastensis*, Lemmon, *Garden and Forest*, x. 184 (1897).
*Picea magnifica*, Gordon, *Pinetum*, 219 (1875).
*Pinus magnifica*, M'Nab, *Proc. R. I. Acad.* ii. 700 (1876).
*Pinus amabilis*, Parlatore, in DC. *Prod.* xvi. 2, p. 426 (in part) (1868).

A tree, attaining in America 200 feet in height and 30 feet in girth. Bark,
buds, and branchlets similar in all respects to those of *Abies nobilis*.

Leaves on lateral branchlets arranged as in *A. nobilis*; but with the median
leaves above not so densely crowded as in that species, portions of the branchlet
being visible from above, whereas in *A. nobilis* the branchlet is completely concealed;
moreover, these median leaves are appressed to the branchlet at their bases for a
shorter distance than in the other species. Leaves longer than in *A. nobilis*, up to
about $1\frac{3}{4}$ inch long, $\frac{1}{16}$ inch wide, tapering gradually to the base, uniform in width
elsewhere; apex rounded, entire; obscurely quadrangular in section; upper surface
with a central ridge and several (often eight) rows of stomata; lower surface with
two bands of stomata, each of four to six lines; resin-canals marginal. Leaves on
leading shoots erect and acuminate, with long rigid points pressed against the stem.
Leaves on fertile branches much thickened, crowded, upturned, acute with short
callous tips.

Staminate flowers dark reddish. Pistillate flowers with rounded scales much
shorter than their oblong pale green bracts, which end in elongated slender tips.

Cones very large, 6 to 9 inches long, 3 to 5 inches in diameter, cylindrical, but
slightly narrowing to the rounded, truncate or retuse apex; purplish-violet when
growing, brown when mature, pubescent. Scales fan-shaped; lamina, $1\frac{1}{4}$ to $1\frac{1}{2}$
inch broad, 1 inch long, upper margin rounded and incurved, the sides gradually
narrowing to a cordate base; claw nearly $\frac{1}{2}$ inch long, narrowly obcuneate. Bracts,
in the usual form of the species, about two-thirds as long as the scale and not
exserted; variable in shape; upper expanded part oval, acute or acuminate,
terminated by a mucro; claw sharply contracted below the lamina. Seeds brownish,
more than $\frac{1}{2}$ inch long, slightly shorter than their pink obovate-cuneate wings.

Var. *shastensis*, Lemmon, *West. Amer. Cone-bearers*, 62 (1895); Sargent, *Silva N. Amer.* xii. 138, t. 620 (1898), and *Trees N. Amer.* 67 (1905).

> Var. *xanthocarpa*, Lemmon, *Third Report*, ex Masters, *Journ. Roy. Hort. Soc.* xiv. 193 (1892), and *Gard. Chron.* xli. 114, figs. 51, 52, 53 (1907).
>
> *Abies shastensis*, Lemmon, *Garden and Forest*, x. 184 (1897); Coville, *Garden and Forest*, x. 516 (1897).
>
> *Abies nobilis robusta*, Masters, *Gard. Chron.* xxiv. 652, f. 147 (1885) (not Carrière).
>
> *Abies nobilis*, Lindley, var. *magnifica*, Masters, *Journ. Linn. Soc.* (*Bot.*) xxii. 193, Pl. 5 (1886).

This differs from the type only in the cones, which have much longer bracts, yellow in colour, rounded or obtusely pointed (not acute), exserted, usually reflexed, and covering about half the outer surface of the scales.

This variety, which is known as the Shasta Fir, occurs on the mountains of southern Oregon, in the cross and coast ranges of northern California and on the southern Sierra Nevada. In Oregon it is met with in the lower parts of the mountains; but in the other localities it only occurs at very high elevations.

It is rare in cultivation in England, or at any rate has been rarely noticed. A tree at the Cranston Nursery, near Hereford, produced cones[1] of this kind in 1878, which were figured[2] by Dr. Masters. Another is growing at Durris Castle, Aberdeenshire, where Mr. Crozier states that intermediate forms between this and *A. nobilis* exist.

## IDENTIFICATION

This species is only liable to be confused with *A. nobilis*; but in large trees, as seen in cultivation, the difference in habit between the two species is remarkable. The formal arrangement of the branches in *A. magnifica*, though difficult to describe, when once seen can seldom be mistaken. The differences in the foliage are given in the Key, p. 718. (A. H.)

## DISTRIBUTION

The most northerly point at which this tree has been found is on the mountains east of Odell Lake in about lat. 44° N. in southern Oregon, where Dr. Coville collected it in 1897, many miles south of where *A. nobilis* occurs; and it is not mentioned among the trees of the Cascade Forest Reserve, so that it really belongs to the Californian rather than to the North Pacific flora. It becomes common on the Trinity Mountains, and on Mt. Shasta is the only fir besides *A. Lowiana*. The tree extends along the entire length of the western slope of the Sierra Nevada, from 6000 to 9000 feet above the sea, and extends to the eastern slope at high elevations.

The northern form has been separated by Lemmon under the name of *A. shastensis*, on account of the bracts which protrude from the scales; being in this respect, as in its geographical distribution, midway between *A. nobilis* and *A. magnifica*;

---

[1] *Gard. Chron.* 1878, p. 343.
[2] *Journ. Linn. Soc.* (*Bot*). xxii. 193, plate v. (1886).

but this character is variable, and I judged from the specimens shown me by Miss A. Eastwood, that the two forms cannot always be defined.

On the west slopes of Mount Shasta the tree occurs higher up than *A. Lowiana*, mixing with it at about 6000 feet, and at 8000 feet it is the only species of fir. It is not on Mount Shasta a very large tree, the biggest that I measured being not much over 100 feet in height, and 15½ feet in girth, the average being 80 to 100 feet high, by 6 to 8 feet in girth. I could not see very much difference in the bark; though *A. magnifica* is known as the red, and *A. Lowiana* as the white fir, but the very much larger cones of the former distinguish it at once. These were borne only near the summit of the trees, and could only be procured by shooting them off with a rifle, or by felling trees on purpose. They were fully formed but unripe in the first week of September. The soil here was very rocky, and drier than that of any mountain which I have ascended in this latitude; and there was little herbaceous vegetation, though the snow is said to lie deep from November until May or June.

## INTRODUCTION

This species was introduced in 1851 by John Jeffrey, who believed it to be *A. amabilis*; and the seedlings were distributed under this name amongst the members of the Scottish Oregon Association. The tree in Scotland is frequently labelled *A. amabilis*, in consequence of this error.

W. Lobb[1] sent seed in 1852, also under the name of *A. amabilis*; but later on the plants were found to differ from that species, and were distributed by Messrs. Veitch as *A. nobilis robusta*.

## REMARKABLE TREES

Though it is quite possible that larger trees exist, which have been mistaken for *A. nobilis*, yet we have identified none in England which at all approach that species in size, and all the best we have seen are in the eastern and southern counties. The largest perhaps is one at Fulmodestone, Norfolk, a handsome and well-shaped tree growing in a damp soil and well sheltered situation, which in 1905 was 61 feet by 5 feet 9 inches, and bore cones near the summit.

At Bayfordbury, on a much drier soil, it has flourished better than *A. nobilis*, and in 1905 was 56 feet by 5 feet 9 inches (Plate 222). Mr. H. Clinton Baker recently measured a good specimen, 61 feet by 4 feet 4 inches, at Flitwick Manor; and another, 60 feet by 4 feet 9 inches at High Leigh, near Hoddesdon. At Grayswood, Haslemere, a tree planted as recently as 1881, measured in 1906 56 feet by 4 feet 11 inches; and at Petworth, in 1905, there was a slender and less vigorous tree 47 feet by 3 feet 1 inch.

At Eridge Park, Kent, a tree planted in 1880 by Count Gleichen was, in 1905, 34 feet by 3½ feet.

In a pinetum close to Presteign, Radnorshire, planted about fifty years ago,

[1] *Hortus Veitchii*, 336 (1906).

now the property of Mr. J. H. Wall, I saw a good specimen of *Abies magnifica* in 1906, which measured 53 feet by 5 feet 7 inches.

The largest reported[1] at the Conifer Conference in 1891 was at Revesby Abbey, Lincolnshire, and then measured only 40 feet by 5 feet.

In Scotland it is more numerous and larger. The late Malcolm Dunn, who had an exceptionally wide experience in the cultivation of conifers in Great Britain, wrote of it as follows in a paper[2] which he sent to the Conifer Conference :— "It is in truth a stately tree and one of the handsomest of all the taller-growing conifers for ornamental purposes. It is one of the very hardiest of the firs, and is seldom affected by spring frost, and the timber being straight, clean-grained, and of good quality, it will no doubt be a useful forest tree." But this latter opinion has not so far received any proof so far as we know, for the tree is, and seems likely to remain, difficult to obtain, and like most of its congeners is slow and costly to raise from seed.

Probably the finest trees in Scotland are one at Durris,[3] Aberdeenshire, which was, in 1904, 80 feet high by 6 feet 6 inches in girth, and when I measured it in 1907 had increased to about 85 feet; and another (Plate 223) at Bonskeid, near Pitlochry, of which Mr. J. Forgan has been good enough to send me a photograph, and which measured, in 1908, 87 feet by 8 feet. When he first knew it thirty-five years ago it was about 12 feet high; it has not produced cones. Mr. Bean[4] noticed in 1906 a tree at Abercairney 70 feet high, and another at Blair Castle 60 feet high.

At Farthingbank, Drumlanrig, there is, growing on clay loam at 650 feet above sea-level, a tree 50 feet by 5 feet 3 inches in 1905, which was planted, according to Mr. Menzies, the forester, thirty-one years previously.

The tree is rare in Ireland, but there is a specimen[5] at Castlewellan, which was 47 feet by 6 feet in 1906; and at Powerscourt, a tree, planted thirty-five years ago, was 57 feet by 6 feet 8 inches in 1906, and is said to bear cones nearly every year.

(H. J. E.)

[1] *Journ. Roy. Hort. Soc.* xiv. 568 (1892).     [2] *Journ. Roy. Hort. Soc.* xiv. 83 (1892).

[3] *A. magnifica* closely resembles *A. nobilis*, but in strong contrast as regards seed-bearing. It does not seem as if the tree is likely to become acclimatised in this respect as, although planted in considerable numbers throughout the policy grounds and plantations, and most of those trees now between fifty and sixty years of age, cones have been produced only on one occasion, and that on only a few trees. The timber when closely grown is closer in texture, richer in colour, and better in quality than *A. nobilis*. Like that species it is impatient of side shade and sheds its branches freely. Constitutionally it is less robust than its near relative, and also less accommodating in its demands on site and soil.—(J. D. CROZIER.)

[4] *Kew Bulletin*, 1906, pp. 264, 267.

[5] Figured in *Garden*, June 28, 1890, p. 591.

## ABIES BRACTEATA, BRISTLE-CONE FIR

*Abies bracteata*, Nuttall, *Sylva N. Amer.* iii. 137, t. 118 (1849); Hooker, *Bot. Mag.* t. 4740
(1853); Masters, *Gard. Chron.* v. 242, f. 44 (1889), and vii. 672, f. 112 (1890); Kent, Veitch's
*Man. Coniferæ*, 493 (1900).
*Abies venusta*, Koch, *Dendrol.* ii. 210 (1873); Sargent, *Silva N. Amer.* xii. 129, tt. 615, 616
(1898), and *Trees N. Amer.* 63 (1905).
*Pinus venusta*, Douglas, *Comp. Bot. Mag.* ii. 152 (1836).
*Pinus bracteata*, Don, *Trans. Linn. Soc.* xvii. 442 (1837).
*Picea bracteata*, Loudon, *Arb. et Frut. Brit.* iv. 2348 (1838); Coleman, *Garden*, xxxv. 12, with fig.
(1889).

A tree attaining in America 150 feet in height and 9 feet in girth. Bark
brown, smooth; becoming, near the base in old trees, slightly fissured and broken
into thick appressed scales. Buds unique in the genus, elongated, fusiform,
broadest near the base, and gradually tapering to a sharp point, about $\frac{1}{2}$ to $\frac{3}{4}$ inch
long, brown in colour, non-resinous; scales thin, membranous, glabrous, loosely
imbricated, obtuse at the apex, shorter at the base of the bud, gradually lengthening
above. Young shoots glabrous, greenish, with slightly raised pulvini and inconspicuous furrows. Base of the shoots usually ringed with the scars of the
previous season's bud-scales, which in most cases all fall off and do not persist in
part, as is usual in other species.

Leaves on lateral branches pectinately arranged, those below spreading
outwards in two sets in the horizontal plane; those above slightly shorter, falcate,
directed outwards and slightly upwards and forwards, forming a shallow V-shaped
depression on the upper side of the branchlet. Leaves, up to 2 inches long, $\frac{1}{10}$ inch
wide, rigid, thin, flat, linear, ending in long spine-like cartilaginous points, never
bifid; widest in the lower third, gradually tapering to the apex, and abruptly
narrowed close to the base; upper surface dark-green, shining, slightly concave in
the lower half and flat near the apex, no definite median groove being formed; lower
surface with two wide white bands of stomata, each of 10 to 12 lines; resin-canals
marginal. Leaves on cone-bearing branches upturned, falcate.

Male flowers, $1\frac{1}{4}$ to $1\frac{1}{2}$ inch long, cylindric, shortly-stalked, surrounded at the
base by numerous lanceolate, fawn-coloured parchment-like scales, similar to
those of the leaf-buds. Pistillate flowers, with oblong scales rounded above and
nearly as long as the cuneate obcordate yellow-green bracts, which end in slender
elongated awns.

Cones, remarkable for the long spiny rigid tips to the bracts, ovoid, rounded
and full at the apex, 3 to 4 inches long, about 2 inches in diameter, glabrous,[1]
resinous, purplish brown. Scales, about 1 inch broad by $\frac{1}{2}$ inch long, almost
reniform; upper margin incurved, with a short obtuse denticulate cusp; claw
obcuneate. Bracts oblong-obovate, adnate to the scale to beyond the middle and

---

[1] Remarkable, as all the other species of Abies have the scales of the cones pubescent.

deciduous with it, terminating in linear, rigid spines, 1 to 2 inches long, which in the upper half of the cone point towards its apex, and in the lower half are spreading and often recurved. Seeds dark reddish-brown, about $\frac{3}{8}$ inch long and nearly as long as their pale reddish-brown shining wings. (A. H.)

## DISTRIBUTION AND HISTORY

*Abies bracteata* has perhaps the most restricted distribution of all the silver firs, as, according to Sargent, it only occurs in a few isolated groves along the moist bottoms of cañons at about 3000 feet elevation on both slopes of the western ridge of the Santa Lucia Mountains in Monterey County, California. The most northerly point where it is now known to grow is in Bear Cañon, twenty-five miles south of the Los Burros mines; the other localities mentioned by Sargent are in the San Miguel Cañon and in a gorge at the head of the Nacimiento river.

The discovery of this tree is assigned by Don, Sir W. J. Hooker,[1] and Sargent to Dr. T. Coulter, who, according to a letter[2] of Douglas to Hooker dated November 23, 1831, arrived at Monterey after he began the letter in question. Douglas also, in a letter[3] dated October 1832, states[4] that he found the tree, which he called *Pinus venusta*, in the preceding March "on the high mountains of California," and that it is never seen at a lower elevation than 6000 feet above sea-level, in lat. 36°, where it is not uncommon.

But Kent says,[5] in a note, that a comparison of the dates shows that Douglas was the first discoverer, which, however, is not proved; as, according to Douglas's own showing, Coulter was at Monterey, near to the place where the tree grows, three months before Douglas found the tree himself. Prof. Hansen[6] also has incorrectly stated the date of Douglas's discovery of this tree as March 1831 instead of March 1832.

William Lobb, when collecting for Messrs. Veitch in 1853, introduced it to cultivation, and in a letter in *Gardeners' Chronicle*, 1853, p. 435, describes it as "the most conspicuous ornament of the arborescent vegetation. On the western slopes, towards the sea, it occupies the deep ravines, and attains the height of from 120 to 150 feet, and from 1 to 2 feet in diameter, the trunk as straight as an arrow, the lower branches decumbent. The branches above are numerous, short, and thickly set, forming a long tapering pyramid or spire, which gives to the tree that peculiar appearance not seen in any other kind of the Pinus tribe. Along the summit of the central ridges, and about the highest peaks, in the most exposed and coldest places imaginable, where no other pine makes its appearance, it stands the severity of the climate without the slightest perceptible injury, growing in slaty rubbish, which to all appearance is incapable of supporting vegetation. In such situations it becomes stunted and bushy. The cones are quite as singular as the growth of the tree is beautiful; when fully developed the scales, as well as

---

[1] *Bot. Mag.* t. 4740 (1853).
[2] *Comp. Bot. Mag.* ii. p. 149.
[3] *Ibid.* 151.
[4] *Ibid.* 152.
[5] Veitch's *Man. Coniferæ*, 497, note (1900).
[6] *Journ. Roy. Hort. Soc.* xiv. 459 (1892).

the long leaf-like bracts, are covered with globules of thin transparent resin. Douglas was mistaken in saying that this tree does not occur below 6000 feet elevation; on the contrary, it is found as low as 3000 feet, where it meets *Taxodium sempervirens.*"

In 1856 another expedition to collect seeds was made by W. Beardsley, who gives a good account of his journey, which is quoted from by Murray.[1]  In the middle of October the seeds were already shed, and Murray says that Mr. W. Peebles, who went for the same purpose on September 17, 1858, found the cones so ripe that when the tree was felled they fell to pieces.

According to Beardsley, the soil on which it grows is "exclusively the calcareous districts, abounding with ledges of white, veined, and grey marble."

## CULTIVATION

*A. bracteata* has never been a common tree in English gardens and, owing to the difficulty of procuring seeds in California, it is rarely to be had from nurseries. It seems to be quite hardy as regards winter cold, but susceptible to spring frosts; and all the good specimens I have seen are in sheltered and rather elevated situations on well drained soil.

The seedlings which I have raised from English-grown seeds have not thriven on my soil, though the tree does not appear to dislike a moderate amount of lime. All the best specimens we know of are in the south and west of England, and in Ireland.   A list of them is given by Kent,[2] and they all are probably of about the same age, being raised from William Lobb's seeds by Messrs. Veitch in 1854.

## REMARKABLE TREES

The finest trees in England are in the valley of the Severn, the largest being at Eastnor Castle (Plate 224), where two are growing.   They were stated[3] by the late Mr. Coleman to have been planted in 1865, and the best of them was 40 feet high in 1889.   It first bore cones in 1888; when I measured it last in 1908, I found it to be 78 feet by 9 feet, and though very healthy and handsome in appearance, the top had become forked.   It bore cones freely in 1900, from which I raised numerous seedlings, but these have grown very slowly, and do not seem able to make roots freely on my soil.

At Highnam Court, Gloucester, the seat of Sir Hubert Parry, there is another fine tree, difficult to measure on account of its situation, but, in 1908, I made it 64 feet by 6 feet 5 inches.   It has several times produced cones, four being borne in 1907, from which seedlings were raised.

At Tortworth Court there is a tree which Lord Ducie believes to have been planted between 1858 and 1862, and in 1908 was 63 feet by 6 feet.   It is growing on old red sandstone, about 250 feet above sea level in a situation much exposed to the south-west wind.

[1] *Edin. New Phil. Journ.* x. 1, pls. 1 and 2 (1859).   [2] Veitch's *Man. Coniferæ, loc. cit.*
[3] *Garden,* 1889, xxxv. 12.

At Nevill Court, near Tunbridge Wells, I measured a tree which, though only 48 by 4½ feet in 1906, is one of the best shaped I have seen, with a very slender spire, as described by Lobb in California.

At Fonthill Abbey there is a tree about 72 feet by 5½ feet, in a sheltered though elevated situation on greensand. At Osborne, in the Isle of Wight, a tree was 60 feet by 5 feet 9 inches in 1908, but this does not appear to be thriving, on account perhaps, of the dry soil.

At Ponfield, Hertford, the seat of P. Bosanquet, Esq., Henry saw in 1906 a tree, very thriving and about 25 feet in height; and in the same district, at High Canons, near Shenley, Mr. Clinton Baker showed me a tree 53 feet by 4 feet which bore about twenty cones in 1907. At Pampisford, Cambridgeshire, there is a tree, about 25 feet high, growing in a sheltered position, and very thriving.

At Monk Coniston, in Westmoreland, the seat of Victor Marshall, Esq., I have seen a tree which has borne cones, and which measured, in 1906, 60 feet by 5 feet.

Several others are mentioned by Kent: at Kenfield Hall,[1] near Canterbury; at New Court, and at Streatham, near Exeter; at Upcott, near Barnstaple; and at Warnham Court, near Horsham. A large tree at Orton Hall, Peterborough, was, before it was cut down in 1905, 59 by 6 feet, but became unhealthy owing to the soil being too heavy.

In Wales, where the species should grow well, I have seen no trees of any size.

In Scotland the only specimens I have seen are at Castle Kennedy and at Cawdor Castle, neither of which are large, and the climate of Scotland generally seems to be too cold for it.[2]

In Ireland Henry has seen specimens at Fota, in the south-west, a fine young tree which, in 1903, was 48 feet by 4 feet; at Castlewellan, in the north-east, another, in 1906, was 35 feet by 3 feet 2 inches; and a smaller one also exists at Glasnevin.

On the continent of Europe this tree is very rare, the only fine one I have seen being a tree at Pallanza in the nursery grounds of Messrs. Rovelli, which, in 1906, was about 70 feet by 7 feet, but not very healthy and bearing no cones.

M. Pardé states that there is a fine specimen in the domain of the National Society of Agriculture at Harcourt (Eure); and I saw a small one in M. Allard's collection at Angers. (H. J. E.)

---

[1] This tree produced cones in 1886. Cf. *Gard. Chron.* xxvi. 85 (1886).

[2] The one specimen now remaining at Durris—between forty and fifty years of age—if it can possibly be taken as a fair example of the growth of the tree in this locality, proves it of little use for planting. It is quite healthy, but its growth is slow in proportion to that of *A. pectinata.*—(J. D. CROZIER.)

## ABIES LASIOCARPA, Rocky Mountain Fir

*Abies lasiocarpa*, Nuttall, *Sylva*, iii. 138 (1849); Masters, *Gard. Chron.* v. 172, ff. 23-27, 32, (1889), and *Journ. Bot.* xxvii. 129 (1889); Sargent, *Silva N. Amer.* xii. 113, t. 611 (1898), and *Trees N. Amer.* 61 (1905); Kent, Veitch's *Man. Coniferæ*, 515 (1900).
*Abies bifolia*, Murray, *Proc. Roy. Hort. Soc.* iii. 320 (1863).
*Abies subalpina*, Engelmann, *Am. Nat.* x. 555 (1876).
*Abies arizonica*, Merriam, *Proc. Biol. Soc. Wash.* x. 115, ff. 24, 25 (1896).
*Pinus lasiocarpa*, W. J. Hooker, *Fl. Bor. Am.* ii. 163 (1839).
*Picea bifolia*, Murray, *Gard. Chron.* iii. 106 (1875).
*Picea lasiocarpa*, Murray, *Gard. Chron.* iv. 135 (1875).

A tree, attaining occasionally 175 feet in height, with a trunk 15 feet in girth, but usually not over 80 to 100 feet high. Bark of young trees smooth and silvery grey; of old trees shallowly fissured and roughened by reddish brown or whitish scales; in some trees becoming corky and white in colour. Buds small, about $\frac{1}{4}$ inch long, ovoid-conical, obtuse at the apex, brownish, resinous; scales embedded in the resin but roughening the surface of the bud by their raised tips. Branchlets swollen at the nodes, those of the first year ashy grey, smooth, and covered with a moderately dense short wavy pubescence. Branchlets of the second year retaining some pubescence, darker grey, smooth, with the bark slightly fissuring.

Leaves on lateral branchlets irregularly arranged; sometimes irregularly pectinate with some of the leaves above and below not directed outwards, but forwards at an angle with the axis of the shoot; usually with most of the leaves directed upwards, those in the middle line above covering the shoot and standing edgeways with their apices almost vertical, a few leaves in the middle line below pointing forwards and downwards. Leaves linear, up to $1\frac{1}{2}$ inch long by $\frac{1}{12}$ inch broad, uniform in width except at the gradually tapering base; apex rounded and either entire or with a slight emargination; upper surface with a shallow continuous median groove, and with four to five lines of stomata on each side of the groove in its anterior half, the lines fewer in number and broken in the basal half; under surface with two bands of stomata, each of six to eight lines; resin-canals median. The stomatic lines above give the foliage a glaucous appearance; the bands below vary very much in whiteness. Leaves on leading shoots closely appressed to the stem with their tips directed forwards, flattened in section, and ending in long slender rigid points. Leaves on cone-bearing branchlets upturned, directed forwards, usually acute and not more than $\frac{1}{2}$ inch long.

Cones sub-sessile, cylindrical; rounded, truncate or depressed at the slightly narrowed apex; 2 to 4 inches long by $1\frac{1}{2}$ inch in diameter, dark purple and tomentose, with the bracts concealed.[1] Scales very variable in size and shape, from $\frac{7}{8}$ inch long by $\frac{3}{4}$ inch wide to $\frac{1}{2}$ inch long by 1 inch wide: lateral margins rounded or with sinuses, usually auricled on each side of the short obcuneate claw. Bract situated at the base of the scale or slightly above it, quadrangular or

---

[1] According to Piper, *Contrib. U.S. Nat. Herb.* xi. 93 (1906), cones on trees growing in the Olympic Mountains have exserted bracts.

orbicular, denticulate, emarginate with a long slender mucro. Seed $\frac{1}{4}$ inch long, with dark purplish shining wings, which vary in length according to the height of the scale which they cover almost completely.

Var. *arizonica*, Lemmon, *Bull. Sierra Club*, ii. 167 (1897); Masters, *Gard. Chron.* xxix. 86, 134, ff. 52, 53 (1901).

> *Abies arizonica*, Merriam, *Proc. Biol. Soc. Wash.* x. 115, ff. 24, 25 (1896); Purpus, *Garten-welt*, v. 4, 26 (1896).

This form occurs in the San Francisco mountains in Arizona, where it is common between 8500 and 9500 feet elevation, and occasionally ascends to 12,000 feet. It is remarkable for the creamy-white thick corky bark of the trunk. As seen in cultivation, young plants differ from the type, in the leaves being emarginate at the apex, whiter beneath, and more regularly pectinate in arrangement. Sargent[1] states that bark equally corky occurs in trees of *Abies lasiocarpa* in other regions, as in Colorado, Oregon, South Alberta, and British Columbia; and, as there is no difference in the cones, he does not assign even varietal rank to the Arizona tree.

The best account of this variety is by Prof. Purpus in *Mitt. D. D. Ges.*, No. 13, p. 47 (1904), who visited the San Francisco mountains in 1901, and introduced the tree to Europe. It seems to be a strictly alpine tree, growing on basaltic and trachytic rocks, where the soil is never quite dry, either scattered or mixed with *Populus tremuloides*, *Pinus flexilis*, *Pseudotsuga Douglasii*, and *Picea Engelmanni*. It attains a height of 60 to 70 feet with a girth of 6 to 9 feet. The bark is very corky and corrugated, in old trees milk-white or silver-grey in colour. It is replaced in these mountains at 7000 to 8000 feet by *Abies concolor*.

This form has only recently been introduced into cultivation. Plants were for sale in the Pinehurst Nurseries, North Carolina, in 1901; and Dr. Masters saw a stock of young plants in Moser's nursery at Versailles in 1903. It is too soon yet to form any opinion as to the suitability of this variety for ornamental gardening.

## IDENTIFICATION

*Abies lasiocarpa* is perhaps most readily distinguished by the conspicuous bands of stomata on the upper surface of the leaf, which separate it clearly from the other species[2] with median resin-canals and long narrow leaves. The following points are also noteworthy :—the irregular arrangement of the leaves, which are usually quite entire at the apex; the ashy-grey pubescent shoots; and the resinous obtuse buds.

(A. H.)

## DISTRIBUTION

This is essentially the alpine fir of the Rocky Mountains and higher ranges on the west coast of North America, and is the most widely distributed fir of the New World, occurring from about lat. 61° N. in Alaska to Arizona and New Mexico. It does not occur in California.[3] In the west it extends to the summits of the Olympic

---

[1] *Silva*, xii. 113.  [2] As *A. sibirica* and *A. sachalinensis*, which it somewhat resembles in general appearance.
[3] U.S. Forest Service, *Sylvical Leaflet* 1, Alpine Fir.

Mountains in Washington, and in the east to the mountains of Idaho, Montana (Plate 225), Wyoming, Colorado, and Utah. Everywhere it grows up to or very near the timber line, and on the shores of Lake Bennett in northern British Columbia descends to 2500 feet. In Colorado it reaches 10,000 feet.

Macoun states that it crosses the Rocky Mountains into the Peace River region, and the country between the Little Slave Lake and the Athabasca River; and that in the Rocky Mountains of Alberta it occurs with *Picea Engelmanni*, but is less common; and in a letter states that it is an enormous tree at Glacier, but becomes dwarfed at higher elevations, ascending to 7000 feet in that region.

Wilcox[1] writes of it as follows :—

"The balsam fir has about the same range as the white spruce (*Picea Engelmanni*) but is less common. At a distance it is hardly to be distinguished from the spruce, but the bark on branches and young trees is raised in blisters which contain a drop or two of balsam. This balsam exudes from the bark wherever it is bruised. At first it is a very clear liquid, regarded by old trappers and woodsmen as a certain cure, when brewed with hot water, for colds and throat troubles. On exposure to the air it hardens into a brittle resin, which the woodsman melts into pitch to seal boxes or mend leaky canvas. The camper-out makes his bed from balsam boughs, as they are more springy and less rigid than those of the spruce."

I saw this tree in perfection in the Paradise valley on the south-west slopes of Mount Rainier in August 1904. An excellent illustration of this locality is given by C. O. Piper in *Garden and Forest*, vol. iv. p. 382, which shows the tall slender spiry habit of the fir. Here it lives in company with *A. amabilis* in the lower part of its range, and with *Tsuga Pattoniana*, and *Cupressus nootkatensis* higher up; growing in small clumps and groves, as shown in the illustration referred to. It seems to be a very slow grower, a tree felled by Plummer being only 15 inches in diameter at 125 years old. The tallest that I measured here was 77 feet by 5 feet 8 inches, but Sargent says that it occasionally attains 175 feet in height (probably in the Olympic Mountains). The seedlings, which I usually found growing on rotten logs, were very slow in growth, and must be often eight to ten years old before their roots reach the soil.

### HISTORY AND CULTIVATION

*Abies lasiocarpa* was discovered by Douglas in 1832, and his specimen, which is the type of *Pinus lasiocarpa* of W. J. Hooker, the first name applied to the species, is preserved in the herbarium at Kew.

Seeds were first collected about 1863, by Dr. Parry in Colorado; but it is not known if any plants raised from these still survive. The first plants raised in the Arnold Arboretum date from 1873, the largest of them being now only 10 to 12 feet in height. Roezl collected seeds in 1874 in Colorado.[2] According to Syme,[3] a small

---

[1] *The Rockies of Canada*, 62 (1900).

[2] Masters, *Journ. Bot.* xxvii. 135 (1889), refers these seeds doubtfully to New Mexico; but there is no doubt that they were collected in Colorado. Cf. Lavallée's article on *Nouveaux Conifères du Colorado et de la Californie*, in *Journ. Soc. Cent. Hort. France*, 1875.     [3] *Gard. Chron.* iii. 586 (1888).

plant of this origin was alive in Perthshire in 1888; but it was only $2\frac{1}{2}$ feet in height, forming a wide spreading bush, though it was growing in rich black loam.

No trees of this species are recorded by Kent; nor were any specimens sent to the Conifer Conference in 1891. It appears to be unsuitable for cultivation in this country. Young trees at Kew, a few feet in height, are stunted and dying. Waterer had a large stock of plants in 1889 in the nursery at Bagshot; but they all did badly and were thrown away, only one or two surviving and showing the same wretched appearance as the young trees at Kew. Plants cultivated some years ago at Glasnevin have since died. Henry, however, lately saw in the Pinetum at Hatfield, Herts, a tree, planted in 1893 when it was about 3 feet high, which is now 20 feet in height and 15 inches in girth. It has thriven well hitherto, but is slightly attacked by knotty disease. The best specimen we have seen is one at Bayfordbury, about 14 feet high, and fairly thriving. A small tree at Ochtertyre bore cones in 1906. The tree appears to succeed better in Germany. I have raised seedlings from cones sent by Prof. Allen in 1905 from Mount Rainier. (H. J. E.)

## ABIES BALSAMEA, Balsam Fir

*Abies balsamea*, Miller, *Dict.* No. 3 (1768); Sargent, *Silva N. Amer.* xii. 107, t. 610 (1898), and
 *Trees N. Amer.* 58 (1905); Masters, *Gard. Chron.* xvii. 422, figs. 57-60 (1895); Kent,
 Veitch's *Man. Coniferæ*, 492 (1900).
*Abies balsamifera*, Michaux, *Fl. Bor. Am.* ii. 207 (1803) (in part).
*Pinus balsamea*, Linnæus, *Sp. Pl.* 1002 (1753).
*Picea balsamea*, Loudon, *Arb. et Frut. Brit.* iv. 2339 (1838).

A tree, commonly 50 to 60 feet in height and 3 to 5 feet in girth, but sometimes larger, with spreading branches, usually forming an open broad-based pyramid. Bark, greyish brown and with numerous blisters; on old trees broken on the surface into small scaly plates. Buds small, globose or occasionally dome-shaped, reddish, shining and resinous. Young shoots smooth, ashy grey, with very short scattered pubescence; on the shoots of the second year some of the pubescence is retained, and the bark fissures slightly between the pulvini. The branchlets when cut have a very resinous odour.

Leaves on lateral branches pectinately arranged, in two sets directed outwards in the horizontal plane; upper leaves of each set shorter than the others, and directed also slightly upwards, thus forming a shallow V-shaped arrangement. Leaves linear, flattened, uniform in width except at the tapering base; rounded and slightly bifid at the apex, up to about 1 inch long and $\frac{1}{20}$ to $\frac{1}{16}$ inch wide; upper surface dark green, shining, with a median continuous groove, and with two or three broken rows of stomata in the middle line towards the apex; lower surface with two narrow, greyish bands of stomata, composed of six to eight lines; resin-canals median. Leaves on cone-bearing branches more or less upturned, stouter and broader than those on barren shoots, acute and not bifid at the apex.

Staminate flowers yellow, tinged with purple. Pistillate flowers with nearly orbicular purple scales, shorter than the serrulate greenish-yellow bracts, which are emarginate above and end in long, recurved tips.

Cones sub-sessile, ovoid-cylindrical, tapering both at the base and towards the round or flattened apex; purple[1] in colour, 2 to 4 inches long, about an inch in diameter. Scales, about ⅝ inch wide and long; lamina fan-shaped, rounded and undulate above, lateral margins denticulate and curving to the truncate or auricled base; claw wedge-shaped. Bracts variable in length, exserted or concealed between the scales; claw oblong; lamina trapezoidal and denticulate, ending in a mucro. Seeds purplish, about ½ inch long; wing about as long as the body of the seed.

In the wild state considerable variation occurs in the habit of the tree, which becomes a mere shrub at high altitudes. The cones vary both in size and in the length of the bracts, which are either slightly exserted, or quite concealed between the scales. Prof. Balfour found on the same tree at Keillour cones both with long and with short bracts.

Var. *Hudsonia*, Engelmann, *Trans. St. Louis Acad.* iii. 597 (1878).

*Abies Hudsonia*, Bosc. ex Carrière, *Conif.* i. 200 (1855).

According to Engelmann this is a sterile dwarf form which occurs above the timber line on the White Mountains in New Hampshire. Whether this is identical with the *A. Hudsonia*, which occurs in cultivation, is uncertain. The latter, according to Sargent,[2] is of unknown origin, but is probably, though it has never produced cones, a depauperate form of *A. balsamea*. It has densely crowded branches, short numerous branchlets, and small broad leaves, about ¼ inch in length; and is a dwarf spreading shrub, only a foot or two in height. It differs from *A. balsamea* in having marginal resin-canals.

Var. *macrocarpa*.[3] This was discovered near the Wolf River, Wisconsin, and raised by Robert Douglas at Waukegan nursery; it is said to be a distinct and beautiful form with longer leaves and larger cones than the type.

## DISTRIBUTION

The balsam fir extends far to the northward in the Dominion of Canada, its northerly limit being a line drawn from the interior of Labrador north-westward to the shores of the Lesser Slave Lake. It occurs in Newfoundland and in the provinces of Quebec and Ontario, and descends in the United States in the west through northern Michigan and Minnesota to northern and central Iowa, and in the east extends through New England and New York, along the Catskill and Alleghany mountains to south-western Virginia. It is common and often forms a considerable part of the forest on low swampy ground, while on well-drained hill-sides it is met with as single trees or small groves chiefly in the spruce forests. It ascends to 5000 feet on the Adirondacks.                                                     (A. H.)

---

[1] In cultivated specimens the cones are occasionally olive-green in colour, and rarely exceed 2 inches in length.
[2] *Garden and Forest*, x. 510 (1897).          [3] *Ibid.* v. 274 (1892) and x. 510 (1897).

## REMARKABLE TREES

The most noted trees of this species in cultivation were those growing in the Keillour Pinetum, Perthshire, now the property of Captain Black of Balgowan. This pinetum was visited by Prof. Balfour[1] in 1895, who found about 30 trees still living out of 200, which were planted in 1831. The largest tree was about 60 feet high with a girth of 5 feet 1 inch at three feet from the ground. There were several others over 4 feet in girth. In 1904, when Henry made a hurried visit to the Keillour Pinetum, where there was much of interest to be seen, he only saw one tree of *A. balsamea*, with the top broken and in a dying state. Mr. W. Causand informed him that in 1903 there was a tree 68 feet by 5 feet.

The finest specimen of which we have any account in Great Britain is recorded in the *Conifer Conference Report, Journ. Roy. Hort. Soc.* xiv. 511, as having grown at Saltoun Hall, East Lothian, the seat of A. Fletcher, Esq., until 1891, when it was swept away by a flood on the river Tyne. This tree was supposed to have been given by Bishop Compton, who introduced the species in 1697, to Bishop Burnet, formerly incumbent of the parish of Pencaitland, and was thus something like 190 years old. It was 68 feet high though it had lost its top, and at ten feet from the ground no less than 7 feet 10 inches in girth, and was said to have been healthy and growing vigorously up till the time of its destruction.

In England we have never seen a tree of any great size or age, the largest being at Bicton, 52 feet by 4 feet 4 inches in 1908; and this species seems to have been neglected and forgotten by modern planters, as it is only twice mentioned in the numerous reports sent to the Conifer Conference.

Loudon states that it arrives at maturity in twenty to twenty-five years, after which it soon dies, though he mentions trees of 30 to 40 feet high as then existing at Syon, Whitton, and Chiswick.

It appears therefore to be of no horticultural value in this country, though if the Saltoun report was correct it may be grown successfully in some parts of Scotland.[2]

In Norway, according to Schübeler, the Balsam fir succeeds better than here. He mentions three at Bogstad near Christiania, planted about 1772, of which the largest was 55 feet by 6 feet 4 inches, and another 8 feet 2 inches in girth; but when I visited this place in 1904 I could not find these trees, and do not know whether they are still living. Hansen[3] states that specimens, about 50 years old and 40 feet high, are to be met with in Danish Gardens.

## TIMBER, RESIN

Sargent describes the wood[4] as being light, soft, coarse-grained, and perishable, and only used for cheap lumber. From the blisters on the bark, a straw-coloured

[1] See *Gard. Chron.* xvii. 422 (1895), which gives an interesting account of this remarkable pinetum.

[2] *A. balsamea* was planted at Durris freely about fifty years ago, the largest trees now being from 40 to 45 feet in height. Timber of good quality, and contains an exceptionally small percentage of water in a green state. I have seen no account taken of the latter fact, but it has been a continual surprise to me in handling timber in a green state.—(J. D. CROZIER.)

[3] *Journ. Roy. Hort. Soc.* xiv. 458 (1892).

[4] H. von Schrenk, in *Missouri Bot. Garden Report*, 1905, p. 117, describes and figures logs of this timber, felled in Maine for pulpwood, which show on cross-section irregular areas, perfectly smooth and shining as if they had been planed.

transparent resin, known as Canada balsam, is collected by Indians and poor whites in the province of Quebec.  This resin, which was formerly largely used in medicine on account of its stimulating action on the mucous membrane, is now chiefly used for mounting objects to be examined under the microscope, for which, and kindred purposes, it is specially suitable by reason of its transparency.      (H. J. E.)

## ABIES FRASERI, Southern Balsam Fir

*Abies Fraseri*, Poiret, in Lamarck, *Dict. Suppl.* v. 35 (1817); Forbes, *Pinetum Woburnense*, 111, t. 38 (1840); Sargent, *Silva N. Amer.* xii. 105, t. 609 (1898), and *Trees N. Amer.* 57 (1905); Masters, *Gard. Chron.* viii. 684, fig. 132 (1890); Kent, Veitch's *Man. Coniferæ*, 509 (1900).
*Pinus Fraseri*, Lambert, *Genus Pinus*, ii. t. 42 (1837).
*Picea Fraseri*, Loudon, *Arb. et Frut. Brit.* iv. 2340 (1838).

A tree attaining in America 70 feet in height and 7 feet in girth, with rather rigid branches, forming an open symmetrical pyramid.  Bark smooth and with numerous blisters in young trees, becoming on older trunks covered with thin appressed reddish scales.  Buds small, broadly ovoid or globose, reddish, resinous. Young shoots smooth, yellowish grey, densely covered with reddish, short, twisted or curved hairs, the pubescence being retained on the older branchlets.

Leaves on lateral branches pectinately arranged, as in *A. balsamea*; linear, flattened, shorter than in that species, rarely exceeding $\frac{3}{4}$ inch long and $\frac{1}{20}$ inch broad, uniform in width except at the shortly tapering base, rounded and bifid at the apex; upper surface dark green, shining, with a continuous median groove and without stomata; lower surface with two broad conspicuously white bands of stomata, each of eight to twelve lines; resin-canals median.  Leaves on cone-bearing shoots upturned, crowded, broader than on barren shoots, rounded and entire at the apex.

Staminate flowers yellow tinged with red.  Pistillate flowers, with rounded scales, shorter than the oblong bracts, which are broad and rounded above, ending in long slender tips.

Cones sub-sessile, ovoid, cylindrical, slightly tapering at the base and towards the rounded or flattened apex, purple, about 2 inches long by $1\frac{1}{4}$ inch in diameter, with the bracts conspicuously exserted and reflected.  Scales as in *A. balsamea*, but wider in proportion to their length.  Bracts; claw oblong; lamina broad, trapezoidal, denticulate in margin and bifid above with a mucro in the emargination.  Seed with wing about $\frac{1}{2}$ inch long; wing purplish, broadly trapezoidal, denticulate in the upper margin, about twice as long as the body of the seed.

### Identification

This species can readily be distinguished from *Abies balsamea* by the different pubescence on the young branchlets and the shorter, more coriaceous leaves, which

Logs with this so-called "glassy" appearance are occasionally rejected; but examination showed that this peculiarity was simply due to the presence of ice, which follows the radial lines on the healed-over branches of the logs.

have broader bands of stomata than in that species—eight to twelve lines in *A. Fraseri*, usually only six lines in *A. balsamea*. The cones differ mainly in the larger bracts, which are much exserted and reflexed over the edges of the scales next below ; whereas in *A. balsamea* the bracts are either concealed, or, if slightly exserted, are never reflexed.  (A. H.)

## DISTRIBUTION

*Abies Fraseri* is very restricted in its range of distribution, being only found in the Alleghany Mountains of south-western Virginia, North Carolina, and eastern Tennessee, where it often forms forests of considerable extent at elevations of 4000 to 6000 feet above sea-level. These forests are usually pure ; but occasionally this species grows mixed with black spruce, birch, and beech. The tree averages about 40 feet in height; it only rarely attains 70 feet.

Sargent in an article[1] on this species gives a good illustration of a forest, at about 5000 feet altitude on the Black Mountain range, a spur of the Blue Ridge in North Carolina; which is very like some forests that I saw when I visited this most interesting region in 1895.

## HISTORY AND CULTIVATION

*Abies Fraseri* was discovered by the Scotch traveller and botanist whose name it bears, John Fraser, in the first decade of the nineteenth century ; and plants of it were first distributed from Messrs. Lee's nursery, at Hammersmith, in 1811. The excellent figure in *Pinetum Woburnense*, was taken from the original tree in this nursery, where it had then attained 16 feet in height, at about twenty-eight years of age.

The tree is short-lived, and the plants of the first introduction are probably all long since dead. According to Sargent,[1] seeds of *A. balsamea*, collected in Pennsylvania and Canada, where specimens are occasionally found, in which the tips of the bracts of the cone are slightly exserted, have been very generally sold as *A. Fraseri*. Seedlings of the Carolina tree were, however, distributed by the Arnold Arboretum a few years prior to 1889. We know of no trees of any size now living in this country. Some seedlings which I brought from N. Carolina in 1895 soon died.

(H. J. E.)

[1] *Garden and Forest*, ii. 472, fig. 132 (1889).

## ABIES RELIGIOSA, Mexican Fir

*Abies religiosa*, Schlechtendal, *Linnæa*, v. 77 (1830); Lindley, *Penny. Cycl.* i. 31 (1833); Seemann,
    *Bot. Voy. 'Herald*,' 335 (1852-1857); Hooker, *Bot. Mag.* t. 6753 (1884); Masters, *Gard.
    Chron.* xxiii. 56, f. 13 (1885), and ix. 304, ff. 69, 70 (1891), and *Journ. Linn. Soc. (Bot.)*
    xxii. 194, t. 6 (1886); Kent, Veitch's *Man. Coniferæ*, 536 (1900).
*Abies hirtella*, Lindley, *loc. cit.* (1833).
*Pinus religiosa*, Humboldt, Bonpland et Kunth, *Nov. Gen. et Spec.* ii. 5 (1817); Parlatore, in DC.
    *Prod.* xvi. 2, p. 420 (1868).
*Pinus hirtella*, Humboldt, Blonpland et Kunth, *loc. cit.* (1817).
*Picea religiosa*, Loudon, *Arb. et Frut. Brit.* iv. 2349 (1838).
*Picea hirtella*, Loudon, *loc. cit.* (1838).

A tree, attaining in Mexico 150 feet in height and 18 feet in girth. Bark [1] greyish-white, rough, divided into small roundish plates.

Buds shortly cylindrical, rounded at the apex, covered with white resin. Young shoots brown on the upper surface, olive green beneath, covered with minute erect pubescence; pulvini prominent. Second year's shoots reddish-brown, smooth, and striate between the pulvini, which are no longer raised.

Leaves on lateral branches, arranged as in *A. Nordmanniana*; but with the median upper leaves much fewer than in that species, covering the upper side of the branchlet, and pointing forwards and slightly upwards; lower leaves in two lateral sets, spreading outwards and slightly forwards in .the horizontal plane. Leaves twisted above the base, linear, flattened, gradually narrowing in the anterior half to an obtuse apex, which is usually entire or rarely minutely bifid; upper surface dark green, shining, with a median groove (usually not continued to the apex) and without stomata;[2] lower surface with two greyish bands of stomata, each of eight to ten lines; resin-canals marginal. The upper leaves are about half the length of those below, the latter about an inch in length and about $\frac{1}{16}$ inch broad. Leaves on cone-bearing branches similar to those on barren branches.

Cones on short stout stalks, 4 inches long, 2 inches in diameter, conical, broadest near the base and gradually tapering to an obtuse and narrowed apex, bluish before ripening, dark brown when mature, the large reflexed bracts being then of a chestnut brown colour. Scales broadly fan-shaped, nearly $1\frac{1}{4}$ inch wide by $\frac{5}{8}$ inch long; upper margin almost entire; lateral margins laciniate and denticulate; base broad with a sinus on each side of the short obcuneate claw. Bract: claw wide, obcuneate; lamina quadrangular, denticulate, emarginate with a short triangular cusp. Seed with wing about $\frac{3}{4}$ inch long; wing broad and $1\frac{1}{2}$ times the length of the body of the seed.

### DISTRIBUTION

This species extends throughout the mountains of Mexico, from near Durango in the Sierra Madre range (lat. 24°), where it was collected by Seemann,[3] to the

---

[1] In this tree, as in the other species with prominent pulvini on the branchlets, the bark of the trunk speedily becomes scaly and like that of a spruce, not remaining smooth for a considerable period, as in the common species of silver fir.

[2] On leaves towards the tip of the shoot, short irregular lines of stomata are present on their upper surface near the apex. Some of these leaves turn their ventral surfaces upwards towards the light.        [3] *Bot. Voy. Herald*, 335 (1852).

mountains of northern Guatemala (lat. 15°), where it was observed by Hartweg[1] and collected by Skinner. It is known to the natives as Oyamel, and occurs mainly in forests at 8000 to 10,000 feet, though it occasionally descends to 4000 feet. It apparently reaches its best development on the Campanario, the highest point of the mountains of Angangueo, a range about 100 miles west of the city of Mexico. Here Hartweg found trees 150 feet in height and 5 to 6 feet in diameter. Parry and Palmer collected it in the province of San Luis Potosi in Central Mexico, and gave its range as from 6000 to 8000 feet. Linden found it on the peak of Orizaba, inland from Vera Cruz, growing between 9000 and 10,000 feet elevation.

Stahl, in Karsten and Schenk's *Vegetationsbilder*, 2 Reihe, Heft 3, gives a good account of this tree, which he found growing near Orizaba between 2600 and 3500 metres above sea-level, and in the higher mountains round the valley of Mexico, in pure forests or mixed with pines, oaks, and alders. He gives no dimensions, and the two excellent figures 17 and 18 taken in the Sierra de Ajusco, near Salazar, at about 9500 feet, show the trees to be smaller there than those which Elwes saw on Popocatapetl.[2]

Dr. Gadow[3] found it growing in the mountains of Omiltelme, at 8000 feet; and describes the trees as "veritable giants, from 5 to 6 feet in diameter, as straight as a mast, and may be 100 feet high."

Humboldt supposed that there were two species, one with glabrous and the other with pubescent branchlets; but Seemann and Hartweg were convinced that this distinction is unfounded; and the type specimen of *Pinus religiosa*, the supposed glabrous form, according to Bolle, has pubescent branchlets.

The branches of the tree, which are very elegant, are used in Mexico for decorating churches at the times of religious festivals.

This species was discovered in 1799 by Humboldt, who saw it near the city of Mexico in two localities, at 4000 feet elevation between Masantla and Chilpantzingo, and near El Guardia at 8400 feet. It was introduced into cultivation in 1838 by Hartweg, who collected for the Horticultural Society of London.

## REMARKABLE TREES

*Abies religiosa* is tender and will not live, except in the warmer parts of these islands, close to the sea coast, where the temperature never falls much below freezing point. Trees planted long ago at Kew and Bayfordbury, do not now survive. Murray mentions[4] in 1876 specimens growing at Woodstock in Kilkenny, Highnam

[1] *Trans. Hort. Soc.* iii. 123, 138 (1848).
[2] I believe that this was the silver fir which clothes the lower slopes of the volcano of Popocatepetl, in Mexico, which I ascended to the limit of vegetation, about 13,000 feet, in March 1888, with my wife and Mr. F. D. Godman. The trees formed in some places dense forests at an elevation of 9000 to 10,000 feet, but though my recollection is that they grew to a great size, we took no measurements, being at the time engaged in collecting birds and insects. In the dry volcanic soil in which they grow we found abundantly *Pinguicula rosea*, one of the most charming ornaments of our greenhouses; and higher up lupins and pentstemons were the most conspicuous plants.—(H. J. E.)
[3] *Through Southern Mexico*, 378 (1908).
[4] *Gard. Chron.* v. 560 (1876).

Court in Gloucestershire, Munches in Kirkcudbright, and Hafodunos in Denbighshire, places which we have visited; but none of these trees can now be found.

There are several trees in Cornwall. Specimens with cones were sent to Kew in 1899 from Trevince, near Redruth, the residence of Mr. E. B. Beauchamp. There is also a tree [1] in Mr. Boscawen's garden at Lamorran, which produced cones in 1890.[2] There is a small tree at Mr. Rashleigh's garden, Menabilly, which was figured in the *Gardeners' Chronicle*;[1] and at Tregothnan, the seat of Viscount Falmouth, there is a large tree which Elwes measured in 1905 as 56 feet by 6½ feet. Though bearing cones, this tree did not seem healthy, and its top was broken by the wind.

A tree at Castle Kennedy is fairly large in size; but it was blown down some years ago, and then replaced in position. It is in consequence very irregular in shape. It produces cones freely, but the seeds are never fertile.

There were formerly two trees at Fota, which differed somewhat in colour of the foliage and hardihood; one [3] has since been blown down. The surviving tree (Plate 226) is a handsome one, though its trunk was broken at about thirty-six feet up, and it has now four leaders: when measured by Elwes, in 1908, it was 66 feet high by 7 feet 3 inches in girth. It is branched to the ground, one very large branch coming off near the base. The foliage is variable in colour, being bluish-green towards the ends of the branchlets, and elsewhere of a light or dark green colour, so that there are three tints visible on the tree. It was bearing in August 1904 numerous cones and male flowers, scattered all over the tree. The cones exude a white resin; and are peculiar, as the scales do not all fall at the same time, some remaining at the base and apex of the axis for two or three years.

The tree does remarkably well on the shores of the Italian Lakes.[4] Carrière says it is killed by frost at Paris; but at Cherbourg [5] there was a tree 30 feet high in 1867. It seems, however, to be very rare if at all existing in France.

(A. H.)

---

[1] *Gard. Chron.* ix. 304, figs. 69, 70 (1891).

[2] This tree was the first in Britain to produce cones, which were exhibited at the Royal Horticultural Society in 1876.

[3] This tree was much more tender than the other, and had the top and some lateral branches killed by frost in the winter of 1880-1881. See Osborne, in *Gard. Chron.* xxiii. 56 (1885).

[4] Sargent, *Silva N. Amer.* xii. 97, *adnot.* (1898). But Elwes saw none that he could identify in the neighbourhood of Pallanza.

[5] Hickel and Pardé, in *Bull. Soc. Dendr. France*, 1908, pp. 206, 224, state that the trees of this species in the neighbourhood of Cherbourg died in the severe winter of 1879-1880; and believe that there are now no living specimens in France.

# PSEUDOTSUGA

*Pseudotsuga*, Carrière, *Conif.* 256 (1867); Bentham et Hooker, *Gen. Pl.* iii. 441 (1880); Masters, *Journ. Linn. Soc.* (*Bot.*) xxx. 35 (1893).
*Abies*, section *Peucoides*, Spach, *Hist. Vég.* xi. 423 (1842).
*Pinus*, section *Tsuga*, Endlicher, *Gen. Pl. Suppl.* iv. Pt. ii. 6 (1847).
*Tsuga*, section *Peucoides*, Engelmann, *Trans. St. Louis Acad.* ii. 211 (1863).
*Abietia*, Kent, Veitch's *Man. Coniferæ*, 474 (1900).

EVERGREEN trees belonging to the tribe Abietineæ of the order Coniferæ. Branches irregularly whorled. Branchlets of one kind; pulvini slightly projecting, persistent, and showing, when the leaves have fallen, an oval scar at their apex. Buds spindle-shaped, acute at the apex, brownish, shining, glabrous; one terminal larger, and one to four lateral and smaller in the axils of the uppermost leaves; scales numerous, imbricated, rounded and entire at the upper margin, increasing in size from below upwards; some of the scales persistent for three or four years at the base of the branchlets, ultimately falling and leaving ring-like scars. Leaves arising in spiral order; but on lateral branches, thrown by a twist of their bases into two spreading ranks; persistent for four to eight years; linear, flat, narrowed at the base; upper surface green and longitudinally furrowed; lower surface with a prominent midrib and two stomatic bands; fibro-vascular bundle single, resin-canals two on the under surface next the epidermis.

Flowers, arising from buds formed in the previous summer, erect, solitary, surrounded at the base by involucral scales. Male flowers axillary, scattered along the branchlets, cylindrical; pedicel short at first, ultimately elongated; composed of numerous spirally arranged short-stalked globose anthers, opening obliquely: connective surmounted by a short spur; pollen-grains globose, without air-sacs. Pistillate flowers, terminal or in the axils of the uppermost leaves, composed of numerous spirally imbricated rounded scales, much shorter than the acutely three-lobed bracts; ovules two on each scale, inverted. Fruit, a woody pendulous cone, ripening in the first season, ovoid-oblong, acute at the apex, rounded and narrowed at the base; peduncle short and stout; scales persistent on the axis after the fall of the seeds, small and sterile towards the base and apex of the cone, rounded, concave, rigid; bracts conspicuous, exserted, oblong, three-lobed at the apex, the middle lobe awn-like and longer than the two lateral lobes. Seeds, two in shallow depressions, which occupy about half the surface of the scale, triangular, without resin-vesicles, winged. Cotyledons, six to twelve, linear; with a prominent midrib and stomatiferous on the upper surface.

The genus comprises three species, two inhabiting western North America, and the third restricted to small areas in Japan and Formosa.

In the absence of cones, they are distinguishable as follows :—

1. *Pseudotsuga Douglasii*, Carrière. Western North America.

Branchlets usually pubescent, occasionally glabrous. Leaves straight, undivided at the apex.

2. *Pseudotsuga macrocarpa*, Mayr. Southern California.

Branchlets covered with short, stiff pubescence. Leaves curved, undivided at the apex.

3. *Pseudotsuga japonica*, Sargent. Japan, Formosa.

Branchlets glabrous. Leaves straight or curved, bifid at the apex.

The latter two species, not being yet introduced into England, will now be briefly dealt with.

PSEUDOTSUGA MACROCARPA, Mayr, *Wald. Nordamer.* 278 (1890) ; Sargent, *Silva N. Amer.* xii. 93, t. 608 (1898), and *Trees N. America*, 54 (1905).

*Abies Douglasii*, var. *macrocarpa*, Torrey, Ives' *Rep.* pt. iv. 28 (1861).
*Abies macrocarpa*, Vasey, *Gardeners' Monthly*, xviii. 21 (1876).
*Tsuga macrocarpa*, Lemmon, *Pacific Rural Press*, xvii. No. 5, p. 75 (1879).
*Pseudotsuga Douglasii*, var. *macrocarpa*, Engelmann, in Brewer and Watson, *Bot. Calif.* ii. 120 (1880).
*Abietia Douglasii*, var. *macrocarpa*, Kent, Veitch's *Man. Coniferæ*, 478 (1900).

A tree usually 50, rarely 80 feet high, with a trunk 3 to 4 feet in diameter. It differs from the common species in the following characters :— Branches comparatively larger and more remotely placed. Branchlets covered with a short, stiff, white pubescence. Leaves, $\frac{3}{4}$ to 1 inch long, resembling those of *P. Douglasii*, except that they are distinctly curved. Buds short and broad, usually not more than $\frac{1}{8}$ inch long. Cones very large, $4\frac{1}{2}$ to 6 inches long ; scales $1\frac{1}{2}$ to 2 inches wide, thick, very concave, puberulous on the outer surface ; bracts, only slightly exserted, short, narrow, with broad midribs produced into short flattened flexible tips. Seeds, $\frac{1}{2}$ inch long, dark brown or nearly black and shining above, pale brown below ; wing $\frac{1}{2}$ inch long.

This species[1] occupies an isolated area in the arid mountains of southern California, at 3000 to 5000 feet elevation, forming open groves or growing in mixture with oak and pines on western and southern slopes. Its distribution extends from the Santa Inez Mountains near Santa Barbara on the coast to the Cuyamaca Mountains on the southern border of California.

PSEUDOTSUGA JAPONICA, Sargent, *Silva N. Amer.* xii. 84, adnot. 2 (1898) ; Shirasawa, *Icon. Ess. Forest. Japon*, text 21, t. 7 (1900).

*Tsuga japonica*, Shirasawa, *Tokyo Bot. Mag.* ix. 86, t. 3 (1895).

This species is not represented by dried material in European herbaria ; but I have seen a specimen[2] recently sent from Japan by Capt. L. Clinton Baker, R.N.

---

[1] A view of a forest of this species is given in *Garden and Forest*, x. 24, f. 5 (1897).

[2] The buds on this specimen were not developed ; but the scales of the previous season's buds remained persistent at the base of the branchlets, and resembled those of *P. Douglasii*.

It is distinguished from the other species by its glabrous branchlets and by its leaves bifid at the apex. The leaves are pectinately arranged, $\frac{3}{4}$ to 1 inch long, $\frac{1}{12}$ to $\frac{1}{20}$ inch wide, straight or curved, yellowish green above, conspicuously white beneath, broadest near the contracted base, and gradually tapering to an acute apex, which is minutely bifid. The cones are small, $1\frac{1}{2}$ to $1\frac{3}{4}$ inch long, 1 inch in diameter; scales few, about twenty in number, more woody in consistence than those of *P. Douglasii*, glabrous externally; bracts strongly reflexed, the central awn-like lobe only slightly larger than the lateral lobes. According to Shirasawa, its discoverer,[1] the tree attains a height of 100 feet and a diameter of 3 feet, and occurs at 1000 to 3000 feet elevation in the mountains of the provinces of Ise, Yamato, and Kii in Japan. It grows in mixed forests, composed mainly of Tsuga, Oak, Beech, Magnolia, and other broad-leaved species. Elwes, when at Koyasan, endeavoured to reach the habitat of this species, but owing to the distance, the heavy rain, and inability to find a guide, was unsuccessful. According to Hayata,[2] this species occurs also on Mount Morrison in Formosa. Its Japanese name is *Togasawara*.

Young plants are reported by Beissner[3] to be in cultivation in Ansorge's nursery, at Flottbeck near Altona, and in the Botanic Garden at Hamburg. Two small branches, recently sent to Kew from Flottbeck and from Herr Langen's nursery at Grevenbroich, are only distinguishable from those of the American species by some of the leaves being bifid at the apex. Apparently in the young stage, the leaves are acute or mucronate and entire, the bifid character only being assumed after two or three years.

Except for its botanical interest this species does not seem likely to have any value in this country.[4] (A. H.)

---

[1] Shirasawa discovered this species in July 1893, on the road between Owashi (in Kii province) and Yoshino (Yamato province), about 10 miles from the coast. He states that the forests in which it occurs are small in area and very inaccessible.

[2] *Tokyo Bot. Mag.* xix. 45 (1905).

[3] *Mitt. Deut. Dendr. Gesell.* 1902, p. 53, and 1906, pp. 84 and 144. Mayr, in *Fremdländ. Wald- u. Parkbäume*, 406 (1906), states that seeds of the Japanese species have never germinated in Europe. The young plants, however, referred to above, are unquestionably this species.

[4] While the above was passing through the press, Mr. H. Clinton Baker writes that he had just received from Pallanza four plants of *P. japonica*, about 2 feet high, which are being planted at Bayfordbury. The buds on these plants are about $\frac{1}{4}$ inch long, shining brown, and without resin; and the leaves are nearly all bifid at the apex.

## PSEUDOTSUGA DOUGLASII, Douglas Fir

*Pseudotsuga Douglasii*, Carrière, *Conif.* 256 (1867); Mayr, *Fremdländ. Wald- u. Parkbäume*, 396 (1906).

*Pseudotsuga Lindleyana*, Carrière, *Rev. Hort.* 1868, p. 152, fig.

*Pseudotsuga taxifolia*,[1] Britton, *Trans. N. York Acad. Sc.* viii. 74 (1889); Sargent, *Bot. Gazette*, xliv. 226 (1907).

*Pseudotsuga mucronata*, Sudworth, *Contrib. U.S. Nat. Herb.* iii. 266 (1895); Sargent, *Silva N. Amer.* xii. 87, t. 607 (1898), and *Trees N. Amer.* 53 (1905).

*Pseudotsuga glaucescens*, Bailly, *Rev. Hort.* 1895, p. 88, fig.; André, *Rev. Hort.* 1895, p. 159; Bellair, *Rev. Hort.* 1903, p. 208, f. 85.

*Pseudotsuga glauca*, Mayr. *Mitt. Deut. Dendr. Ges.* 1902, p. 86, and *Fremdländ. Wald- u. Parkbäume*, 404 (1906).

*Pinus taxifolia*, Lambert, *Pinus*, i. 51, t. 33 (1803) (not Salisbury).

*Pinus Douglasii*, D. Don, in Lambert, *Pinus*, iii. t. (1837).

*Abies taxifolia*, Poiret, in Lamarck, *Dict.* vi. 523 (1804).

*Abies mucronata*, Rafinesque, *Atlant. Journ.* 120 (1832).

*Abies Douglasii*, Lindley, *Penny Cycl.* i. 32 (1833); Loudon, *Arb. et Frut. Brit.* iv. 2319 (1838).

*Picea Douglasii*, Link, *Linnæa*, xv. 524 (1841).

*Tsuga Douglasii*, Carrière, *Conif.* 192 (1855).

*Tsuga Lindleyana*, Roezl, *Cat. Conif. Mex.* 8 (1857).

*Tsuga taxifolia*, Kuntze, *Rev. Gen. Pl.* ii. 802 (1891).

*Abietia Douglasii*, Kent, Veitch's *Man. Conif.* 476 (1900).

A tree, attaining in the moist climate of the Pacific coast 250 to 300 feet in height and 40 feet in girth; but in the dry regions of the interior and at high altitudes rarely more than 100 feet high and 10 feet in girth. Bark of young stems thin, smooth, shining, grey; on older trunks, 2 to 12 inches in thickness, corky, divided by deep longitudinal furrows into broad oblong scaly ridges. Young branchlets usually pubescent, occasionally glabrous. Buds $\frac{1}{4}$ to $\frac{5}{8}$ inch long. Leaves $\frac{3}{4}$ to $1\frac{1}{4}$ inch long, straight, rounded or obtuse, rarely acute at the apex; variable in colour, the stomatic bands beneath either dull grey or conspicuously white.

Cones, 2 to $4\frac{1}{2}$ inches long; scales thin, slightly concave, rounded or slightly prolonged at the apex, about $\frac{3}{4}$ inch wide; pubescent on both surfaces; before ripening bluish below, purple towards the apex and bright red on the closely appressed margins, the bracts being pale green; scales and bracts brown when ripe. Bracts variable in length; the three-pointed apex always, however, extending beyond the scale, usually appressed, but occasionally reflexed. Seeds, about $\frac{1}{4}$ inch long, reddish brown and shining above, paler and with whitish spots below; wings longer than the body of the seed, dark brown, rounded at the apex.

### VARIETIES

1. Var. *glauca*, Beissner, *Nadelholzkunde* 419 (1891), Colorado Douglas fir.

In the interior of the continent, the Douglas fir, growing in a dry climate at

---

[1] According to the rules of botanical nomenclature adopted by the Vienna Congress of 1905, *P. taxifolia* is the correct name for the species, as pointed out by Sargent, in *Bot. Gaz.* xliv. 226 (1907); but we prefer to use *P. Douglasii*, the name which is universally in use amongst foresters and arboriculturists.

high elevations in the Rocky Mountains, through Montana, Colorado, Utah, Arizona, New Mexico, and Mexico, is a smaller tree than the form which occurs in the moist climate of the Pacific coast region. It bears small cones, 2 to 3 inches in length, which in rare cases have the bracts reflexed, but resemble in all essential characters, except size, the cones of the coast form. The leaves are usually thicker in texture and are very glaucous beneath; but the bluish tint visible on the upper surface of the leaves, which is supposed to be characteristic, while common in certain localities, and in others occurring on scattered individual trees, is no more constant than the similar coloured variation which is met with in trees like *Picea pungens* and *Cedrus atlantica*. Mayr has separated the Rocky Mountain form as a distinct species, *P. glauca*; but the differences, being rather physiological than morphological, do not entitle it to rank as more than a variety. The main difference lies in the rate of growth and the hardiness of the tree, when seeds of it are raised in countries remote from its native habitat.

Dr. C. C. Parry discovered this variety of the Douglas fir in the outer ranges of the Rocky Mountains in 1862; and in the following year seeds were sent to the Botanic Garden of Harvard College, from which plants were raised, that have proved perfectly hardy and vigorous in growth in New England. In the northeastern States the Pacific Coast form, whether introduced by seeds collected in Oregon or produced by trees growing in England, has not proved hardy.

The exact date of the introduction of the Colorado Douglas into Europe is uncertain; but it appears to have been unknown in 1884, when the first edition of Veitch's *Manual* was published, and was described as a distinct variety by Beissner in 1891. Seeds were apparently sent from Mexico by Roezl in 1856, and plants[1] raised from these on the continent do not seem to differ from the Colorado Douglas.

According to the experiments of Johannes Rafn, of Copenhagen, the germination of the seed of Douglas fir from Colorado is quicker and much better than that from the Pacific coast.[2]

In England young plants of the Colorado Douglas[3] have ascending branches, and are more narrowly pyramidal in habit than the Oregon Douglas, which has wide-spreading horizontal branches. Owing to its slowness of growth, the Colorado variety has short internodes between the branches, which give it a bushy appearance. The blue tint of the foliage can scarcely be relied on as a distinctive character, as it is variable in intensity and often disappears with age. The leaves are usually thicker, but do not differ in length or shape from those of the Oregon Douglas, the sharp-pointed apex being characteristic of both forms in the young stage. The young branchlets of the Colorado variety are often either quite glabrous or show only a few minute hairs under the lens, whereas those of the other form are distinctly pubescent. In wild trees, judging from herbarium specimens, this distinction does not occur.

---

[1] *Pseudotsuga Lindleyana*, Carrière, raised from Mexican seed sent by Roezl, and *P. glaucescens*, Bailly, also probably from Mexican seed, belong to var. *glauca*, and bear cones with strongly reflexed bracts.

[2] *Trans. Roy. Scot. Arbor. Soc.*, xvi. 408 (1901).

[3] The Colorado Douglas in cultivation in England has been supposed by Schwappach (cf. Richardson in *Trans. Roy. Scot. Arbor. Soc.* xviii. 195, with figure) to be *Pseudotsuga macrocarpa*; but there is no evidence to support this opinion.

The Colorado Douglas displays in cultivation well-marked peculiarities, which are mentioned in detail on pages 825, 826.

2. Lemmon[1] has described several other wild varieties, as var. *suberosa* from Arizona and New Mexico, var. *elongata* from the base of Mount Hood in Oregon, and var. *palustris* from swamps in the Lower Columbia Valley.

3. A considerable number of cultivated varieties have been distinguished by Carrière and Beissner, most of which are not worth mentioning, as their distinctive characters are trifling and inconstant. Fastigiate and pendulous forms are known, but are rarely met with.

Var. *Stairii*, with yellowish foliage, originated[2] at Castle Kennedy.

Var. *Fretsii*, Beissner,[3] is very peculiar in the foliage, as the leaves are short and broad, only $\frac{1}{2}$ inch in length, very obtuse at the apex, and resembling those of *Tsuga Sieboldii*. This originated in the seed-bed, and was sent out by Messrs. Frets & Sons of Boskoop, Holland.                              (A. H.)

Other varieties occur in cultivation which, though very distinct in habit, are not, in my opinion, worth naming. Among the best of the pendulous forms is one at Bury Hill, Dorking, the seat of R. Barclay, Esq., which in 1908 was 88 feet high.

I noticed in October 1907, near Boldrewood in the New Forest, on the north side of the drive, two trees, one of which was a typical Oregon Douglas fir with drooping branches, and leaves very silvery on the under side. Another close by had much narrower, stiffer, and darker foliage, and denser branches, the leaves much less silvery, and the cones closely packed near the summit of the tree in a manner unusual in this species. The former tree measured 70 feet by 6 feet 10 inches, the latter 66 feet by 5 feet 2 inches. Others of the latter type, standing near the gate leading into Mark Ash, bore no cones at all.

Near Eggesford House, in the higher walk, I saw a Douglas fir tree, so distinct in habit that it might be easily mistaken for another species. It had thin greyish foliage, pendulous branchlets, and very few cones, and measured 80 feet by 5 feet 7 inches, whilst ordinary Douglas firs planted close by were much thicker in proportion. I believe that by selecting such trees as seed-bearers we may ultimately succeed in obtaining distinct races which, for economic planting, will be much more valuable than trees of unknown origin.                              (H. J. E.)

## DISTRIBUTION

The Douglas fir has an extremely wide distribution in western North America, extending from north to south over 33° of latitude, between the parallels of 55° and 22°, and ranging from the Pacific coast to east of the Rocky Mountains. It occupies practically all of this vast territory except the higher elevations of the mountains and the desert and prairie regions of lower altitudes, where the rainfall is slight. It is the dominant tree of the great western forest, always growing in mixture with other

---

[1] *West American Conebearers*, 57 (1895).

[2] This variety is fully described in *Gard. Chron.* 1871, p. 1481. I have seen the original, which is now a small unhealthy looking tree; as are all those we have seen elsewhere. The best, perhaps, is a large dense bush rather than a tree, growing in Wood's nursery, near Buxsted, Sussex.—(H. J. E.)

[3] *Mitt. Deut. Dend. Gesell.* 1905, p. 74, f. 8.

conifers, which have a much more restricted distribution. In Montana it is associated with the western larch : in California it encroaches on the redwood belt; in south-western Oregon it is mixed with the Lawson cypress; while in the rest of the great forest of this state, and of Washington and northern Idaho, *Thuya plicata* is usually its constant companion. The various silver firs, hemlocks, and the Sitka spruce also take part, in different localities, in the mixture of coniferous trees in the Douglas forest. Towards the edges of the prairie regions and in the drier parts of the mountains, the Douglas fir gradually gives place to *Pinus ponderosa*, which is the characteristic tree of dry soils, where a very moderate rainfall prevails.

The northern limit of the Douglas fir extends from near the head of the Skeena River, latitude 54°, in the coast range of British Columbia to Lake Tacla in the Rocky Mountains, latitude 55°, reaching its most easterly point near Calgary in Alberta. In the coast range, the tree grows at some distance inland north of latitude 51°; while south of this line it is common on the coast of the mainland and in the island of Vancouver ; and in this region, and in Washington and Oregon, between the western foothills of the Cascades and the sea, it is most abundant and of its largest size. It attains its maximum development, 300 feet in height, in Vancouver Island and on the northern slopes of the Olympic Mountains in Washington, where the rainfall is excessive ; whereas, on the Cascades and in the interior of the continent, it rarely exceeds 150 feet in height. It is common, but only of moderate size, in the forests of northern Idaho and of western Montana,[1] ascending to 6000 feet.

The Douglas fir extends southwards along the Rocky Mountains, in the Yellowstone Park in Colorado, where it grows between 6000 and 11,000 feet altitude ; in Utah, to the east of the Wasatch range ; in northern and central New Mexico and northern Arizona, where it is common between 8200 and 9000 feet, being rare and of small size in the southern parts of these two states, where it ascends to 6000 or 7000 feet ; in the Guadalupe Mountains of western Texas, where it is abundant; and it spreads into Mexico, along the Sierra Madre range of Chihuahua and the mountains of Nuevo Leon, reaching its most southerly point near the city of San Luis Potosi.

In California it extends southward in the coast mountains[2] as far as Punta Gorda in Monterey county, but is not abundant, and is rarely over 150 feet in height ; inland it extends to the Sierra Nevada,[3] where it grows to a large size and ascends to 7000 feet. It does not occur in the arid tracts of Nevada and Utah, which lie between the Sierra Nevada and the Wasatch ranges. (A. H.)

So little seems to be known by British foresters as to the conditions under which the tree grows in America, that though I quite agree with the preceding account, it may be as well to add some of my personal experience of the tree as I saw it on my last journey in 1904. In the Blackfoot valley of Montana it is associated

[1] At Whitefish, Montana, an average tree, growing with the western larch, was 140 feet in height and 8 feet in girth, and showed 245 annual rings ; the sapwood, $\frac{3}{4}$ inch wide, containing 45 rings ; the bark was $2\frac{1}{2}$ inches in thickness.

[2] Jepson, in *Flora Western Mid. California*, 19 (1901) says that it is frequent in the Santa Cruz mountains; but is not known in the Mt. Diabolo and Mt. Hamilton ranges, or in the Oakland hills.

[3] Sargent in *Garden and Forest*, x. 25 (1897), says that it does not extend in the Sierras, south of the head of King's river, or within 100 miles of the territory occupied by *P. macrocarpa*. Jepson (*op. cit.* 20), makes its southern limit on the Sierras, about the head-waters of Stevenson Creek, which is not far from the head of King's river.

with *Larix occidentalis*, on the damper and shadier slopes of the mountains, at 4000 to 6000 feet, giving place to *Pinus ponderosa* in drier and sunnier situations, and never attains, so far as I could see or learn, more than 140 to 150 feet in height.

In Washington and British Columbia it is not seen in the dry country east of the Cascade range, but appears as soon as the forest begins to thicken near the watershed ; and on the western slopes of the mountains, from about 6000 feet downwards, is almost everywhere, except in swampy land, the dominant tree of the forest, attaining 200 to 300 feet in height from sea level to about 2000 feet.

It grows usually in mixture with *Thuya plicata*, *Tsuga Albertiana*, *Picea sitchensis*, and *Abies grandis* ; sometimes with a smaller proportion of *Pinus monticola*, and in drier situations with *Pinus ponderosa* ;[3] but in all the coast forests which I saw in Oregon, Washington, and British Columbia, including Vancouver Island, it outnumbers all the other conifers, except where forest fires have destroyed it, and its place is being to some extent taken by the hemlock, whose seeds seem able to germinate and grow in denser shade and in deeper humus than the young plants of the Douglas fir can endure.   Wherever the soil becomes too dry and rocky for hemlock and Thuya, the Douglas fir is able to grow, climbing up to the dry ridges and sunny slopes until it meets the more alpine species of conifers.   Its habit and size vary according to the soil and situation ; but I never observed any trees even in the most open situations, whose branches extended so far from the trunk as they do in English parks and gardens, and it does not attain anything like its full size unless it has a deep soil, a sheltered situation, and has been drawn up in youth by the struggle for existence, which prevails everywhere in the forest.

I saw a section of bark in the Washington State exhibit at the St. Louis Exhibition, taken from a tree cut at M'Cormick in Lewis Co., Washington, in the spring of 1904 ; which was said by Mr. Baker, who was in charge of it, to have been 390 feet high.   The same tree was recorded, however, in a Washington newspaper as having been 340 feet high and 42 feet in circumference (probably at three to four feet from the ground), and above 300 years old.   The tree is said to have contained 79,218 feet board measure, equal to above 8000 cubic feet, quarter-girth measure. The discrepancy in the account of the height and that given me by Mr. Baker may arise, in part, from the tree in falling, having jumped some distance from its stump.

Another tree even more remarkable, though not so large, was cut by Mr. Angus M'Dougall of Tacoma for the Chicago Exhibition in 1893.   This grew in Snohomish Co., Washington, and measured on the stump only 4 feet in diameter.   In falling it broke off at a height of 238 feet, where it measured 17½ inches in diameter, and was nearly free from branches to a height of 216 feet, which length was sent to Chicago.

The largest tree I have ever seen myself, which is said to be perhaps the largest known in Vancouver Island, grows by the roadside at Mr. P. Barkley's farm at Westholme, about 40 miles north of Victoria and 4 miles south of Chemainus Station.

---

[1] In the Bow river of Alberta it grows mixed with aspen (*Populus tremuloides*), and cottonwood (*P. balsamifera*).— Wilcox, *The Rockies of Canada*, p. 65.

What its height may originally have been is impossible to say, as it is broken off at about 175 feet. This tree has a very swelling base, which does not show so well as I could wish in the photograph (Plate 227). At the ground it measures 21 paces in circumference. Above the swelling, at about 6½ feet, I made it 41 feet 5 inches in girth. Assuming this tree to be 24 feet in girth at 100 feet high, and to have had a top at all in proportion to its girth, it must have contained 7000 to 8000 cubic feet of timber, or even more. The soil in which it grows is a deep fertile loam, and the timber standing in the valley is some of the finest in the island. Plate 228, from a photograph also taken in Vancouver Island, gives an idea of the forest, and shows on the right the trunk of a typical Douglas Fir; on the left, a trunk of *Thuya plicata.*

In the eastern part of the Washington Forest Reserve, Mr. Martin W. Gorman found this species up to 6000 feet, and measured a tree growing at 5500 feet, 132 years old, which was 18¾ inches diameter on the stump, with the bark 3 inches thick. Another tree at 1200 feet elevation, 244 years old, was 43 inches in diameter, with bark 6 inches thick. In the dry region the tree ranges from 70 to 120 feet high and from 20 to 50 inches diameter. He remarks that the species resists fire better than any other conifer of this region, and bears fertile cones at an earlier age than any other, a tree of only twelve years old having well developed cones.

Observations on the rapid growth of Douglas fir at various ages in its own country are given by Mayr.[1] In southern Oregon, on the best sandy loam, with a rich humus, he measured Douglas firs 130 feet high at eighty years old; a fallen stem, 100 feet high, contained 135 cubic feet.

The wood of the Douglas fir is known in the European, South African, and Australian markets as Oregon pine or Oregon fir, on the Pacific coast of North America as red fir or yellow fir, in Utah, Idaho, and Colorado as red pine, and in California is sometimes incorrectly called spruce or hemlock. It produces probably a larger quantity of commercial lumber than any other conifer in the new world, or at least on the Pacific coast; and is likely to continue the principal source of supply for most purposes, as the white pine (*P. Strobus*) of the New England states and Canada, and the long-leaved or pitch pine of the southern states become scarcer; and as its timber is likely in the future to become an important article of trade in Europe, both as an imported and home-grown product, I think it may be useful to give some particulars of the way in which the immense sawmills of Oregon, Washington, and British Columbia are managed.

First, as regards the growth of the timber, Prof. Sheldon, the Oregon State forestry expert, has published in the *Oregon Timberman*, May 1904, a valuable paper, which entirely confirms my own much more limited observations, and goes to show that the two forms locally known as red and yellow fir are not in any way distinct, but are simply the result of different conditions of growth.

When the trees grow in an open space, and have the annual rings, as is usually the case in youth, pretty far apart, they may attain at the butt 16 to 18 inches diameter at forty years. In such trees the thickness of the sap-wood is from 2 to 3

[1] *Fremdländ. Wald- u. Parkbäume*, 398 (1906).

inches, and the thickness of the bark, which under such conditions is comparatively smooth and greyish in colour, is about ½ inch. The timber of such trees would be known as red fir. When the tree, however, becomes crowded by its neighbours, and its girth increment is much slower, all the energy of the tree being devoted to upward growth, the rings become much closer, and trees of fifty to sixty years of age may be only 1 foot in diameter. The bark in such cases is much thinner, and the quality of the timber from the point at which the slower growth began much better, so that it would be classed as yellow fir when sawn up. Prof. Sheldon gives figures showing sections of such trees, his Plate 6 showing the influence of light, room, and nourishment on the growing tree. The tree from which this section was made was 143 years old- with a diameter of only 16½ inches. For 116 years it had stood in a crowded forest with large trees 4 and 5 feet in diameter all round it. Twenty-seven years ago the large trees were felled, and the growth immediately became much more rapid. The sap-wood in this case is 3½ to 4 inches and the bark 1½ inch thick. He says, "The result of this study is to conclude that the rapid growth of Oregon fir in the open produces red fir, and the subsequent growth when the trees begin to crowd each other produces yellow fir. Trees grown in dense clumps crowded all their life produce solid yellow fir. The growth of the upper portion of the tree may show larger annual growths in the centre than are found near the butt of the same tree. This is of interest in accounting for the immense height of the Oregon fir in many places, as trees 300 to 350 feet high are found in the forests of Oregon and Washington."

I asked experienced loggers whether they could distinguish red from yellow fir as they grew, and my impression was that they could not, though they said a very few blows of the axe would soon show the difference in the hardness of the wood. With the object of finding out the age at which the tree comes to maturity, I measured the rings of several trees recently felled at the logging camps which I visited. I am much indebted to the managers of these mills, for the facilities which they gave me to see the whole operations of a modern west coast lumberman. Among them Mr. Bradley of the Bridal Veil Company, Oregon; Mr. Browne, president of the St. Paul and Tacoma Sawmills, Tacoma, Washington, and his logging contractor, Mr. M'Dougall; Mr. Palmer of the Chemainus Mills, Vancouver Island; and Mr. Kenneth Ross, manager of the Big Blackfoot Lumber Company, Montana, were all most obliging and hospitable.

I found that the average age of mature trees 4 to 6 feet in diameter on the stump is 300 to 500 years. At an age of from 400 to 500 years, and possibly much sooner in some cases, the trees begin to decline in health, and some of those felled are more or less hollow. In all cases the annual rings for the first fifty to seventy years are very much thicker than for the next 300 years, the best trees having from four to five rings to the inch at first, and afterwards as many as fifteen to twenty. The better class trees are clear of branches up to about 120 to 150 feet, and in such cases produce wood free from knots, or "clear lumber" as it is called in the trade. Such clear lumber, however, even when a large number of trees are rejected by the fellers, does not exceed 15 to 30 per cent of the total product, and is worth a much higher price than the more or less knotty lumber known as " merchantable."

The business of lumbering which has been carried on for many years on a very large scale is, on the Pacific coast, as in most parts of North America, conducted in a way which, though perhaps necessary in order to meet the severe competition for price which everywhere prevails, would shock the feelings of any European forester, on account of its wastefulness and the absolute disregard which is paid to the future of the forest; which is in most cases abandoned to fire, as soon as the soundest, cleanest, and most accessible trees have been extracted.

A tract of land having been first surveyed, and its probable contents roughly estimated by the "cruiser," on whose judgment in selecting the best field of operations much of the success of the business depends, is purchased or leased from the owner on the basis of so much per thousand feet board measure. This estimate runs in most cases from 20,000 to 70,000 feet per acre, and as far as I could judge is rarely more than half, and often much less than half, of the actual contents, which in favourable situations amounts to as much as 300,000 to 500,000 feet per acre.

Unless the timber to be felled is near the sea,—in which case it is on Puget Sound often slid direct into the salt water, made up into rafts, and towed by steamers to the sawmill,—the next operation is to build a railroad up the valley to bring the logs from the forest to the sawmill. Sometimes the mill is in the forest itself, and a wooden flume of many miles in length is built, by which the sawn boards can be floated down to the nearest railway station. Sometimes the logs themselves are floated to the mill, where a large enough river exists; or a combination of railway, river, and flume may have to be adopted as the distance from the mill or station increases. The cost of extracting the logs from the forest and bringing them to their shipping point, governs the value of the growing timber, which is rapidly becoming less and less accessible as the best areas are cut over.

When the means of transport are completed, a "skid road" or a temporary tramway is built right up to where the trees grow, and powerful movable donkey engines are used, which are able, with a steel-wire rope, to drag logs of 40 to 50 feet long to a distance of 1000 yards or more from where they fall. Felling then commences and is managed as follows :—The most experienced man in the gang, having marked the trees to be felled, cuts a deep notch into one side at 4 to 6 feet from the ground, after carefully considering which way the tree should fall, so as to run least risk of lodging, or of breaking in falling. Both the undercutting and the sawing which follows, are done on spring boards fixed into a notch cut into the butt at 3 to 4 feet from the ground. When the two fellers, who sometimes make the notch themselves, have got within 5 to 6 inches of it, they insert large iron wedges in the sawcut, carefully watching the top of the tree to see where the wedges should be driven, so as to fell the tree with least danger to themselves and the log. After a few blows on the wedges the tree begins to lean and the men jump clear, calling out to warn others who may be near. There is some risk of large branches being torn off the falling tree or adjacent trees, and many accidents occur.

When the tree comes down, it is cross-cut by other men paid at a lower rate than the fellers, into such lengths as seem best. The smaller end of the log is then bevelled off and two deep notches cut, into which a pair of iron claws are fixed, and attached to the wire rope of the donkey engine. A signal is then given by wire from the men in charge of the log to the engineer, who commences winding up the rope, and with frequent stoppages caused by the log being jammed among stumps and other obstructions, it is at last dragged either to a prepared skid road, where another donkey engine hauls it to the loading point, or direct to where the trucks are able to load it. The loading is managed by building a rolling stage of heavy timbers down which the logs can be slid, or up which they are rolled by a donkey engine on to the trucks. Sometimes a dam is built and a pond formed, into which all the logs are dragged and rolled out on to the trucks. In fact there is no end to the ingenuity of the logging contractor in devising mechanical means for handling these great logs, often 4, 5, and 6 feet in diameter, with the least expense and trouble. Many logs which to an inexperienced eye would be thought valuable, are left either because they would cost more than they are worth to get out, or because they are more or less faulty; and in all cases that I saw, the work is done without the least regard to the younger trees, or to the future. Sometimes half the trees are left standing and as much is left after felling as is taken. The price per 1000 feet at the sawmill is the one governing idea.

When the logs reach the sawmill they usually go into a pond, from which they are hauled as required up an inclined plane to the saw bench. In the largest and modern mills the band-saw has replaced the gang-saws formerly used, and works at an incredible speed, saving a great deal of wood which was formerly eaten up by the saw. Some of the band-saws are double-edged; and after taking off the slabs and squaring the log, it is then converted into whatever sized lumber is wanted; the best quality being cut into vertical grained decking or flooring, 4 to 6 inches wide.

The ingenious arrangements by which everything in these great mills is arranged so as to save manual labour, must be seen to be appreciated. I found many of the men employed were Japanese, who are said to be excellent workmen and to possess both nerve and pluck.

When the boards are cut, the best are sorted out and sent to the drying kiln where they are dried for four or six days in order to prepare them for planing, tonguing, and grooving; which is usually done in another part of the same establishment by machinery, before the finished wood is put in cars for transport to the interior.

Much of it now goes to the middle states, and a great deal to South Africa, China, and Australia; but whenever very large-sized balks, masts, or piles are wanted, the Puget Sound mills are called on to fill the order, because no others in the world can supply timbers of such great size at so cheap a rate. Logs of 24 inches square and up to 100 feet long are regularly quoted.

The Douglas Flagstaff,[1] in Kew Gardens, came from Vancouver Island, and was

[1] Cf. *Journ. R. Hort. Soc.* xiv. 452 (1892).

presented in 1861 by Mr. Edward Stamp. It is 159 feet high, about 12 feet being underground, and is about 4½ feet in girth at ground-level. It weighed 4 tons 8 cwt., and was about 250 years old. In the British Museum of Natural History there is a section cut in 1885, 7 feet 7 inches in diameter, including bark, on which 533 annual rings may be counted. There is also in the Timbers Museum at Kew a fine section, 8 feet in diameter, cut from a tree on Puget Sound.

A technical report on the strength, weight, and structural value of Douglas timber is given by Hatt in *U.S. Bureau of Forestry*, Circular No. 32 (1904), from which it appears that the possibility of obtaining long and large pieces, combined with the exceptional strength and stiffness of the material, compared with its very moderate weight, renders it an ideal timber for structural purposes, and durable on exposure to weather.

In a report on the Forest Products of the United States for 1906 (issued March 1908)[1] I find that this species now comes second in the quantity of timber produced, being only surpassed by "yellow pine," under which heading are included all the various pines of the south and east except white and Norway pine (*P. Strobus* and *P. resinosa*). The quantity cut in 1906 was estimated at 5 billion feet, valued at 70 million dollars, of which the state of Washington yielded 68.5 per cent, Oregon 27.2 per cent, and all the other states together less than 5 per cent. The increase in production was very rapid in the last few years, and the average value had increased from 8.67 dollars per 1000 in 1899 to 14.20 dollars in 1906.

I am informed by Mr. R. S. Kellogg of the United States Forestry Bureau, Washington, that on the Pacific coast all masts except the smallest, and on the east coast the largest masts, are made of Douglas fir, which is transported overland from the Pacific coast.

It is the opinion of Lieut.-Commander Williams of the Bureau of Construction and Repair, U.S. Navy Dept., that there is practically no difference in the strength of Douglas fir and long-leaf pine (*P. palustris*); the latter, however, is considerably heavier. This appears to be now generally recognised by yacht-builders in Europe who use Douglas fir in preference to any other timber for the masts of racing yachts.

A letter on the timber of this tree in *Gardeners' Chronicle*, 1862, p. 452, gives the results of experiments made at Cherbourg by M. Serres on twelve specimens of squared mast timber sent from Vancouver, which showed that in strength it was almost equal to Florida pitch pine, and stronger than Baltic or Canadian pine. The weight of a compound mast made up of pitch pine in the centre and Baltic or Canada pine on the outside was about 12,200 kilos., whilst a solid mast of the same dimensions, made of Douglas fir, weighed only 8900 kilos. The cost of material and workmanship of the latter was very much less.

Mayr's comparison[2] of the wood as grown in various parts of Europe, with that grown in America, and also with that of silver fir, spruce, and larch, is well worth studying; but the age of the trees was insufficient to make the comparison conclusive.

[1] *U.S. Dept. Agr. Forest Service Bull.* 77.  [2] *Fremdländ. Wald- u. Parkbäume*, 399, 400 (1906).

With regard to the future of the Douglas fir forests it is very hard to say to what extent or for what period the present supply will last. Axe and fire are certainly destroying them at a great rate, but the reproduction all over the coast region is so good, and the danger of fire in dense young growths of trees so small, that many places cleared twenty to forty years ago are already covered with healthy young trees; and though the size and quality of these will probably never equal those of the virgin forests, yet there is no reason why, with reasonable care, the forests should be devastated as they are now. On the drier mountains of the interior, the danger of destruction is greater; and it seems to me that whilst Douglas pine is the dominant tree of the coast region, *Pinus ponderosa*, owing to its thicker bark and greater adaptability to dry soils and climates, will replace it in the interior.

## INTRODUCTION

The Douglas fir was discovered by Menzies at Nootka Sound in 1797. Seeds were, however, first sent home by Douglas in 1827, from which plants were raised by the Horticultural Society of London and distributed throughout the country. According to a note by Mr. Frost[1] the tree at Dropmore, which is usually considered to be the oldest in England, was raised from seed sown by himself in the winter of 1827-1828.

## CULTIVATION

The best account[2] of the cultivation of Douglas fir yet written is by Mr. Crozier, forester on the Durris estate in Kincardineshire, who has paid special attention to this tree, and is one of the most experienced foresters in Scotland. He prefers to collect home-grown seed, and considers that much may be done to improve the type of the tree commercially, by selection of the best varieties as seed-bearers; and states that the production of seed in good years is enormous, no less than 15,000 cones having been counted on an outlying specimen tree about 40 years old.

The seed ripens about the beginning of October, when the cones should be gathered without delay before the seed escapes. After storage in a dry loft through the winter, the cones are exposed to sun heat, which causes them to shed the seed. In the beginning of May the seed is sown in beds 3 to 3½ feet wide, one pound being allowed to every 8 or 10 yards. The seedlings are transplanted at two years old, and Mr. Crozier prefers to plant them out in the month of April one year later by notching, or if the ground is liable to be covered with bracken or herbage, by pitting in plants a year older.

So far as my own observations go this tree will not grow well on clay or on the oolite formation, but it thrives on greensand, and on sandstone of the Llandovery group at Tortworth. If desired to grow to a large size, it should be planted in a well-sheltered situation, where the soil is of sufficient depth and fertility to keep the trees growing for a long period, but in exposed situations the tops are ruined by the

[1] *Gard. Chron.* 1871, p. 1360.          [2] *Trans. Roy. Scot. Arb. Soc.* xxi. 31 (1908).

wind. All attempts to grow this tree into timber on bare, exposed, or barren downs and hillsides will, I believe, prove futile.[1]

The Colorado or glaucous variety has been so much spoken of, and is recognised so universally in cultivation as a distinct form, that we must speak of its peculiarities in full. It is usually supposed to be known by its colour, which is variable in all races of the tree; and I know of a case in which colour alone was considered by a forestry expert, to be sufficient to condemn as seed-bearing parents, a large number of vigorous healthy trees of great size, which were certainly of Pacific coast origin.

The Rocky Mountain forms, of which the Colorado one may be taken as typical, are constitutionally able to endure a continental climate; namely, one characterised by extremes of summer heat and winter cold; whilst the coast form is less hardy, though it will endure the extremes of climate in most parts of Great Britain, and is a very much larger, faster-growing, and, from a forester's point of view, more valuable tree.

They are at Colesborne equally liable to suffer from late spring frost after growth has commenced; but Mayr, whose experience of both is considerable, says that the Colorado form in Germany, does not suffer like the other, from the freezing of the immature shoots in autumn and early winter; and wherever this is a common cause of injury to the coast form, the mountain form should be tried instead. Such places, however, are rare in England; and on this subject I cannot do better than quote the opinion of Mr. Crozier. In a letter to me he says, "That there are two well-defined forms no one with practical experience of the tree will deny, but whether that known as 'Colorado' is confined to the state of that name seems doubtful. As a timber tree, however, my experience convinces me that in the north of Scotland at least it is a failure, and whatever advantages it may possess over the Oregon variety in its nursery stages, is really of no moment, as after a trial of between thirty and forty years, under the most favourable conditions of soil, shelter, etc., it has failed to make timber on this estate; while the Oregon variety, under much less favourable conditions, has never failed to make good headway. The cone[2] also differs from that of the Oregon variety in some important respects, being much smaller, with the bracts a great deal longer and reflexed."

" I made a further experiment with this tree some years ago, and may give you the dimensions of average specimens at the present time of Oregon and Colorado Douglas and Norway spruce, grown under exactly similar conditions side by side. The age of the Colorado Douglas and Norway spruce is twelve years, while the Oregon Douglas is ten years from sowing.

|  | Height. | Three last years' growth. | Girth at 6 inches high. |
|---|---|---|---|
| Oregon Douglas | 15 feet 6 inches | 8 feet 10 inches | $9\frac{1}{2}$ inches |
| Colorado Douglas | 10 ,, 11 ,, | 5 ,, 5 ,, | $6\frac{1}{4}$ ,, |
| Spruce | 8 ,, 10 ,, | 4 ,, 5 ,, | $5\frac{1}{4}$ ,, |

[1] According to Mr. Bean, in *Kew Bulletin*, 1906, p. 268, this species is used as a hedge plant at Monzie Castle, and answers the purpose very well, being dense and well-furnished.

[2] The cones on cultivated trees are very variable. Cf. *Gard. Chron.* xxviii. 12 (1900).

These have been planted seven years, and though for a time the Colorado held their own with the spruce they are now being left behind."

" We have raised some millions of the Oregon variety and find it sufficiently hardy for all practical purposes.   It does frequently make a second growth in the nursery stages, and these may be killed back, but the damage done in this respect is not serious.   After being planted out and established in the plantation, they are capable of bearing a greater degree of exposure than the Norway spruce, and may be seen on the lower spurs of the Grampians easily beating the latter.   In the treeless district of Buchan it does not do well, but neither does any other tree; but for general planting in Scotland, and with ordinary precautions, it is quite valuable."

" A member of an old firm of nurserymen informed me that it is about fifteen years since the Colorado variety first began to be sent to this country in quantity, and they only found out the mistake after the seedlings came up.   To speak of the Colorado as 'glaucous' and the Oregon as the green variety would be incorrect, as both vary in colour.   The Colorado may be found of all shades from green to a rich glaucous, while the Oregon runs from a dark bluish tint to a light green."

A most striking instance of the different rate of growth of the two trees may be seen in Dr. Watney's avenue at Buckhold in Berkshire, where Oregon Douglas about $3\frac{1}{2}$ feet high were planted in the winter of 1882-83, in trenched ground on a gravelly soil with some clay, underlaid at a depth of 10 to 12 feet by chalk.   Five of the best of these average in 1908 59 feet 8 inches in height by 4 feet in girth.   The largest was 65 feet by 5 feet 3 inches, showing $2\frac{1}{2}$ feet of annual height increase for twenty-four years.   Colorado Douglas (so-called) planted on the same land at the same time, were, when I saw them, not above half this size.

In the Great Bear plantation, on the same estate, planted October 1895, and steam cultivated 15 inches deep, Dr. Watney has measured six average Colorado Douglas, planted about 3 feet high, now 13 feet by $6\frac{2}{3}$ inches; six average Scots pine, planted about $1\frac{1}{2}$ feet high, now 18 feet by 12 inches; six average larch, planted about 2 feet high, now 19 feet 7 inches by $9\frac{1}{4}$ inches.   According to his experience the Colorado have many small branches which extend but a short distance from the stem, whilst the Oregon are distinguished by wide-spreading branches set much farther apart on the stem.   He says that the latter is the fastest-growing tree he knows, whilst the former is probably the most useless of all the common conifers he has grown; and yet he is told by a leading nurseryman that about one-third of the seed he buys produces plants which are apparently of the Colorado variety.   These trees are sold and planted somewhere, to the great ultimate loss and disappointment of the unwary planter.

The Douglas fir is usually healthy and little liable to insect or fungus attacks. However, of late years, a fungus, *Botrytis Douglasii*, Tubeuf, which is known as the Douglas fir blight,[1] has caused considerable danger to young trees growing in nurseries.   The leaves, especially those on the upper shoots, wilt and fall off; and

[1] Fisher, Schlich's *Man. Forestry*, iv. 461 (1907).

the plants frequently die. There is an illustrated article on this fungus in the *Journal of the Board of Agriculture* for June 1903.

I am informed by Capt. the Hon. R. Coke that in January 1907 there was a bad attack of this fungus, on two-year-old plants in the nurseries at Weasenham, Norfolk, and on some trees of the same age which were planted out in the previous autumn. He was advised at Kew to burn all the affected plants, and spray the remainder with "Violet Mixture." [1] About 25 per cent of the infected plants died or were removed as worthless; the remainder outgrew the disease, and are now (June 1908) looking well, though the fungus has not entirely disappeared. Capt. Coke adds that, after trying the so-called Colorado variety, he will plant no more of them; and that as seedlings of the Oregon variety vary a good deal, he prefers those which show a tendency to stop growing in time to ripen their leader.

The seeds are liable to be destroyed by the larva of an insect,[2] *Megastigmus spermotrophus*, which has been introduced into Europe from Oregon. The eggs are laid by the insect in the young cones, and one larva develops in each seed and destroys it. This pest has been observed at Mariabrunn, and has done great damage in Denmark; and during 1905 and 1906 was so serious at Durris in Aberdeenshire, that no seed was worth collecting there.

## REMARKABLE TREES AND PLANTATIONS

The largest tree that we have heard of in Europe, is at Eggesford, in Devonshire (Plate 229). This tree must be as old as any existing, for it was reported[3] in 1865 to be then about forty years old and 100 feet high. This, however, was an exaggeration, as three years later it was recorded[4] by Mr. A. Spreadbury, as being 93 feet by 12 feet at three feet from the ground. I measured it carefully in company with Mr. Asprey, agent to the Earl of Portsmouth, in April 1908, and found it, by the mean of two measurements from opposite sides, to be 128 feet by 18½ feet. About four feet from the ground two very large spreading branches come off, which at two feet from the trunk are 6 feet 9 inches and 5 feet in girth. At 30 feet from the ground, the stem is still 13 feet 5 inches round, and at 100 feet it girths 3 feet 3 inches; so that it must contain about 700 feet of timber. It grows on a lawn facing east, a little above the river Exe, on a soil which is evidently deep and fertile; and if the top is not broken may become a much larger tree, though it has only increased 35 feet in height in forty years.

The largest tree in the grounds at Endsleigh was reported by Mr. R. G. Forbes to be, in 1906, 100 feet high, with a quarter-girth of 26 inches in the middle; but in remeasuring it by climbing in 1908, he informs me that it is only measurable to a height of 87 feet. The quarter-girth over bark at 43½ feet is 26½ inches. Allowing

---

[1] This is composed of sulphate of copper, 2 lbs.; carbonate of copper, 3 lbs.; permanganate of potash, 3 oz.; soft soap, ½ lb.; rain water, 18 gallons.

[2] Cf. *Gard. Chron.* xxxix. 57 (1906), *Trans. R. Scot. Arb. Soc.* xix. 52 (1906), and *Journ. Board Agriculture*, xii. 615 (1906), where an article on the insect with figures is given by Dr. R. Stewart MacDougall.

[3] *Trans. Scot. Arb. Soc.* iii. 80.      [4] *Gard. Chron.* 1868, p. 1189.

2¼ inches for bark, its contents are therefore 87 feet by 24¼ inches, making 355 cubic feet, instead of 469, as stated in *Quarterly Journal of Forestry*, i. 107 (1907).

At the time of the Conifer Conference in 1891, a tree[1] at Dropmore was stated to have been then 120 feet by 11 feet; but I measured this tree in 1905 and could not make it more than 107 feet by 11½ feet, a considerable part of the top having been, as I was told, broken off by the wind. I measured it again carefully in June 1908, when it was 110 feet by 12 feet in girth.

At Walcot there is a very large tree planted in 1842, of which the Earl of Powis gave me a series of measurements. The first taken in 1860 was 74 feet by 7 feet; the second in 1872, 85 feet by 8 feet 10 inches; the third in 1892, 107 feet by 12 feet 9 inches; the fourth in April 1906, taken with a theodolite, was given as 122 feet by 15 feet 6 inches; all the girths taken at 4 feet. The cubic contents were 393 feet. I measured what I believed to be the same tree carefully from both sides, in March 1906, and made it 114 feet by 14 feet 2 inches at 5 feet, and noticed that the top had been somewhat broken. Thus it is evident in both these trees that after they had attained about sixty years old, the height increased much more slowly.

There are two trees at Powis Castle, one of which on the rabbit bank, near the park gate, I made from 112 to 115 feet by 11 feet 10 inches (this is the mean of two measurements from opposite sides as the tree[2] leans a good deal), and the other in a thick plantation, close to a pond, which, though I cannot, owing to its position, be confident that it is over 130 feet, may be 5 feet or more higher, and is more likely to increase in height than any Douglas fir that I have seen. It is only 9 feet 5 inches in girth and quite the finest timber tree of the sort I know in England.[3]

There is a tree at Highclere which, in 1903, was about 100 feet by 13 feet 8 inches. At Barton, Suffolk, a tree planted in 1831 measured in 1904, 107 feet high by 10 feet 1 inch in girth, and, while beautifully clothed to the base, was rather thin at the top with a divided leader. At Bury Hill, near Dorking, are perhaps the oldest and largest trees in Surrey, which, as Mr. R. Barclay told me, were planted by his father about 1832. The largest in 1908 measured 104 feet by 12 feet 2 inches, and appeared to have lost its leader recently.

At Albury, Sussex, there are two trees, which in 1904 measured 95 feet by 6 feet 2 inches and 82 feet by 8 feet 3 inches. At Cassiobury, Herts, there is another, which, according to the label, was planted in 1830, and had attained in 1904, 99 feet in height and 11 feet 3 inches in girth. This has now lost its leader, and has remarkably pendent branches, with leaves conspicuously white

---

[1] The tree at Dropmore, raised from seed, sown in the mid-winter of 1827, was planted out in 1829, and has shown the following growth :—

| Measured | 1837 | 1843 | 1846 | 1851 | 1853 | 1860 | 1862 | 1867 | 1868 | 1871 |
|---|---|---|---|---|---|---|---|---|---|---|
| Height in feet | 18 | 40 | 48½ | 62½ | 65 | 78 | 85 | 93 | 95 | 100 |

Cf. *Gard. Chron.* 1843, p. 808; 1846, p. 661; 1851, p. 246; 1853, p. 343; 1860, p. 854; 1867, p. 808; 1868, p. 465; 1871, p. 1360.

[2] Lord Powis had this tree measured in 1908, by a man climbing, as 127 feet by 12 feet 1 inch. I cannot account for the difference.

[3] I measured this tree again in July 1908, and having found a spot from which I could see the top, am confident that it is more than 130 feet high. It had increased 5 inches in girth in two years.

beneath. At Fulmodestone, Norfolk, Sir Hugh Beevor measured, in 1904, a tree 98 feet high by 8½ feet in girth. Henry saw a tree there in 1905, which was 82 feet by 9 feet 4 inches.

Many other trees which approach if they do not exceed 100 feet in height, may be found in the southern and western counties.

At Endsleigh, in Devonshire, which was visited by the English Arboricultural Society[1] in August 1906, there is a very fine plantation of Douglas fir in Gunoak Wood, of which careful measurements were made by Mr. R. G. Forbes, forester to the Duke of Bedford, in November 1906, from which it appears that the three largest trees in this plantation measure as follows :—

> No. 16. 120 feet high by 11 inches quarter-girth = 100 cubic feet.
> No. 23. 100    ,,    13    ,,    = 117    ,,
> No. 30. 110    ,,    13    ,,    = 129    ,,

Mr. E. C. Rundle, agent for the property, writes to me as follows : " The forester says that the trees must not be taken as a full crop, for there is space on the ¼ acre for forty trees instead of thirty-two. As to their age I believe they must be over fifty years, probably fifty-five, though an old man remembers their being planted. The quarter-girth was taken over bark at half the length of the tree, and an inch to the foot would be sufficient allowance. They are growing in an exposed position, but in the middle of a wood on high ground, and the soil is not at all good." The total contents of the thirty-two trees is 2857 cubic feet, an average of rather over 89 feet per tree. If 357 feet is deducted from the total for bark and small tops, it will leave a result of 10,000 feet per acre.

At Woburn, in a plantation called "The Evergreens," on a very light sandy soil, Mr. Mitchell, forester to the Duke of Bedford, showed me a plantation made in 1882, well sheltered by surrounding trees, and wrote me the following particulars :—

"The number of trees planted was 160, of which 132 are now left. I thinned them a few years ago, taking out only dead and suppressed trees. The area of land is as nearly as possible 2 chains square, and includes a few old Scotch and spruce fir. I measured the trees in three classes, as follows :—

> 72 trees : 50 feet by 6 inches quarter-girth = 900 cubic feet.
> 40   ,,   55   ,,   6½ ,,    ,,    645   ,,
> 20   ,,   50   ,,   4   ,,    ,,    111   ,,
>                            1656   ,,
> Deduct for bark at 8 per cent   .   136   ,,
> Total contents of timber .   1520   ,,

This works out at 3800 cubic feet per acre at twenty-six years after planting," say thirty years from seed. I may add that though these trees were planted close enough to kill all their lower branches, yet none of these had fallen, and a good many of the stems showed the same want of straightness which is so often evident in this species in England, and which I attribute to the unripeness of their sappy leaders before winter.

In 1904 the Earl of Ducie showed me a plantation of Douglas fir on a steep bank called Ironmill Wood near Tortworth, which, though of insufficient area to give the best results, is a good illustration of the growth of this tree on sandstone (Plate 231). The plantation was made in 1868, and was therefore thirty-six years old when I saw it. The area, as measured by Mr. Harle, agent to Lord Ducie, was 1 acre 28 perches; the number of trees standing was 238; their average height was about 80 feet; and their average cubic contents I estimated at slightly over 20 feet, making a total of about 5000 feet per acre. Mr. A. P. Grenfell, who visited the same place in the same year, made a more careful estimate based on the measurements of the trees standing on $\frac{1}{10}$ of an acre, and came to the conclusion that the total volume, with allowance for bark, was 5250 feet, which gives an annual average increase of 150 cubic feet per acre, no allowance being made for thinnings.

Mr. G. F. Luttrell of Dunster Castle, Somersetshire, showed me, in August 1906, a plantation of Douglas fir which he made in 1880 on a piece of waste land, which was growing only furze, on gravelly soil close to the rock, which is on the Old Red Sandstone formation. In the following December he had this carefully measured, with the following result:—Broom Ball Wood, area 3 roods 10 perches, planted entirely with Douglas fir at about 10 feet apart. Number of trees now standing, 264. Total contents, allowing half an inch for bark, 2491 cubic feet. Of these, 158 trees contain less than 10 cubic feet each, and only 7 contain above 20 cubic feet, the largest tree measuring 42 feet timber length and 10 inches quartergirth, equal to a volume of 29 cubic feet. The actual height of the tallest was 73 feet, of the shortest 48 feet.

The trees are valued as timber by Mr. Luttrell's forester at 6d. a foot, which amounts on the estimated quantity to £62 : 6s., equal to £76 : 13 : 6 per acre. Deducting from this sum, the expenses of planting and fencing, £6 an acre in 1880, equivalent in 1906, at 4 per cent. interest, to  .      .      .      .    £16  12   6
and the annual deferred rent at 5s. an acre, from 1880 to 1906, equivalent to  .      .      .      .      .      .      .      .      11   0   0
                                                                    _____
                                                                    £27  12   6
the balance, £49 : 1s., represents the actual profit per acre   .      .      49   1   0
                                                                    _____
                                                                    £76  13   6

It seems to me that the price of 6d. per foot for trees of this size is somewhat excessive, as those of less than 10 feet are hardly fit for anything but pitwood or rough fencing; but the value of the trees over 10 feet might be somewhat higher.

From the appearance of this plantation, in which many of the smaller trees were already suppressed and not likely to grow much more, it seemed to me that either a heavy thinning or clean felling was the proper thing to do, but this must depend on the local demand for timber of this size and quality. And if the small area, exposed position, and inferior agricultural quality of the land be taken into consideration, there can be no doubt that this has been an unusually profitable investment, and one which would fully justify planting Douglas fir on a large scale in this district. Mr. Luttrell states that where there is sufficient room and light the trees reproduce freely from seed.

In Scotland there are many Douglas firs exceeding 100 feet in height, but we cannot say which is actually the largest; and if we did, it would not hold good for many years to come. The tallest recorded at the Conifer Conference[1] in 1891 was at Lynedoch, on Lord Mansfield's property in Perthshire, which was then reported to be 92 feet by 12 feet, but had a fork at 60 feet from the ground[2] (Plate 230). Another tree at the same place, is the parent of the seedlings planted at Scone and Taymount, and was only $72\frac{1}{4}$ feet by 11 feet 2 inches, though planted in 1834.

One of the oldest trees is in the grounds of Scone Palace, and bears the inscription "raised from the first seed, brought by David Douglas in 1827, planted 1834." In 1850 it was transplanted to its present position, and this has doubtless checked its growth. It measured, in 1904, 96 feet high by 10 feet in girth. Its foliage is conspicuously white beneath.

At Drumlanrig, in Dumfriesshire, there is also an original tree, which was sent by David Douglas to his brother, who was clerk of works at Drumlanrig about 1832. It is growing in shallow gravelly soil near the top of a hill, overlooking a glen, and in 1904 measured 90 feet high by 11 feet 4 inches in girth.

Mr. R. Macleod of Cadboll sends me the measurements of four trees taken in 1907 by Mr. C. E. Cranstoun at Corehouse, near Lanark, as follows :—

|  | Height. | Girth at 5 feet. |
|---|---|---|
| No. 1. | $70\frac{1}{2}$ feet. | 12 feet 5 inches. |
| ,, 2. | 83 ,, | 10 ,, 7 ,, |
| ,, 3. | 85 ,, | 12 ,, 4 ,, |
| ,, 4. | 92 ,, | 7 ,, 6 ,, |

He adds that these were raised from the first seed sent to Scotland by Douglas; and that he finds by repeated measurements of several trees, that their rate of girth increase is about 2 inches per annum.

At Durris, in Kincardineshire, the original and largest tree, planted about seventy-two years ago, has now reached 114 feet by $12\frac{1}{2}$ feet, and contains over 300 feet of timber. At Buchanan Castle, Stirlingshire, Mr. Renwick measured[3] in 1900 an original tree 85 feet by 13 feet 2 inches. He informs us that the girth in 1908 is 14 feet $2\frac{1}{2}$ inches.

At Murthly Castle there are probably more large trees of this species than anywhere in Scotland, the plantation below the castle being especially fine, and also the avenue called the Dolphin Walk, where the trees, planted about 8 yards apart,

[1] *Journ. Roy. Hort. Soc.* xiv. 537 (1892).

[2] An accurate measurement of this tree, made in January 1908 by Mr. A. T. Kinnear, makes it 108 feet high by 13 feet 9 inches at 5 feet. The main stem up to the fork, 60 feet from the ground, contains 415 feet, and the two tops together, 48 feet, making the whole 463 cubic feet. [3] *Trans. Nat. Hist. Soc. Glasgow*, vi. 256 (1900).

average about 90 feet by 8 feet, and grow at the foot of a bank, in deep sand with pebbles in it, which looks like an old bank of the Tay, which is not far off.   In *The Garden* for 19th May 1900, some particulars are given of the trees here.   One, planted in 1847, measured on 11th August 1892, 86½ feet by 8 feet 10 inches, and on 24th March 1900, 97 feet 4 inches by 9 feet 10 inches.   A great many others were of about the same size.   This proves the diminishing rate of increase, both in height and girth, after forty to fifty years' growth, even when the lower branches remain. Mr. Fothringham states that all these measurements were taken by sending men or boys up the trees, and not with a dendrometer.   He adds that the temperature[1] in February 1895 was for several days below zero, and on one night went down to − 11°.

There is probably no plantation in Great Britain about which so much has been written as the Taymount plantation on the estate of the Earl of Mansfield, in Perthshire.    It lies about seven miles north of Perth, one mile from Stanley Station, and may be seen from the Highland Railway, which passes close to the east of it. The plantation covers eight acres of flat land, which is locally known as "till," two feet of light loam over red clay, and which may be worth for agricultural purposes 12s. to 15s. per acre.   This plantation was first fully described in the *Gardeners' Chronicle* of 10th, 17th, and 24th November 1888, by Dr. Schlich, than whom there can be no higher authority.   It was planted by the late W. M'Corquodale in the spring of 1860, with Douglas firs, two-year seedlings, two years transplanted, at 9 feet by 9 feet apart, with larch four years old, between every two firs, and an additional line of larch between every two rows, so that the trees stood 4½ feet apart, and each acre contained 538 Douglas and 1613 larch.   The latter were gradually thinned out, and were all removed by 1880.   The first thinning of Douglas took place in 1887, when about half the trees had already disappeared, 277 per acre only remaining.   Of these 75 per acre were cut, leaving 202 per acre.

Dr. Schlich made a careful estimate of a sample plot measuring $\frac{4}{10}$ of an acre of average appearance, and had a tree felled to ascertain its actual contents ; and from these data came to the conclusion that the total per acre was 3738 cubic feet of wood over 3 inches diameter, exclusive of top and branches, which gives an annual increment of 133 cubic feet per acre.   But this estimate being the gross volume, when reduced by about one-fourth, makes the quarter-girth measurement, as adopted in English practice, to be 2934 cubic feet.

After inspecting a sample area of Scots pine in the same district, Dr. Schlich goes on to say, "If grown in a well-stocked, overcrowded wood, and in localities of equal quality, Douglas fir is not likely to produce more solid wood, during the first thirty or forty years, than the larch, and probably also not more than Scotch pine."   He then goes into careful estimates of the probable future increase of the Douglas, based on data taken from America, where Dr. Mayr found that in the most favourable localities in the Cascade Mountains the average height of mature Douglas

---

[1] At Balmoral, where there are 25,000 to 30,000 trees, planted in the 'eighties, on a northern aspect at 1000 to 1200 feet altitude, Mr. Michie informs me in a letter that this severe frost, when the temperature fell to − 17½°, did no harm to the Douglas fir.

firs on the best soil was 213 feet, with a diameter of 6½ feet, whilst in Montana it only reached an average height of 148 feet, with a diameter of 2.6 feet, thus showing what an immense influence the soil and rainfall have on the growth of this tree.

From a cross section of Douglas fir grown in Washington and then in the museum at Cooper's Hill, Dr. Schlich remarks "that the rate of growth indicated in this section, up to thirty years old, resembles that of an average tree in the Taymount plantation in a striking degree, as follows : diameter of average tree at Taymount at 4½ feet, 12 inches ; diameter of thirty years' growth on the section from America, 11.9 inches.

After visiting a second growth area of pure Douglas fir on Ladds farm, about four miles from Portland, Oregon, which was believed to be of about fifty years' growth ; I came to precisely the same conclusion, and though I had not then seen Dr. Schlich's article, I wrote in my journal at the time, that the trees in Oregon were very similar in density to those at Taymount, but decidedly cleaner and better grown, and having regard to their greater age and better soil, they might average 100 feet by 4 feet, and I estimated their cubic contents at something like 6000 feet per acre.

When I first visited Taymount, in April 1904, I determined to estimate it for myself, without regard to what others had done. I therefore paced an area of 100 yards long by 50 yards wide in what I thought a fair average of the whole planta-tion, and found that there were on it ninety-nine trees of the first size, and fifty trees of the second. I did not reckon a number of other trees, which were so small, crooked, or poor, that they could not have been sold profitably with the better ones ; and, judging from a fallen tree which I was able to measure accurately, which was 55 feet long by 10 inches quarter-girth, equal to 38 cubic feet, came to the conclusion that the total volume of saleable timber at forty-four years after planting, or forty-eight years from the seed, did not much exceed 5000 feet per acre.

Sir Hugh Beevor visited Taymount in the autumn after I was there, and made an estimate in a different way by taking three different areas of ¼ acre each, and measuring everything on those areas. He found 96 trees of 12 inches quarter-girth and upwards at six feet from ground; 44 of 10 and 11 inches ; 44 of below 10 inches ; and estimated the total contents per acre at 6226 cubic feet.

I revisited Taymount in September 1906 in order to compare it again with what I had seen since in America and in England. I measured twenty trees in the fifth row and twenty in the tenth row from the bank on the east side of the plantation nearest to the high road. I found that their average girth over bark at 5 feet was slightly under 4 feet, the largest being 7 feet 10 inches and the smallest 2 feet 3 inches. I estimated the average timber length of these trees at 60 feet, and the quarter-girth, under bark at half this length, at 8 inches. If this is approximately correct, their average contents would be 26 feet 8 inches, and their total per acre something like 5400 feet, which very closely agrees with my previous estimate, allowing for the increase of two years.

A very different estimate was made by Dr. Somerville in a paper on " Exotic Conifers in Britain," which was printed in the *Journal of the Board of Agricul-*

*ture*, December 1903, and of which an abstract appeared in *Transactions of Royal Scottish Arboricultural Society*, xvii. 269.   This was based on measurements made in June 1903, by the late Mr. Pitcaithley, forester to the late Earl of Mansfield, who selected two typical areas of $\frac{1}{10}$ acre each, on which he counted and measured the trees, of which he found eighteen on one and twenty-five on the other area, and accurately measured the cubic contents of two trees, one of which contained 46.76 cubic feet and the other 39.49 cubic feet measured down to 3 inches diameter. Dr. Somerville, assuming Mr. Pitcaithley's measured trees to be average ones, brings out the total cubic contents per acre by quarter-girth measure as 7977 cubic feet, and comparing this with Dr. Schlich's estimate of 2934 cubic feet made fifteen years previously, comes to the conclusion that the average increase per acre in that period was no less than 336 cubic feet per annum.

This in my opinion is a mistaken calculation, and if compared with the annual increment of 150 feet per acre per annum in Lord Ducie's plantation and the results of the measurement of Gunoak wood, both on better land than that at Taymount, we must hesitate to accept it as even approximately correct.[1]

The important point to consider is how long these trees will continue to maintain their rapidity of growth, and what will be the value of the timber?  My own belief is that they fall off in their rate of increase; that the larger ones will continue to suppress and starve out the weaker ones, as they have already done to a great extent; and that the timber of Douglas fir grown in the country will never compare in quality or value with the imported timber, which, it must always be remembered, is from 200 to 300 years old, and selected both in the forest and the mill from a very much larger quantity.

Dr. Schlich writes me as follows : — "As to the quality of the Douglas fir timber, I merely quoted what the late Mr. M'Corquodale told me.   Since then I have paid some attention to the subject and noticed that in timber from young Douglas firs there is a considerable difference between spring wood and summer wood; hence I am sure, and in this I agree with you, that only trees of considerable age will yield timber equal in quality to that of larch, if at all."

There are other causes, which tend to make the production of clean, straight timber difficult, in many situations and on many soils in this country, and which should be considered by all who contemplate planting this species largely for profit.

The first is its tendency to form large and spreading branches, which it shows in a very marked degree.   In order to prevent this, the trees must be crowded to an extent which is only possible with success on soils of unusual depth, or on slopes composed of rock which is sufficiently disintegrated to allow the roots to penetrate deeply; in which case they may clean themselves when they attain a height of 60 to 80 feet; though I have never seen any in England which have naturally cleaned

[1] After this was in print I sent it to the Earl of Mansfield for his opinion, and am informed that in 1908 a careful measurement was made by his forester, Mr. A. T. Kinnear, of the whole of Taymount plantation, which now contains 1536 Douglas firs on the whole area = 192 trees to the acre.   These contain 51,456 cubic feet (under bark) or 6432 cubic feet per acre, being an increase of 3498 cubic feet per acre since it was measured by Dr. Schlich in 1888, equal to about 134 feet per acre per annum since it was planted, the rate of increase from 1888 to 1908 being about 175 feet.   The largest tree is 93 feet high, and contains 118 cubic feet.

their trunks.   In default of these conditions recourse must be had to pruning, which entails considerable expense, and must be repeated at frequent intervals.

The second is the tendency which I have observed in so many places for the trees to ripen their leading shoot prematurely in dry summers, and to make a fresh start in the autumn when wet warm weather sets in.   The result is that the second shoot is weak, immature, and usually becomes crooked either from frost, wind, or the settling of birds on it.   A double lead is then often produced, and the result is seen in many plantations, in the more or less crooked stems,[1] or in forks, which must seriously depreciate the value of the timber when brought to the sawmill.

A third is the effect of gales on the leading shoots, which owing to their great length and weakness, seems greater than on any other conifer, especially as owing to the rapid growth of the tree it overtops other species with which it may be mixed.   Even if the tops are not broken they become crooked, and often forked, in places exposed to wind, and the taller the trees become the more they are liable to this source of injury.

For these reasons it seems to me that the most profitable way of utilising Douglas fir, is to cut it at a comparatively early age, and utilise the wood for pit timber and estate purposes, for which purposes I am disposed to class it as superior to spruce or silver fir and inferior to larch.

In Ireland the Douglas fir grows very fast, and has attained in many places a large size.   The late Lord Powerscourt planted at Powerscourt in 1865, with his own hands a tree which measured in 1904 100 feet in height and 9½ feet in girth. There are good trees at Fota, Queenstown, 84 feet by 9½ feet in 1903 ; at Carton, 81 feet by 7½ feet in the same year ; at Stradbally Hall, Queen's County, 86 feet by 8 feet 3 inches in 1907 ; at Coollattin, Wicklow, 85 feet by 9 feet in 1906.   At Coollattin there are a few natural seedlings,[2] and several trees bear cones profusely ; but the forester has not been able to raise plants from their seeds, doubtless owing to the cones being attacked by the insect which has done so much damage at Durris in Kincardineshire.   At Castlewellan, Co. Down, there are fine trees, about 80 feet in height, which I could not measure on account of heavy rain when I was there in 1908.   One measured by the Earl of Annesley in August 1908 is 79 feet by 10 feet, but lost 12 feet of its top in a gale in 1902.

The late Mr. John Booth of Berlin was a great admirer of this tree, and for many years advocated its planting in Germany, where it is now beginning to be looked upon as one of the most valuable forest trees.   The result[3] of an experiment made by the late Prince Bismarck, on his estates at Sachsenwald near Hamburg, was sent me by Mr. Booth just before his death, and may be summarised as follows :—

An area of 1.16 acre, the soil being a coarse, somewhat loamy, diluvial sand, was planted in 1881, half with four - year old Douglas, 5 feet apart, and half with spruce, 4 feet apart.   In 1906, the Douglas plot consisted of 869 trees, measuring 3300 cubic feet of timber ; while the spruce plot, 1335 trees, only

---

[1] This defect is clearly seen in the Taymount plantation.

[2] Natural seedlings were seen by Henry, also at Dereen and at Powerscourt.

[3] Published in detail in *Zeitschrift für Forst- und Jagdwesen*, 1906, p. 8, of which a translation appeared in *Trans. Roy. Scot. Arbor. Soc.* xx. 104 (1907).

measured 1700 cubic feet.   The market value of the timber, which could be used for poles and pit-props, worked out at about 1000 marks for the Douglas, and about 360 marks for the spruce.   Thus, growing on the same soil, the Douglas, as compared with the spruce, had yielded about twice the amount of timber, with about three times the value.   I visited this plantation in August 1908, and measured two of the largest trees, which were 74 feet high by 3 feet 8 inches, and 2 feet 7 inches ; but the average was considerably less.   I noticed that the lower branches, though dead for several years, were not falling off; and that many of the trees showed the same irregularity in straightness that I have noticed elsewhere.   My impression was that unless heavily thinned, a large proportion of the trees would soon be suppressed by their more vigorous neighbours, and that such close planting was neither economic nor desirable.

## TIMBER

I have said so much about the timber of this tree in its own country that it only remains to speak of its probable future value here, and as this subject has been ably dealt with in a recent paper by Mr. J. D. Crozier,[1] I cannot do better than summarise his opinions.

He agrees with me that we cannot hope to compete with the imported timber in size, age, or quality, and thinks that in a young state it is not so dense in fibre or so tough as larch of the same age.   " For standing in contact with soil, and for such purposes as gate-making, fencing, etc., where the ability to stand wear and tear is a desideratum, it is inferior to larch, but there are many other purposes for which it is infinitely superior, and for the supply of which an infinitely greater volume of timber is required.   For constructive purposes of all kinds it is especially suited, and owing to the beauty of its grain and the ease with which it can be worked, it is valuable for the finished work of interiors.   The timber stains well, and when varnished, takes on and retains a beautiful gloss.   Outlying and badly-grown trees, when sawn up are liable to warp, but this defect is not apparent when dealing with trees of clean straight growth ; and with home timber more freedom may be used in regard to nailing.   In a younger state it has been tried and found useful as curing-barrel staves and headings, and for box wood, for which in this locality there is an unlimited demand."

"What the most profitable length of rotation may be is a question which will have to be determined by trade demands, but to provide timber of a class fitted for house construction, any period short of 100 years need not, I feel convinced, be contemplated, and on deep rich soils, probably other ten or twenty years will require to be added to that period."

"As a pitwood tree the Douglas fir is well adapted, and is deserving of consideration wherever crops cultivated for that purpose are found to pay.   Crowded together in pure plantation, by the time they have reached their thirtieth year, they will be found capable of yielding an amount of pitwood almost incredible to those who have not seen the tree so grown.   For this purpose the planting should not be done at more than 3 feet apart."

(H. J. E.)

[1] *Trans. Roy. Scot. Arb. Soc.* xxi. 31 (1908).

# CASTANEA

*Castanea*, Adanson, *Fam. Pl.* ii. 375 (1763); Bentham et Hooker, *Gen. Pl.* iii. 409 (1880); Dode,
    *Bull. Soc. Dendr. France*, 1908, p. 140.
*Fagus*, Linnæus, *Gen. Pl.* 292 (in part) (1737).
*Casanophorum*, Necker, *Elem. Bot.* iii. 257 (1790).

TREES or shrubs, belonging to the order Fagaceæ. Bark furrowed. Buds all axillary, no true terminal bud being formed, as the tip of the branchlet falls off in early summer, leaving a small circular scar close to the uppermost axillary bud, which prolongs the branchlet in the following year. Buds alternate, arranged on the long shoots in two ranks; scales numerous, two or three of which are visible externally, lowest pair lateral and each composed of two connate stipules, next pair each corresponding to a stipule and with or without a leaf-rudiment, following pairs of single stipules each covering a young leaf; all the single stipules accrescent and marking in their fall the base of the shoot with ring-like scars.

Leaves deciduous, alternate, simple, stalked, dentate with slender glandular teeth, penninerved, each lateral nerve ending in a tooth. Stipules ovate or lanceolate, scarious, deciduous, their scars visible in winter on each side of the leaf-scars, which show three groups of bundle-dots, and are placed on prominent pulvini, from which decurrent lines descend along the branchlet.

Flowers monœcious, strong-smelling,[1] fertilised by the wind, unisexual, in slender elongated erect catkins, of which those arising in the axils of the lower leaves of the branchlet open early and are entirely composed of male flowers, while the catkins arising in the axils of the upper leaves are shorter and bear female flowers at their base and male flowers on their upper part, the latter not opening until after the female flowers have been fertilised. Staminate flowers three to seven, in a cyme in the axil of a bract, and surrounded by minute bracteoles; calyx campanulate, deeply divided into usually six segments, stamens twice or thrice as many as the calyx lobes; filaments filiform; anthers two-celled, dehiscing longitudinally; ovary aborted. Pistillate flowers, sessile, solitary or two to three together, placed within an involucre of closely imbricated scales, subtended by a bract and two bracteoles; calyx-tube urn-shaped, divided above into six short lobes; ovary adnate to the calyx tube, six-celled, each cell containing two ovules, surmounted by six simple styles, which are exserted out of the involucre.

---

[1] Cf. Kerner, *Nat. Hist. Plants*, Eng. trans. ii. 200 (1898), concerning the nature of the odour of the flowers.

Fruit, ripening in one season, one to three nuts, ovoid, plano-convex or compressed, enclosed in an involucre, which is tomentose within and is covered externally with branched spines fascicled between deciduous scales, the nuts escaping by the ultimate splitting of the involucre above into two to four valves. Nut crowned by the styles, marked with a scar at the base, its shell lined with tomentum. Seed usually solitary, occasionally two to three in each nut, the aborted ovules, two to eleven in number, remaining at the apex of the seed. Albumen absent. Cotyledons thick, fleshy, undulate, sweet, farinaceous, remaining under ground on germination.

The genus[1] is confined to the warmer parts of the northern temperate zone, and much difference of opinion exists as to the various forms[2] which are met with. Formerly only two species were recognised, viz. *C. sativa* and *C. pumila*; but the former, widely spread over North America, Europe, and Asia, exists in certain well-marked geographical forms, which it is convenient to treat as distinct species. A small shrub, occurring in North America, near the coast in the South Atlantic states and in Louisiana and Arkansas, is considered by American botanists to be another distinct species, *Castanea alnifolia*, Nuttall, and will not be further alluded to. Four species have been introduced into cultivation and are distinguished as follows :—

I. Leaves without stellate tomentum, acute at the base.

  1. *Castanea dentata*, Borkhausen. N. America. See p. 856.

    Leaves tapering at the base, long acuminate at the apex, green and glabrous beneath, pendulous. Petiole glabrous.

II. Leaves with stellate tomentum, rounded or cordate at the base.

  2. *Castanea sativa*, Miller. Europe, N. Africa, Asia Minor, Caucasus, Persia. See p. 839.

    Leaves green beneath, always showing some trace at least of tomentum, not pendulous, coarsely serrate. Petiole and young shoots scurfy pubescent.

  3. *Castanea crenata*, Siebold et Zuccarini. China, Japan. See p. 854.

    Leaves green beneath, tomentum variable in quantity, shallowly and crenately serrate, the teeth often reduced to bristle-like points. Petiole, young shoots, and midrib densely pubescent with short hairs.

  4. *Castanea pumila*, Miller. N. America. See p. 857.

    Leaves silvery white and always tomentose beneath. Petiole and young shoots strongly pubescent.

---

[1] In *Castanea*, the leaves are deciduous, no terminal bud is formed, and the fruits ripen in one season. In *Castanopsis* the leaves are persistent, a terminal bud is present, and the fruits ripen at the end of the second season.

[2] Dode enumerates twelve species, some of which are alluded to in our accounts of *C. crenata* and *C. pumila*.

## CASTANEA SATIVA, Spanish or Sweet Chestnut

*Castanea sativa,* Miller, *Dict.* ed. 8, No. 1. (1768).
*Castanea vulgaris,* Lamarck, *Dict.* i. 708 (1783); Willkomm, *Forstliche Flora,* 428 (1887);
    Mathieu, *Flore Forestière,* 325 (1897).
*Castanea vesca,* Gaertner, *Fruct.* i. 181, t. 37 (1788); Loudon, *Arb. et Frut. Brit.* iii. 1983 (1838).
*Castanea Castanea,* Karsten, *Pharm. Med. Bot.* 495 (1882).
*Fagus Castanea,* Linnæus, *Sp. Pl.* 997 (1753).

A tree, attaining over 100 feet in height and an immense girth. Bark of very young stems smooth and olive green, soon becoming greyish white, after fifteen to twenty years gradually changing into a thick brown bark, which is deeply and longitudinally fissured. Young branchlets green, covered with a minute scattered pubescence above, and with longer hairs near the base; in the second year grey, glabrous.

Leaves (Plate 202, Fig. 11) not pendulous, oblong-lanceolate; broad, unequal, rounded and often auricled at the base; acuminate at the apex; with about twenty pairs of parallel nerves, raised on the under surface of the blade, each ending in a triangular tooth, which is prolonged into a long fine point; upper surface dark green, shining, covered with minute scattered pubescence; lower surface lighter green, with dense appressed stellate pubescence.[1] Petiole scurfy pubescent, ½ to 1 inch long. Stipules ⅜ inch long.

Nut, variable in size, abruptly and shortly acuminate at the apex, usually three in each involucre, in wild trees.

An elaborate description of the fruit is given by Lubbock.[2] The cotyledons are fleshy, occupying nearly the whole of the seed, undulate, and interlocking with each other at the margins. When sown, the pericarp, owing to the swelling of the cotyledons, splits in the soil at the apex, so that the shoot and rootlet emerge, the cotyledons remaining enclosed in the pericarp and being gradually absorbed. The germination thus resembles that of the oak; and the young stem similarly bears several scales (two to six in number) below the primary leaves, which resemble in shape those of the adult plant and bear deciduous stipules.

### IDENTIFICATION

In summer the leaves are unmistakable and can only possibly be confused with certain species of oak, like *Quercus serrata* and *Q. castaneæfolia,* which have, however, very different buds. From the other species of the genus, it is distinguished by the characters given in the key.

In winter the following characters (Plate 200, Fig. 1) are available :—Twigs stout, reddish brown or olive green, shining, conspicuously angled, glabrous for the most part but showing remains of glands and pubescence towards the base, which is conspicuously ringed by the fall of the previous season's bud-scales. Leaf-scars

---

[1] This pubescence often wears off, so that the leaves are glabrescent or even glabrous, when gathered in summer.
[2] *Seedlings,* ii. 537 (1892).

obcordate or semicircular, with three groups of bundle-dots, and set parallel to the twig on prominent pulvini, distichous on the long shoots. Stipule-scars long, linear. Buds ovoid, slightly rounded and not acute at the apex, those nearest the apex of the twig the largest; three scales visible externally, first scale small and short, second scale longer, both glabrous and ciliate; third scale clothed with appressed pubescence and appearing at the apex of the bud.

The twigs and buds of the chestnut resemble those of the lime tree. The pith affords a good mark of distinction, being greenish and five-rayed in Castanea, and whitish and round in Tilia.

In France, single trees have been noticed[1] in several localities, which bore catkins entirely formed of pistillate flowers. Such trees, according to Dode,[2] bear a large quantity of fruit; but the presence in the neighbourhood of a tree with staminate flowers is necessary for fertilisation. Mr. Lynch informs me that an isolated tree in a garden at Cambridge never bore fruit, until branches, with staminate flowers from another tree, were laid upon it; but it is uncertain whether this tree bore only pistillate flowers, or whether its own pollen was ineffective. Dode also mentions[2] a tree in the department of the Loire which never bore fruit, as its catkins were entirely composed of staminate flowers.

The number of seeds in the nut is also variable; and a single chestnut with three seeds has been known to germinate and produce three plants.[1]

### VARIETIES

The chestnut varies very little in the wild state, though the amount of pubescence which occurs on the leaf is remarkably different in many specimens. At the Scientific Committee meeting of the Royal Horticultural Society on 6th November 1900, some remarkable leaves were shown, consisting of but little more than the midribs, which had issued from the stump of a tree that had been felled; and it is possible that some of the narrow-leaved varieties originated in this way.

Schelle[3] enumerates nineteen varieties, which have been obtained in cultivation. Seven of these are forms with variously coloured and variegated leaves, viz.— *argentea*, *marginata*, *argenteo-marginata*, *argenteo-variegata*, *aureo-maculata*, *aureo-marginata*, and *aureo-variegata*. These are sufficiently explained by their names; and of those we have seen *aureo-marginata* is the best.[4]

Var. *heterophylla*.[5] Leaves variable in shape, some with irregularly-shaped teeth and occasional deep sinuses, others repand in margin and with few teeth.

Var. *aspleniifolia* (var. *laciniata*). Leaves with long narrow teeth, ending in long subulate points.

---

[1] Clos, in *Bull. Soc. Bot. France*, xiii. 96 (1866).

[2] Dode, in *Bull. Soc. Dendr. France*, 1908, p. 147.          [3] *Laubholz-benennung*, 63 (1903).

[4] There are small trees of the silver and golden variegated forms at Aldenham which are very handsome and well worth growing. A curious purple-leaved variety is described on p. 852.—(H. J. E.)

[5] At Verrières, near Paris, there is a tree, 28 feet high and 5 feet in girth, which has a few branches with normal foliage, all the others bearing leaves deeply and irregularly lobed. These two different kinds of branches bear fruit, which reproduces, when sown, seedlings with the form of foliage from which the nuts have been derived. Cf. *Hortus Vilmorinianus*, 56 (1906). There is a fine specimen of this variety at Murthly Castle; and Mr. Renwick has sent us specimens from a large tree at Finlayson, Renfrewshire, a few of the leaves of which are of the *heterophylla* type.

Var. *cochleata*. Leaves small, irregularly cut, hollow or with swellings in the middle.

Var. *prolifera*. Some of the leaves, usually the uppermost ones, remaining whitish tomentose beneath.

Var. *glabra*. Leaves thin and shining.

Var. *rotundifolia*. Leaves small, not exceeding 2½ inches in length, oval in shape.

Var. *pendulifolia*.[1] Branches pendulous.

Many varieties of the fruit, which are propagated by grafting, are cultivated in France and Italy. In France, the name *marron* is given to the best varieties, in which the fruit is large, globular, broader than long, and usually single in the involucre. According to de Candolle,[2] the Romans in Pliny's time already distinguished eight varieties, but it is impossible to discover from the text of this author whether they possessed the variety with a single kernel. Olivier de Serres[3] in the sixteenth century praises the chestnuts, *Sardonne* and *Tuscane*, which produced the single-kernelled fruit called the *marron de Lyons*. He considered that these varieties came from Italy; and Targioni[4] states that the name *marrone* or *marone* was employed in that country in the Middle Ages (1170). In England, the cultivation[5] of special varieties of the chestnut for its fruit is so little in vogue that it is not even mentioned in a late and comprehensive book on fruit culture, *The Fruit Garden*,[6] by Bunyard and Thomas.[7]

## DISTRIBUTION

The chestnut occurs wild throughout the whole of southern Europe, in Algeria, Tunis,[8] Asia Minor, the Caucasus, and northern Persia. It has not been found in the Himalayas where there are several species of *Castanopsis* occasionally known in India[9] as chestnuts, and is replaced in China and Japan by a closely allied species.[10]

Its northern limit in Europe is difficult to trace with accuracy, as the original area of distribution has been much extended by cultivation since the time of the Romans; and it has become naturalised in many parts. According to Willkomm the northern limit runs along the edge of the Jura and is continued through Switzerland to the south Tyrol, Carinthia, Styria, and Hungary, where it reaches Pressburg

---

[1] Lavallée, *Arb. Segrez.* 113. t. 33 (1885).
[2] *Origin of Cultivated Plants*, 353 (1886).
[3] *Théâtre de l'Agric.* p. 114.
[4] *Cenni Storici*, p. 180.

[5] Hogg, in *Fruit Manual*, 224 (1875), says that the chestnuts produced even in the southern counties are so inferior to those imported from Spain and the south of France, that no one would think of planting the chestnut for its fruit alone. He mentions two varieties, *Devonshire Prolific* and *Downton*, which succeed in hot seasons. Lord Ducie, however, informed Sir W. Thiselton-Dyer that he had once sent a sack of chestnuts to Covent Garden market, which realised £3; and was asked to send more, as they were the first on the market. [6] In Country Life Library, 1904.

[7] W. A. Taylor, in Bailey, *Cycl. Amer. Hort.* i. 296 (1900), enumerates and describes seventeen varieties of the European chestnut which are in cultivation in the United States.

[8] Battandier et Trabut, *Flore de l'Algérie*, 819 (1888); wild in forests of Edough near Bône in Algeria, and in Tunisia near Aïn-Drahm. Though cultivated near Tangiers and Tetuan it has not yet been found wild in Morocco. Cf. Ball, in *Journ. Linn. Soc. (Bot.)* xvi. 666 (1878).

[9] The chestnut has been planted at Bashahr, in the Punjab, where trees fifteen years old are 30 feet high and 4 feet in girth. *Kew Bull.* 1897, p. 113.

[10] The chestnut has been erroneously supposed, mainly on philological grounds, not to be a native of Europe, but to have been introduced at an early period from Asia Minor. The best discussion on this subject is by De Candolle, in *Geog. Bot.* ii. 688 (1855). A learned paper on the classical names of the oak and chestnut by H. L. Long appeared in Loudon, *Gard. Mag.* 1839, pp. 9-20. Dr. Bettelini's excellent account of this tree in *Flora Legnosa del Sottoceneri*, pp. 83-112 (1904), should also be consulted. He describes sixteen varieties, cultivated for their fruit in Switzerland and Italy.

in the west and the Bihar Mountains in Transylvania.    According, however, to Fliche,[1] it is not truly wild in any part of France nor even in Corsica, as it never forms part of the real forests and is generally found either as coppice or as isolated trees planted by man rather than as a true forest tree.

In France it is common in Provence, Dauphiné, the Cevennes, Perigord, Limousin, and all the central plateau, and it fruits abundantly in the environs of Paris. As in England, it was long supposed that there were large forests of chestnut in ancient times, and it is popularly believed that the severe winter of 1709 caused their destruction in the region of the Loire.   This is, however, an error, and the wood supposed to be chestnut, occurring in ancient churches and other buildings at Troyes, Reims, Sens, Chartres, and in Notre Dame at Paris has been conclusively proved to be oak.[2] The chestnut in France is rarely cultivated in high forest, as the timber is very liable to shake and to rot at the heart, so that sound pieces of considerable size are rarely obtained.    It is, however, often cultivated as coppice, for use as vine props and hoops for casks.    Mathieu mentions a tree growing near Sancerre in the department of Cher, which is 30 feet in girth and appears to be perfectly sound.    Mr. Chaumette[3] saw a chestnut in 1851 near Evian in Savoy, which had a girth of 54 feet, was 85 feet high, wide-spreading and well-shaped, but the trunk was perfectly hollow.

The chestnut is truly wild in Spain,[4] and appears to attain there a greater development than in any other country.   In the north, as in the provinces of Galicia, Asturias, and Biscaya, it constitutes forests of great extent, growing in company with *Quercus Toza*, *Q. sessiliflora* and *Q. pedunculata* or occasionally with the beech, and ascends from sea-level up to 2500 feet.    It abounds in the mountains near Avila; and between Baños and Bejar there are vast woods in which it occurs mixed with *Quercus Suber*.    It also occurs in the mountains of Toledo and of Estremadura and in the Sierra Morena.    In the northern parts of Navarre and Aragon, it ascends in the Pyrenees to 3000 feet.    In the extreme south of Spain, it no longer descends to sea-level, but forms a zone between 2700 and 5400 feet altitude in the Serrania de Ronda and the Sierra Nevada; and small woods also occur on the Alpujarras.    The chestnut is also common in Portugal, and various localities are mentioned for it by Colmeiro.

In Italy the chestnut occurs throughout the Apennines and also in Sicily, rising to 3000 or 4000 feet elevation; and pure woods are found, especially in Tuscany.    The most celebrated tree of this species, the *Castagno di Cento Cavalli*, growing on Mount Etna, was visited by Brydone[5] in 1770, who found it to be a hollow shell, which looked rather like a group of five trees growing together than a single tree.    Brydone made its girth 204 feet.    This ruin has lately been seen by Mr. Druce[6] of Oxford, who found four distinct parts still remaining, three of which

---

[1] Cf. Fliche, in *Bull. Soc. Bot. France*, liv. 132 (1907), concerning the recent discovery of chestnut charcoal at a prehistoric station in the department of Dordogne.   Dr. Christ, in *Flore de la Suisse*, suppl. 48 (1907), discusses the question of the distribution of the chestnut, and now agrees with Engler (*Ber. Schweiz. Bot. Ges.* xi. 81) that it is not truly wild in Switzerland, either in the Jura or in the Alps.

[2] Mathieu, *Flore Forestière*, 328, 329 (1897).          [3] *Phytologist*, iv. 71 (1851).

[4] Cf. Willkomm et Lange, *Prod. Fl. Hispanicæ*, i. 246 (1861); and Willkomm, *Forstliche Flore*, loc. cit.

[5] Brydone, *A Tour through Sicily and Malta*, i. 119 (1790).          [6] Cf. *Pharmac. Journ.* Feb. 27, 1904, p. 258.

looked like mighty trees, though not over 70 feet in height. It still fruits freely, and bears on its branches several bunches of the southern species[1] of mistletoe. Besides this great tree, there are four other enormous trees on Mount Etna, mentioned by Parlatore and Tornabene, viz. the *Castagno della Nave*, 22 metres in girth; the *C. della Navota*, 18.7 metres; and the two *C. di Santa Agata*, 22.6 and 26.3 metres, all sound and much more beautiful than the *C. di Cento Cavalli*.

The chestnut forms a part of the forests in the south of Germany, but is not indigenous, being introduced, it is supposed, by the Romans, as in Alsace, where it forms large woods, ascending to 2000 feet, on the slopes of the Vosges, and in the plain, as around Sulzmatt and Rohrbach. Along the foot of the Vosges in Alsace, chestnut coppice, treated on a fifteen years' rotation, is very common, the wood being used for vine-props. The chestnut is cultivated largely in southern Germany as a fruit tree, and as an ornamental tree in parts of north Germany, where in favourable situations, as near Brunswick and at Blankenburg, it ripens its fruit perfectly.

It is planted in southern Sweden and on the coast of Norway between Christiania and Christiansand, and occasionally ripens its fruit. According to Schübeler, it exists in Norway as a bush as far north on the coast as lat. 63°.

In Austria it is commonly planted, as in Bohemia and Moravia, while farther south it is supposed to be often wild. There is a remarkable wood of chestnut, on the domain of Mokritz in lower Carniola, which lies between 500 and 1500 feet elevation. In Carinthia, the chestnut constitutes 10 per cent of the mixed forest on the Neuhaus estate, ascending to 1800 feet; and at Bleiburg it is still a fine tree at 3100 feet elevation.

On the eastern side of the Adriatic,[2] from Fiume to Castelnuova, the chestnut forms a part of the forest, which is composed mainly of oak and laurel; while in the interior it is a considerable element in the oak forests of western Bosnia and Croatia. It occurs also mixed with the beech in Croatia, Bosnia, Herzegovina, and Montenegro. Wilkomm speaks of grand woods of chestnut in southern Hungary, Slavonia, Croatia, and Dalmatia; and mentions large wild forests in the Etsch valley in the Tyrol. Velenovsky[3] states that in the western Balkans, not far from the town of Berkovitza, there are extensive woods of chestnut, which are apparently wild, and have an undergrowth of the common hazel. Elsewhere in Bulgaria the chestnut appears to be planted, and is not a common tree.

The chestnut[4] is very common in the mountains of Greece, and is met with also in the islands of Keos, Naxos, and Crete. It occurs either solitary or gregariously, and in some parts of the mountains forms extensive woods.

In Macedonia,[5] Thrace, Albania, and Bithynia, the chestnut often forms the lower border of the deciduous forest, at 1200 to 3000 feet, occurring above the region of evergreen shrubs; but here and there it descends to sea-level. Chestnut woods occur on Olympus, in the peninsula of Mount Athos, and on Mount Kortiach near Salonica.

---

[1] This appears to be, judging from an imperfect specimen kindly sent by Mr. Druce, *Viscum laxum*, Boissier et Reuter. Cf. Nyman, *Consp. Fl. Europ.* i. 320 (1878).

[2] Cf. Beck von Mannagetta, *Veg. Verh. Illyrischen Ländern*, 147, etc. (1901).

[3] *Flora Bulgarica*, Suppl. i. 254 (1898).     [4] Halácsy, *Consp. Fl. Græcæ*, iii. 125 (1904).

[5] Grisebach, *Fl. Rumelica*, i. 339 (1843).

In Asia Minor the chestnut has been found wild in northern and western Anatolia; but it appears to be absent from the Taurus and Lebanon. In the Caucasus[1] it is found throughout the whole territory, and also in the Talysch, up to 6000 feet elevation; and it extends into north Persia.

A large chestnut grew in Madeira, on the estate of Count Carvalhal, at Achada, 23 kilometres from Funchal, and was reported by M. Joly[2] to have been about 160 feet in height with a girth at 3 feet 4 inches from the ground of 38 feet 8 inches. It was burnt down three years ago, and no trace of it now exists. The chestnut is not indigenous[3] in Madeira, although formerly many large planted woods existed there, most of which have disappeared.

The chestnut was probably introduced into England by the Romans. Charcoal, supposed to be of chestnut, was discovered by Mr. H. N. Ridley[4] associated with palæolithic implements and the bones of the rhinoceros in a brick-earth pit between Erith and Crayford in Kent. Mr. Clement Reid[5] has not found any evidence corroborating the possibility of the tree being a native of Britain in prehistoric times, and Mr. Ridley's specimen may be capable of some other explanation.

The tree[6] is mentioned in Anglo-Saxon literature as the *cisten* or *cyst-beam*. The modern name *chestnut* is a shortened form of *chesten-nut*, the fruit of the *chesten*, the early English name of the tree, representing the old French *chastaigne*, from the Latin *castanea*. King Henry II., in a grant to the Abbey of Flaxley in the Forest of Dean, says:[7] "*de eadem foresta dedi eis decimam castanearum mearum*"; and it is probable that the chestnuts here referred to were cultivated at this early time for their fruit and not for their timber.

Natural seedlings[8] are common in the southern counties, as in Kent, Surrey, Sussex, and Hampshire; and the chestnut may be considered to be naturalised in some places. Briggs states that it is naturalised in Cotehele wood near Plymouth; but as Bromfield remarks, it does not spread over waste places in the way that oak and pine commonly do.                                                                 (A. H.)

## CULTIVATION

The Sweet or Spanish Chestnut, as it is usually called, is on soils and situations which suit it one of the largest trees in England, and both from an ornamental and an economic point of view one of the most important of exotic hard-woods.

It is most at home in the southern counties, for though hardy in almost any part of Great Britain, it loves a warm soil and a warm summer climate, but will grow to a large size where the rainfall is as much as 60 inches per annum.

---

[1] Radde, *Pflanzenverbreit. Kaukasusländ,* 182 (1899).       [2] *Note sur un Chataignier Colossal.*
[3] Cf. Vahl, in Engler, *Bot. Jahrb.* 1905, p. 307.       [4] *Journ. Bot.* 1885, p. 253.
[5] *Origin British Flora,* 146 (1899).
[6] Cf. Murray, *New English Dictionary,* ii. 329 (1893). The village of Cheshunt does not take its name, as has been supposed by Ducarel and others, from the chestnut. Skeat, in *Place Names of Hertfordshire,* 37 (1904), proves that Cheshunt is a corruption of Cestrehunt, derived from Anglo-Saxon *ceaster,* a camp, and *hunta,* a huntsman.
[7] Ducarel, in *Phil. Trans.* 1771.
[8] There are numerous natural seedlings in Windsor Park, especially amongst the tall pines near Virginia Water. They are also common in Norfolk, at Fulmodestone and at Hargham.

With regard to soil, the chestnut is rather fastidious, as, though it will exist for a time, it rarely thrives on soils of a chalky or limy [1] nature, and will not grow in stiff clay or in peaty soil.

All the largest I have seen are on greensand or old red sandstone; and when cultivated for coppice-wood, which is probably its best economic use, it requires a better soil and climate than any other tree usually so treated. It is propagated by seed, which ripens in the southern counties abundantly in good seasons, though the fruit is inferior in size and quality to what is imported from Spain and France. The largest nuts should be chosen and kept dry in sand until spring, as they are devoured by mice, and if sown in autumn are liable to rot if exposed to much frost and wet. They should be transplanted when one year old and kept rather crowded in the nursery until they are 5 to 6 feet high, as they are liable to become very bushy if they have room to spread. They are not difficult to transplant, if grown in light soil, but must not be left more than two years before transplanting. [2]

A remarkable instance of the grafting of the chestnut on the oak [3] was shown me in the Botanic Garden of Dijon in France by M. Genty, the professor of botany there. The history of this tree is given in full by M. A. Baudot, in a pamphlet published at Dijon in 1907, from which I gather that in 1835 some acorns of the pedunculate oak were sown by M. Meline, five of which were grafted in 1839 with scions from the chestnut. Three of the grafts failed to take, another was injured by wind, the fifth pushed a shoot in the first year about 4 feet long, and grew so vigorously that it is now nearly 40 feet high with a girth of 4½ feet. The tree bore small fruit in 1852; and in 1903 some were sown, which germinated and produced three young plants, of which two are now planted out in the garden at Dijon, and a third was sent to M. M. de Vilmorin at Les Barres.

The varieties of the chestnut grown for fruit are usually grafted in French nurseries, but are rarely planted in England at present so far as I have seen.

As coppice-wood the chestnut is principally found in the hop-growing districts of Kent, Sussex, and Hants, where, until wiring was introduced, it was one of the most valuable products of English woodland, being cut at intervals of 8 to 12 years and realising frequently £2 to £3 per acre per annum. But now, though still more valuable than ash or hazel, it has fallen so much in price that these coppices are not as carefully managed as they used to be; and the split poles, which are so largely used for fencing, are said to be imported from France. In such coppices the stools are at 5 to 6 feet apart, because the thinner a hop pole is in proportion to its height the

---

[1] Fliche and Grandeau (*Ann. Chimie et Physique*, 1874, p. 354) proved by experiments, that the presence of a considerable amount of lime in the soil causes the chestnut to languish or to die, as too little iron is absorbed by the tree, and the normal function of the chlorophyll is deleteriously affected. Alphonse de Candolle, in *Nuovo Giorn. Bot. Ital.* x. 228 (1878), states that the chestnut is never found growing in Switzerland on limestone, and that in places where it is believed to occur on limestone, careful examination shows that the roots are surrounded by siliceous soil. However, he brings forward evidence to show that in the climate of south-eastern Europe, as in Hungary and Istria, the chestnut is occasionally found thriving on pure limestone.

[2] Sir H. Maxwell recommends sowing the best foreign nuts, but these produce seedlings which in my nursery are much more tender when young, than those raised from smaller English-grown seed, and when required for timber trees I should prefer the latter.

[3] M. Trabut, in his pamphlet, *Le chataignier en Algérie*, published as bulletin 37, by the Department of Agriculture in Algeria, states that he saw at the Villa Thuret in Antibes, a fine chestnut, which had been grafted on *Quercus Mirbeckii*.

more valuable it is, the young hop shoots, according to Cobbett, disliking a thick pole to twine round.

At Welbeck the chestnut is considered by Mr. Michie,[1] forester to the Duke of Portland, to be the most profitable tree to grow on sandy soil, as it grows much faster than oak and realises about 1s. 2d. per foot at a much earlier period. He showed me a plantation on Tressless Hill thirty-eight years old in 1903, in which the trees averaged about 65 feet high by 3 feet in girth, and stood about 150 to the acre. He said that they should not be grown without underwood, because in severe winters the unprotected trunks were liable to be cracked by frost near the ground.[2]

We have no exact records of the amount of timber per acre that may be produced by the chestnut when grown for timber in England, but I think that in the south on good land it would probably be greater than that of any other tree. One very remarkable case is a grove of 34 chestnuts and 9 oaks by the drive leading to Bicton House, Devonshire, which average about 100 feet high, by 6 to 7 feet in girth in the middle of the grove, and 9 to 12 feet on the outside (Plate 232). I estimated that this area was about half an acre, and the cubic contents of the timber on it about 5000 feet. At my request the late Mr. Mark Rolle had it carefully measured and wrote me on December 19, 1903, that the exact area on which the trunks stood was 1 rood 32 poles, though, of course, the branches extended over much more. The cubic contents were 7300 feet and the age of the trees about 150 years. We may therefore take at least 10,000 feet per acre as the result here.

Another very striking instance of the same character is a grove called "The Chestnut Tole"[3] in Mr. Ashley Dodd's park at Godinton, Kent, where a great number of fine trees, having clean boles of 50 to 70 feet high by 8 to 10 feet in girth, grow mixed with ash. One of the chestnut trees was 86 feet to the point where the branches began, and I think that the timber in this grove would produce as great

---

[1] Mr. Michie has sent the following note :—

"Sow seed in March, collected from sound, healthy, straight-growing trees, forty-five to fifty-five years of age, as I find that seeds from trees of that age produce stronger seedlings than seed from younger or older trees, or than foreign seed. At one year old I lift the seedlings, shorten the tap-root, and plant in nursery lines. Care must be taken to plant in fresh, sweet soil, as the root is very liable to malformation if in contact with fresh manure. In the following year cut them down to within one inch of the ground, which will cause them to throw out a strong and straight stem from 2 to 3 feet long ; after which, at three years old, they can be planted out with safety. Without this treatment before planting out, they generally require cutting off close to the surface, which is not always desirable in the planted area, owing to rank grass, bracken, etc., which smothers the young shoots.

I am greatly in favour of pure chestnut woods, very little thinning, and the encouraging of as much undergrowth as possible, especially on the outsides of plantations, to prevent cold and frosty winds blowing through. At sixty years of age the trees should stand no more than 16 feet apart, which equals 170 per acre, and taking them at the low average of 50 cubic feet per tree, means £425 (at 1s. per foot) when the crop is realised.

The above crop can be grown on a sandy soil, which is of little value for ordinary agricultural purposes ; for instance, in Birklands Wood, adjoining Budly Forest, where the soil is very sandy and light, oaks covering an area of about 100 acres, although from sixty to eighty years of age, are long and slender, and contain on an average not more than 6 cubic feet of timber each ; whereas some Spanish chestnuts, planted less than sixty years ago, contain fully eight times as much timber as the oaks.

On this estate the timber is used for making gates, gate-posts, and all kinds of fencing; also for window-sills of farm buildings, etc. Timber merchants buy it to supply the Sheffield trade (strickle handles, etc.), and also to put in the inside of threshing machines, for coffin boards, etc. The timber should be slowly and thoroughly dried before being used.

[2] The same thing has occurred at Kew ; and, as Sir W. Thiselton-Dyer pointed out, the cracks occur on the south side, and are the result of too rapid thawing by the sun.

[3] Tole seems to be a local name for a clump of trees standing on the crown of a hill.

a quantity per acre and of better quality, than the grove at Bicton. But, however attractive such plantations may be from an ornamental point of view, there is no doubt that the timber is worth much more if cut young; and, as a matter of fact, most of the old chestnut trees in the south of England are so shaky that a great part of their timber is only fit for firewood or fencing.

The chestnut is a good avenue tree in those parts of England where the soil and climate suit it, and there are fine avenues at several places. One of the best known to me is at Cowdray Park where there is an avenue about a mile long, commencing at the bottom of the hill, where the trees are very large, and running up to an elevation of 500 feet or more. According to Loudon this avenue contained 300 trees. Another very fine one at Thoresby is supposed to have been planted by Evelyn, many of the trees in which are about 20 feet in girth. I noticed here that the spiral twist in the trunk of the chestnut is variable in direction. Of three trees standing together in this avenue, one was twisted from left to right, one from right to left, and one had no twist at all; but this twisting of the trunk is commonest on light sandy soil and usually indicates shaky timber.

Another fine avenue of chestnuts is at Newhouse Park, on the property of Sir Robert Newman, near Mamhead, Exeter. This is 24 yards wide, with the trees 12 yards apart, which seems to be the correct distance for this tree in an avenue, as it requires more room than the lime or elm. These trees average about 15 feet in girth and are 70 to 80 feet high. The largest that I measured was 18 feet 8 inches in girth.

### REMARKABLE TREES

The number of large chestnut trees is so great that it is quite possible we may omit some of them, but there is no doubt that the most celebrated, and perhaps the oldest planted tree in England, is the Tortworth chestnut, which has been frequently described, and is figured by Strutt, plate xxix., and by Loudon, p. 1988. Strutt says that in 1766 it measured 50 feet in circumference at 5 feet from the ground, had a stem 10 feet high to the fork, and had three limbs, one of which was at that time 28½ feet in girth. It was said by Sir R. Atkyns, in his *History of Gloucestershire*, p. 413, to have been growing in King John's reign, and to have been " 197 yards in compass." It has since been mentioned and described by almost every writer on trees, but I am informed by Lord Ducie that a good deal of its history is more or less mythical. At present it is by no means a beautiful tree, and so much of its original trunk is decayed, that no measurement is of much value. I think that no one would recognise the existing tree as having formed the subject of Strutt's plate; but notwithstanding its age it still produces nuts, from which several trees have been raised and planted.

Another very large and celebrated chestnut, also figured by Strutt, plate xiii., and by Loudon, p. 1989, grew at Cobham Hall,[1] Kent, and must have been a finer tree than the one at Tortworth. It measured in 1822, according to Strutt, 29 feet in

[1] The finest chestnut now existing in this park grows at Ashenbank, and measured, in 1906, 93 feet in height and 13 feet 10 inches in girth, with a good bole 40 feet in length.

girth at the narrowest part 3 feet from the ground, 33 feet at 12 feet up, and 40 feet at the point where the trunk divided. It was "called the four sisters, from its four branching stems closely combined in one massive trunk," though the figure does not show this clearly. It has now entirely decayed.

Another historic tree, the "Monmouth Tree,"[1] at White Lackington, in Somerset, was destroyed by the severe storm of Ash Wednesday in 1897. It was reputed by tradition to have been the tree under which the Duke of Monmouth had a famous banquet in 1680. It was 25 feet in girth with a total height of only 49 feet, and had a very venerable appearance. Lord Petre measured in 1758 in Writtle Park, three miles from Ingatestone in Essex, a chestnut 45 feet in girth at 5 feet from the ground.[2]

In Waldershare Park, Kent, the seat of the Earl of Guilford, there are some remarkably fine chestnuts, the largest in girth being 23 feet 3 inches, but not a well-shaped or tall tree. The finest, in my opinion, is a tree 112 feet high with a straight and clean bole 50 feet long by 15 feet 2 inches at 5 feet, and carrying its girth well up. I estimated the contents of the first length alone at 50 feet by 36 inches quarter girth, making 450 feet of clean timber.

Fredville Park, the seat of H. W. Plumptre, Esq., in the same district of Kent, contains some splendid chestnuts, the largest of which is about 80 feet by 26 feet 3 inches. Another is called the Crows' Nest, from the fact of its having a platform, with benches and a table large enough to seat about twenty people, built in the crown at about 12 feet from the ground and reached by a ladder.

An immense but very ill-shaped chestnut tree dividing at 5 feet into three main limbs grows at Sunninghill Lodge, near Ascot, the seat of Percy Crutchley, Esq., of which a photograph was shown by him at the Lincoln Exhibition of the Royal Agricultural Society in 1907. This tree was carefully measured in 1816 by T. Luff, who estimated its contents at 716 cubic feet. A measurement made June 15, 1907, by M. C. Squires, gives its contents as 1282 feet, an allowance for bark of $1\frac{1}{2}$ to 2 inches being made.

The finest chestnuts growing near London are those in Kew Gardens, the largest of which measures 75 feet high, and 20 feet 10 inches in girth. These were probably planted early in the eighteenth century.

In Herts, there is a large chestnut at Lockleys Park near Welwyn, which the Hon. Arthur Bligh informs us is 21 feet in girth; and at Broxbournebury, Mr. H. Clinton Baker measured a tree in 1908, 65 feet by 23 feet 9 inches.

At Betchworth Park, part of the Deepdene estate, near Dorking, Surrey, there are many splendid chestnuts,[3] the finest though not the largest round, being 21 feet 5 inches in girth and 90 feet in height. For girth alone I know of few trees in England equal to one measured here by Henry, which, though its bole is only 8 feet long, is $26\frac{1}{2}$ feet in girth at the narrowest point.

---

[1] Cf. H. Norris in *Proc. Somerset Archæological Society* (1897), where a figure of the tree is given.

[2] Ducarel, *Phil. Trans.* 1771.

[3] An interesting article on the chestnut trees in Betchworth Park appeared in *Gardeners' Chronicle*, 1841, p. 4. At that date there were about 80 trees, all of large dimensions. Dr. Aikin, in *Monthly Magazine* for 1798, mentions the rows of old chestnut trees in this park.

At Petworth Park, Sussex, there are several very fine chestnuts, of which one measured by Sir Hugh Beevor in 1904, was no less than 118 feet high by 19 feet in girth, with a trunk clean to about 70 feet, and estimated to contain 800 feet of timber. It grows on the west side of the drive on the west side of the park, about two miles from the house. Another in a clump close to the house I found to be about 100 feet high by 21 feet 9 inches in girth.

At Steventon, North Devon, there is a very large tree in the garden, which Mr. Barrie measured as follows in 1890:—height 86 feet, bole 22 feet 6 inches, girth 16 feet 11 inches, spread 100 feet in diameter, contents 833 feet.

At Tyberton Court, Herefordshire, near the place where the big oak formerly grew,[1] and on soil heavier than the chestnut usually likes, there is a very fine twin tree, which looks as if two stems had started together from the same root. At the base the two measure 31 feet round and are about 95 feet high, one trunk being 20 feet, the other 17 feet 6 inches in girth.

At Highnam Court, Gloucestershire, there is a chestnut stool, which girths 32 feet at 3 feet from the ground, giving off four great stems 80 to 90 feet in height.

At Croft Castle, Herefordshire, there is a row of fourteen trees which were described in the *Transactions of the Woolhope Society*, 1871, p. 306, where their respective girths are given, and average about 17 feet, the two largest being then 20 feet 3 inches and 20 feet 5 inches. They seemed, when I saw them in 1904, to be long past maturity.

Below Warwick Castle, on the banks of the Avon, there is a chestnut having a large branch resting on the ground, where it has taken root and thrown up a large vertical stem, the only instance of self-layering[2] I have seen in this tree. The trunk in 1907 measured 16 feet 3 inches in girth. This tree is figured in *Gardeners' Chronicle*, 1873, fig. 222.

In Ashridge Park there are many fine chestnuts, one of which has its trunk covered with great burrs and is 24 feet in girth. At Chatsworth there is a chestnut tree of which Mr. Robertson, forester to the Duke of Devonshire, has been good enough to send me a photograph. He makes it 86 feet high, with a bole 45 feet by 15 feet 10 inches, and the cubic contents about 700 feet.

At Harleston, near Althorp, on Lord Spencer's property, are some immense chestnuts growing in a field near the church, on rich red sandy soil, the survivors of a row of which many were blown down many years ago. The largest measures 90 feet by 22 feet 6 inches, and was estimated by Mr. Mitchell, now forester at Woburn, to contain 1200 feet of timber (Plate 233). Another of about the same height has a bole 27 feet by 21 feet 6 inches and contains about 887 cubic feet.

If the length of clean trunk be considered, I have seen no chestnut equal to one at Thoresby (Plate 234), which has been drawn up in a thick wood of beech trees called Osland, and has a clean bole as straight as possible, 70 feet long by 11 feet 3 inches in girth, and a total height of about 110 feet. This was planted about 1730, and is on a sandy soil overlying the Bunter beds of New Red Sandstone.[3]

---

[1] Cf. vol. ii. p. 310.      [2] Henry saw a self-layering tree at Riccarton. Cf. p. 851.

[3] For further particulars of this remarkable plantation see our article on the oak, p. 322.

At Euston Park, Suffolk, the property of the Duke of Grafton, Mr. Marshall showed me, in 1905, in a wood called Barnham Springs, a remarkable growth of chestnut from a stool cut forty-two years previously. Sixteen straight stems about 60 feet high, and 2½ feet in girth, had sprung up from the outer edge of the stump, and collectively measured 30 feet in circumference. This growth seems to show how such trees as the one on Mount Etna have been originally formed, as in another 50 or 100 years these stems will probably seem like one tree. At Merton Hall, Thetford, Norfolk, a chestnut, planted about 1660, is 87 feet high, with a clean bole, 40 feet in length and 11 feet 4 inches in girth.

At Shrubland Park, Ipswich, the property of Lady de Saumarez, there are some very large chestnut trees in the grounds. The largest of these, according to Mr. Taylor, measures at ground line 47 feet; at 3 feet 31 feet; at 6 feet 27½ feet. Having had its top blown off some years ago, it is now only 55 feet high.

The finest existing chestnut, if height and girth together are considered, that I have seen, is a tree in a valley called Mackershaw, near Studley Royal, which seems to be the one figured by Loudon, p. 1986, of which he gives the height as 112 feet and the girth at 1 foot as about 23 feet. When I measured this splendid tree in 1904 I made it 112 feet by 20 feet at 5 feet from the ground, and it seemed to be in perfectly sound condition.

At Rydal Hall, Westmoreland, there is a very fine tree, girthing 26½ feet at 5 feet and 37 feet at the ground, of which the owner, Mr. S. H. le Fleming, has kindly sent me a photograph (Plate 235).

In Wales the finest tree I have seen is one which grows just outside the garden at Dynevor Castle, and measured in 1908 about 113 feet by 16½ feet, with a clean bole about 30 feet high. A photograph which was taken proved unsuccessful owing to its being surrounded by other trees.

Notwithstanding its southern origin, the chestnut grows with great vigour in many parts of Scotland, and, according to Loudon,[1] who quotes from Walker's *Essays*, p. 29, the first exotic tree planted north of the Tweed was a chestnut, of which in 1760 a part of the trunk remained, at Finhaven, an ancient seat of the Earls of Crawford. This was measured in 1744, and, as attested before two justices, was 42 feet 8½ inches in circumference close to the ground.

The largest tree I have seen myself is in the Cherry Park, near the stables at Inveraray Castle, and measures about 77 feet by 20 feet, with a bole about 16 feet long. This tree was said in the *Old and Remarkable Trees of Scotland* to have been in 1867 the largest in Scotland, though one at Tyninghame was as tall; and there are two fine ones, both over 16 feet in. girth, at Ardkinglas, in the same neighbourhood. Lord Kesteven informs us that there is a chestnut 25 feet in girth, growing at Stonefield, near Tarbert, Argyllshire. At Kirkconnell, south of Dumfries, Henry measured in 1904 a fine tree, 73 feet high and 18 feet in girth, with a bole of 25 feet. At Kirkmichael House, Ayrshire, a tree measured 18½ feet in girth in 1892; and at the Auld House, near Glasgow, two trees, about 60 feet high in 1904, were 16 feet 3 inches and 14 feet 11 inches in girth respectively.

[1] *Arb. et Frut. Brit.* i. 34, 90 (1838).

At Castle Menzies in Perthshire there are several very large trees, perhaps over 300 years old, one of which, in the washing-green, is about 20 feet in girth. Another in the park at Murthly, though not remarkable for height, has a trunk about 15 feet high and 19 feet 7 inches in girth, twisting from left to right. At Dupplin Castle, Perthshire, there are some very fine trees in a sheltered dell below the castle. One of these has a short bole no less than 21 feet 4 inches in girth; another is about 70 feet high by 17 feet 9 inches in girth. Many other large chestnuts in this county are recorded by Hunter; but as a rule they are remarkable rather for their age and girth than for their height, which rarely exceeds 80 feet in Scotland.

Sir Herbert Maxwell says[1] that the tallest recorded in Scotland is at Marchmont House, Berwickshire, which in 1878 measured 102 feet by $14\frac{1}{2}$ feet, with a bole of 32 feet; but Sir Archibald Buchan-Hepburn tells us that a tree at Yester House, Haddingtonshire, is 112 feet high by 18 feet 8 inches in girth, according to careful measurements, taken in 1908 by Lord Tweeddale's forester.

The chestnut at Riccarton, near Edinburgh,—which was described and figured in 1829 by Monteath, *System of Draining*, 209, as an old tree remarkable for layering, had two stems in 1905, one 17 feet in girth and the other, very decayed, 12 feet in girth, both giving off branches which had layered and become independent trees.

From Castle Leod in Ross-shire Mr. Wotherspoon sends me a photograph of a tree, which is probably the largest of the species existing so far north. It is 76 feet high and girths 28 feet close to the ground, 21 feet 4 inches at 5 feet, with a bole 14 feet long.[2]

In Ireland the chestnut thrives remarkably well, and, growing fast, might in many places be cultivated for its timber. At Fota the chestnuts in a plantation much exposed to the strong winds from the sea, withstood without injury the severe gale of 1903, when many other species were blown down.

The most remarkable chestnut in Ireland is the famous tree at Rossanagh, Wicklow, which was planted, according to Colonel Tighe, who has the family records, in 1718 (Plate 236). This tree is of the large spreading type with a short bole which divides into three mighty limbs. The girth of the main stem close to the ground was in 1903 49 feet, at 3 feet up $27\frac{1}{2}$ feet, and at 5 feet $29\frac{1}{2}$ feet. The height of the tree is about 80 feet, the spread of the branches being 100 feet in diameter. The three limbs girth respectively 12 feet 8 inches, 11 feet 2 inches, and 10 feet.

A very fine tree is growing at Powerscourt which was 84 feet high in 1905, with a good trunk carrying its full girth up to 18 feet and giving off the first branch at 20 feet up. It was $28\frac{1}{2}$ feet in girth at the ground, and $22\frac{1}{2}$ feet at 5 feet up.

At Clonbrock, Co. Galway, there is a tree growing on limestone, planted in 1801. It was 8 feet 6 inches in girth at 3 feet up in 1871, and 12 feet 9 inches in 1904. The chestnut grows at Clonbrock, where rhododendrons refuse to grow; and in the case of the tree just mentioned there is undoubtedly a large proportion of lime in the forest soil on which it stands. At Shannongrove, near Limerick,

---

[1] Green's *Encycl. Agric.* i. 373 (1907).

[2] This is probably the same tree which Loudon mentions, p. 2001, as growing at Castle Send (*sic*) in Cromarty.

there are some large trees, more remarkable for girth than for height, one being 24½ feet round at two feet from the ground.

At Rostrevor House, Co. Down, the seat of Colonel Sir J. Ross of Bladensburg, a chestnut, about 25 feet high, is remarkable for the large size and colour of the young leaves, which were purplish when I saw them in July, and are said to turn copper colour in autumn. This variety is of unknown origin, and I have seen nothing like it at Kew or elsewhere.

## TIMBER, MISCELLANEOUS PRODUCTS

A great deal has been written as to the use of chestnut wood for the beams and roofs of ancient buildings, both in England[1] and France, but it is now pretty generally admitted that most of the supposed chestnut wood is really that of the oak, which it slightly resembles.[2] This subject has been so well discussed by Loudon (pp. 1787, 1989, and 1992) that I need not further allude to it; but the properties and uses of the wood were apparently much better known formerly than now, and Mr. N. Kent, in 1792, wrote an excellent paper on the subject from which Loudon quotes largely (p. 1993). The pith of it all agrees with what I have been able to learn from various practical men—that the wood when young is as good or better than oak (because it has much less sapwood) for fencing, gate-posts, piles, and hop-poles; but that if allowed to become more than 3 to 4 feet in girth it is so apt to be shaky, that its value rapidly diminishes, and very old trees are usually only fit for firewood.

The timber in some cases remains quite sound to a great age and becomes mottled and streaked with dark brown like brown oak. I found the butt of an old tree of this nature, in a small timber yard in Wilts, where it had been lying seventeen years without any use being found for it. I had it cut into boards, from which the stiles and rails of an overmantel, and the frames of some doors have been made; and these, when polished with oil, were both in grain and colour of remarkable beauty. But even after this long period the wood was not dry, and shrank considerably after it was cut up, so that care must be taken not to put such wood together in a hurry.

Mr. T. Roberts, forester to the Earl of Egmont, at Cowdray, informs me that chestnut is used on that estate for joists, window-sills, door-jambs, and other purposes, and is found to last quite as long as oak and to be much easier to work up; he also thinks it less liable to insect attacks than oak (presumably sappy oak). But the trees when a hundred years old are all more or less inclined to be shaky,

---

[1] Sir George Birdwood in *Reports on the Cultivation of the Spanish Chestnut*, p. 9, note (India Office, 12th March 1892), states that the late Mr. T. Blashill, who was architect to the London County Council, pointed out in a letter to the *Times*, that the only instance he knew of chestnut wood in English mediæval carpentry is that of the chancel screen of the church, formerly of the Knights of St. John, at Rodmersham in Kent. The Rev. A. H. J. Massey, Vicar of Rodmersham, tells me, however, that the chancel screen is a modern one of oak, with portions of an ancient screen of chestnut wood worked into it; but the screen separating the Lady Chapel from the chancel, is composed entirely of chestnut wood.

[2] Mr. Blashill, in *Sessional Papers of the Royal Institute of Architects*, No. 12 (1877-78), has finally settled any lingering doubts which may exist. On the question of oak or chestnut in old timber roofs, he says that in some specimens of English oak, particularly in the variety called *sessiliflora*, the medullary plates are very thin and wide apart, and such specimens are often mistaken for chestnut, but a very clean transverse section will always render the plates visible. Though usually lighter than the rest of the wood, they are often dark, and such specimens have also been mistaken for chestnut. He goes on to say that the clean grain and pleasant working of chestnut make it very suitable for joinery, and there is no fear of its durability

which prevents their being cut into small scantlings. Mr. Weale tells me that it is used extensively in London for making coffins instead of oak. For making hoops, poles of chestnut are considered the best; and the wood is also largely used for making wine casks in France and Spain.

A section from the butt of a chestnut tree, said to be two hundred years old, was shown by the Marquis of Exeter at the Forestry Exhibition of the Royal Agricultural Society at Lincoln in 1907. This tree measured 7 feet in diameter at the butt and was fairly sound and free from shakes. I am informed by Mr. Danson that this tree was grown on a clay soil overlying ironstone, with a north-west exposure, about 230 feet above sea level.

At Shobdon Court, Herefordshire, the seat of Lord Bateman, I saw the trunk of a large chestnut, measuring 19½ feet in girth, lying on the ground. It was quite sound, with the exception of two small ring-shakes, and by counting the rings I found that it was 207 years old.

Such poles as are too thick for hop-poles make, on account of their durability, one of the best forms of park fencing that I know, of which many instances are quoted by Loudon. It is said that a park fence, erected in 1772 by Mr. Windham of Felbrigg, of oak and chestnut thinnings, was taken down in 1792, when the chestnut was found as sound as when put down, while the oak was so much wasted at the ground level that it could not be used again without support.

The Earl of Ducie exhibited at the Stroud show of the Gloucestershire Agricultural Society in June 1907 specimens of fencing posts made from chestnut, planted by himself in 1855, and cut in 1885, which had been in use for twenty-two years, and were still quite sound.

The walking- and umbrella-sticks, which are known in the trade as "Congo sticks," are saplings of the chestnut, which are easily manipulated when growing, the knots or markings for which these sticks are valued being produced by lacerating the bark through to the wood. They were formerly obtained from the north of France, but are now almost exclusively produced near Carlstadt in Croatia.[1]

The fruit of the chestnut is so well known that I need say little about it, and though in the colder parts of England it is often so small as to be of little use for human food, it is eaten by pheasants and deer. The large chestnuts eaten at dessert are imported, and are known under the name of "marrons" in France where they are preserved in sugar and form a very favourite sweetmeat.[2]

being equal, and probably superior to that of any wood (presumably he meant English wood) except oak. He spoke of a large bridge having been built about 1858 of chestnut timber, over the river Wye at Hoarwithy near Hereford. The bridge after nineteen years was taken down in a crippled condition, which he attributed partly to the design of the bridge, and partly to the decay of the timber at the numerous joints where water could lodge. Yet the great bulk of the wood was perfectly sound; and seemed to show that for ordinary work not subject to damp, the timber may be very useful. Although he could not admit its occurrence in ancient roofs, it might be very suitably used in preference to deal or pitch pine; and in church furniture it would probably, in course of time, take a colour which would be far better than that of the stained woods now so much used.

[1] *Kew Bulletin*, 1899, p. 53.

[2] In some parts of Spain and Italy, and in the south of France, chestnuts are ground into flour; and in the form of cakes, soup, and porridge, form a considerable part of the food of the poorer classes during winter. Specimens of chestnut flour and cakes are exhibited in the museum at Kew; and in *Kew Bulletin*, 1890, p. 173, an analysis of the flour is given by Professor Church, who considers that it is easily digestible and probably useful as food for children. Further interesting particulars concerning the use of the chestnut are given in *Reports on the Cultivation of the Spanish Chestnut* (India Office, 12th March 1892).

A tanning material,[1] extracted from chestnut bark, is prepared near St. Malo in France, and is largely exported to Belgium and to Glasgow.   It is said to be used to modify the colour produced by hemlock extract obtained from *Tsuga canadensis*.[2]

(H. J. E.)

## CASTANEA CRENATA, JAPANESE CHESTNUT

*Castanea crenata*, Siebold et Zuccarini, *Abh. Akad. Muench.* IV. iii. 224 (1846);   Schneider, *Laubholzkunde*, i. 804 (1906).

*Castanea japonica*, Blume, *Mus. Bot. Lugd. Bat.* i. 284 (1850).

*Castanea vesca*, Gaertner, var. *pubinervis*, Hasskarl, *Cat. Hort. Bog. Alt.* 73 (1844).

*Castanea vulgaris*, Lamarck, var. *japonica*, A. DC. *Prod.* xvi. 2, p. 115 (1864);   Shirasawa, *Icon Forest. Japon*, text 63, t. xxxiv. ff. 14-25 (1900).

*Castanea vulgaris*, Lamarck, var. *yunnanensis*, Franchet, *Journ. de Bot.* 1899, p. 196.

*Castanea sativa*, Miller, var. *acuminatissima*, von Seeman, in Diels, *Flora von Central China*, 287 (1901).

*Castanea pubinervis*, Schneider, *Laubholzkunde*, i. 158 (1904).

A tree, usually smaller in size than the European species, but occasionally attaining large dimensions.   It is probably only a geographical form of that species, but can readily be distinguished and may be kept separate, as it probably differs in growth and in cultural requirements.

The leaves are borne on shorter petioles, but resemble those of the common chestnut in shape, being rounded or cordate at the base and having about twenty pairs of nerves; but they are smaller in size and have much shallower serrations, with very long and fine spine-like points.   The main difference lies in the pubescence,[3] which is short and dense on the young branchlets, on the petioles, and on the midrib of both sides of the leaf.   In the common chestnut this very distinct pubescence is either absent or replaced by a scurf, very different in appearance.   The catkins of the eastern tree are more slender and the fruits of wild trees smaller than in the common species.   *Castanea crenata* also comes into flower, when still very young, and often bears fruit when quite a small shrub.

In China *Castanea crenata*[4] occurs wild, mainly in the mountains of the central provinces, as a tree about 40 feet in height; and is nowhere abundant, and so far as I have seen never forms woods of any extent.

[1] A similar extract, prepared from the wood of the chestnut, is largely manufactured in Corsica.   Mr. Southwell, Vice-Consul at Bastia, gave me some interesting particulars about this industry, when I visited Corsica in December 1906.   There are four factories near Bastia, which produce about 25,000 tons of extract annually.   The bark is not employed in Corsica, as the dark colour of the extract produced from it is objectionable.   Four tons of wood yield about one ton of extract.   The wood is cut into chips, which are soaked under pressure in hot water, which extracts all the tannin and some of the colouring matter.   The resulting liquor is concentrated *in vacuo*.   Practically the whole of this extract is used in England and Germany for sole-leather.   Mr. Southwell informed me that certain trees in Corsica had brown-coloured wood, which produced an unsaleable extract.   He had found by experiment that this brown colour in the wood is due to the presence of iron in the soil.—(A. H.)               [2] *Kew Bulletin*, 1893, p. 229.

[3] The pubescence over the lower surface of the leaf is similar to that of the European tree, and is very variable in quantity and persistence.

[4] The large chestnut tree occurring wild in China is considered by Dode to be distinct from the Japanese tree, and has been named by him *C. Duclouxii* and *C. Fargesii*, in *Bull. Soc. Dendr. France*, 1908, pp. 150, 158.

The low grassy hills of the Yangtse valley and the hills in Chekiang are often covered in places with a scrubby growth of chestnut bushes, scarcely ever over 5 feet in height. This is a distinct species,[1] and corresponds in many respects to *C. pumila* of America, the branchlets and petioles being covered with a dense, bristly pubescence, and the fruits extremely small, usually three in each involucre. This has been supposed to be *Castanea mollissima*, Blume,[2] an imperfectly known species.

The Chinese have distinguished from the most ancient times two kinds of chestnut, classically known as the *li* and the *erh*. The former, now known as the *pan-li* is the cultivated tree, the latter, known as the *mao-li*, is the wild form of the species, which produces remarkably sweet small fruit. These have been noticed by many observers, as by Abel[3] at Tatung on the Yangtze, by Père David[4] at Kiukiang, and by Fortune[5] near Ningpo, who introduced the small-fruited chestnut into England[6] in 1853; but we are unacquainted with any trees raised at that time. Similar small-fruited chestnuts are known in Japan, and were exhibited in London[7] in 1873.                                                                    (A. H.)

The chestnut is widely distributed in Japan where it is called "kuri," from Kiusiu and Shikoku, through the greater part of the mountain forests of Hondo, and in the plains as far north as central Hokkaido. It is usually mixed with other deciduous trees, but in some places forms pure forests of small area. Its wood is preferred for railway sleepers to any other timber, but is not much valued for building purposes. Though, according to Sargent,[8] it does not attain any great size, yet I measured an old tree in the Atera valley which was 15 feet in girth (Plate 237).

The tree is commonly seen on dry and barren hillsides in the form of coppice, which is cut every few years for firewood. It is also cultivated for its fruit, and several large-fruited varieties are grown which Sargent[8] says are equal in size to the best in southern Europe, and are largely consumed as food in the towns, and also exported from Kobé to San Francisco. These varieties are more precocious than the European tree, bearing abundant fruit when only 10 or 12 feet high, and he recommends their introduction from Aomori in the north of Hondo, as being more likely to endure cold winters than the French or Kobé varieties.

The Japanese chestnut was introduced into the United States[9] about 1891, and Rehder[10] states that it is shrubby and usually begins to fruit when about six years old. It has proved hardy as far north as Massachusetts. So far as we know it has not yet been introduced into England.                                                (H. J. E.)

[1] The shrubby chestnut of China is considered by Dode, in *Bull. Soc. Dendr. France*, 1908, pp. 151, 152, 153, to constitute three new species, *C. hupehensis*, *C. Seguinii*, and *C. Davidii*.

[2] *Mus. Bot. Lugd. Bat.* i. 286 (1850). Cf. Diels, *Flora von Central China*, 288 (1901).

[3] *Narrative of a Journey in China*, 165 (1818).          [4] *Plantæ Davidianæ*, i. 277 (1884).

[5] *Residence among the Chinese*, 51, 144 (1857).          [6] *Gard. Chron.* 1860, p. 170.

[7] *Ibid.* 1875, p. 270.          [8] *Forest Flora of Japan*, 69 (1894).

[9] Cf. Bailey in *Amer. Garden*, May 1891, who gives a description and figure of the tree; and *Garden and Forest*, viii. 460 (1895). W. A. Taylor, in *Cycl. Amer. Hort.* i. 294 (1900), who enumerates nineteen varieties of the Japanese chestnut, which have been introduced of late years into North America, gives the date of the first introduction as 1876.

[10] *Cycl. Amer. Hort.* i. 257 (1900).

## CASTANEA DENTATA, American Chestnut

*Castanea dentata*, Borkhausen, *Handb. Forstbot.* i. 741 (1800); Sargent, *Silva N. Amer.* ix. 13, tt. 440, 441 (1896), and *Man. Trees, N. Amer.* 220 (1905).
*Castanea vesca americana*, Michaux, *Fl. Bor. Amer.* ii. 193 (1803); Loudon, *Arb. et Frut. Brit.* iii. 1984 (1838).
*Castanea americana*, Rafinesque, *New Pl.* iii. 82 (1836).
*Castanea vulgaris*, γ. *americana*, A. De Candolle, *Prod.* xvi. 2, p. 114 (1864).
*Castanea sativa*, var. *americana*, Sargent, *Garden and Forest*, ii. 484 (1889).
*Fagus Castanea dentata*, Marshall, *Arbust. Amer.* 46 (1785).

A tree attaining in America 100 feet in height. Bark dark brown, and divided by shallow irregular fissures into broad flat ridges, separating on the surface into small thin appressed scales. Young branchlets with minute scurfy pubescence above, and with long hairs near the base; glabrous and grey in the second year.

Leaves (Plate 202, Fig. 13), pendulous, oblong-lanceolate, gradually tapering and unequal at the base, long acuminate at the apex, with about twenty pairs of parallel nerves, raised on the under surface, each ending in a triangular tooth, which is prolonged into a fine point; upper surface dull, dark green, glabrous; lower surface lighter green, glabrous, or with minute scattered hairs, thin but firm in texture. Petiole, $\frac{1}{2}$ to $\frac{3}{4}$ inch long, glabrous. Stipules, ovate-lanceolate, acute, puberulous, about $\frac{1}{2}$ inch long.

Nut,[1] usually much compressed, $\frac{1}{2}$ to 1 inch wide, gradually acuminate at the apex; two to three fruits together in each involucre.

This species is distinguished from the European one by the leaves being always cuneate and never cordate at the base, and never having any stellate tomentum, the under surface being either glabrous or covered with minute glandular hairs.

In winter it is readily distinguishable from *C. sativa* by the glabrous twigs and the more pointed ovoid buds, which have glabrous ciliate scales as in that species. The buds are smaller than in *C. sativa*, being only about $\frac{3}{16}$ inch long. In the specimens seen the twigs are much more slender, with very minute lenticels and small semicircular leaf-scars. (A. H.)

In America the chestnut is a common tree, and has a wide range from New England and southern Ontario southward along the Alleghany Mountains to central Alabama and Mississippi, and westward to Michigan, Indiana, central Kentucky, and Tennessee. So far as I have seen it does not attain so large a size as the European species, though Sargent says it occasionally reaches 100 feet in height.[2] The largest I saw was a fine old tree on the lawn of Mr. Nathaniel Thayer's house at Lancaster, Mass., which was 80 feet by 13 feet 6 inches, and though rather decayed at the top, where its branches were supported by iron stays, had produced suckers from the root, 40 feet high.

[1] The seedling of this species is described and figured by Rowlee and Hastings, in *Bot. Gazette*, xxvi. 351, fig. 18 (1898).
[2] In *U.S. Forest Service, Circ.* 71 (1907), a leaflet on the cultivation of this species, it is stated that the tree has been known in the region of its best development to reach a height of 120 feet. Throughout the greatest part of its range, it is much smaller, with an average height of 80 to 100 feet, and a diameter of from 2 to 4 feet.

Emerson mentions a tree at Bolton, Mass., which in 1840 was 15½ feet in girth at 6 feet, with an unbranched trunk 24 feet long; and another, on the road to Sheffield, which was 21 feet in girth at 4 feet from the ground. He states that though near the coast it does not ripen fruit so well, yet that in the interior when growing in sunny places it yields abundance of sweet and delicious nuts; and according to Sargent these, though smaller than European chestnuts, are superior to them in sweetness and flavour, and are sold for food in the eastern cities.

In *Garden and Forest*[1] there are several pictures of the chestnut in America, one representing a large tree at Dauphin in Pennsylvania, which is about 6 feet in diameter. Another represents a young forest in West Virginia about forty years old, showing good natural reproduction. A tree on a farm belonging to D. M. Ridgely, near Dover in Delaware, is noted for its excellent fruit, and it has been propagated, the chestnuts being known as Ridgely or Dupont chestnuts.

In the *United States Bureau of Forestry Bulletin No.* 53 (1904), R. Zon gives an interesting account of the chestnut tree in Maryland, where it is an important timber tree, being used for railway sleepers, telegraph poles, and fencing. It is usually coppiced, and Zon states that the sprouts usually come from the root collar, only 10 per cent. arising from the top of the stump. He has never seen any sucker shoots. The capacity of sprouting from the stool is retained to an advanced age, over 100 years. The tree in America usually becomes unsound at about 100 years old.

The American chestnut has rarely been tried in cultivation in Europe, and though not likely to succeed so well as the common species, there are thriving young trees at Kew.

Emerson states that the timber is one of the best native woods on which to lay mahogany veneers; and Mr. Weale informs me that it is now imported into England both in logs and boards; but the demand is not very great. It is used by builders as a substitute for oak, and by cabinetmakers. It carves well, and as it fumes readily, is a favourite wood with makers of antique furniture. In the log its value is from 1s. 6d. to 1s. 9d. per cube foot in Liverpool. In the board it is worth from 2s. to 2s. 6d. After conversion it cannot be distinguished from the English-grown chestnut. Hough states that the wood is rich in tannin, which is extracted and used for tanning purposes.                                        (H. J. E.)

## CASTANEA PUMILA, Chinquapin

*Castanea pumila*, Miller, *Dict.*, ed. 8, No. 2 (1768); Loudon, *Arb. et Frut. Brit.* iii. 2002 (1838); Sargent, *Silva N. Amer.*, ix. 17, tt. 442, 443 (1896), and *Trees N. Amer.*, 221 (1905).
*Fagus pumila*, Linnæus, *Sp. Pl.* 998 (1753).

A tree, rarely attaining in America 50 feet in height and 9 feet in girth, usually much smaller. Bark light-brown, slightly furrowed and broken on the surface

[1] *Garden and Forest*, ix. 114, f. 12, 234, f. 34 (1896), and vii. 484 (1894). Cf. also *Ibid.* x. 372, f. 48 (1897).

into loose plate-like scales. Young branchlets covered with numerous long erect hairs, becoming grey and glabrous in the second year.

Leaves (Plate 202, Fig. 12) smaller than in *C. sativa*, rarely as much as 5 inches long, not pendulous, oblong-oval, base unequal and rounded or tapering, apex acute, with about 15 pairs of nerves, which end in triangular serrations, tipped by short spine-like points; upper surface dull dark green, minutely pubescent; lower surface greyish white and densely tomentose. Petioles short, $\frac{1}{4}$ inch long, pubescent. Stipules about $\frac{1}{4}$ inch long, pubescent, those of the two lowest leaves broad, ovate, acute, on the middle leaves ovate-lanceolate, towards the top of the branch linear.

Nut ovoid, rounded at the slightly narrowed base, gradually narrowed and pointed at the apex, $\frac{3}{4}$ to 1 inch long, $\frac{1}{3}$ inch broad; only one fruit in each involucre, which opens generally by two or three valves. The fruit,[1] which is ripe in America in September, is delicious in flavour, and is occasionally gathered for market.

*Castanea pumila* is distinguished from the other species by its smaller leaves, which remain densely whitish tomentose underneath and have fewer nerves. In winter it is distinguished from the common chestnut by the twigs being slender and having a scattered loose pubescence, especially marked towards their apex. The buds are ovoid, not acute at the apex, minute, about $\frac{1}{8}$ to $\frac{3}{16}$ inch long, with both the first and second scales appressed-pubescent and ciliate. The leaf-scars and stipule-scars are smaller than in *C. sativa*.

*Castanea pumila*,[2] according to Sargent, occurs on dry, sandy ridges, rich hillsides, and the borders of swamps, from southern Pennsylvania to northern Florida and the valley of the Neches River, Texas. It is usually shrubby east of the Alleghany Mountains, becoming a tree west of the Mississippi River, and is most abundant and largest in size in southern Arkansas and eastern Texas.[3] The wood[4] is similar to that of *C. crenata*, with very thin sapwood, and is used for fences, posts, railway sleepers, etc.

According to Loudon it was introduced in 1699 by the Duchess of Beaufort, but it is extremely rare in cultivation, the only specimens which we have seen being small shrubs at Kew, which, however, seem perfectly hardy.

There are two specimens at Verrières,[5] near Paris, the smaller of which has a curiously twisted stem and resembles in appearance a dwarf Japanese tree. The other has two stems, each about 28 inches in girth and about 18 feet high, and produces fruit regularly and often in great abundance.[6]                    (A. H.)

---

[1] Hough, *Trees of N. States and Canada*, 137 (1907).

[2] According to Taylor, in Bailey, *Cycl. Amer. Hort.* i. 295 (1900), this species commonly throws up root-suckers.

[3] *Castanea neglecta*, Dode, in *Bull. Soc. Dendr. France*, 1908, p. 155, said by Dode to occur in the eastern part of the United States, is apparently only distinguishable from *C. pumila* by its larger and less pubescent leaves. It is possibly, as this author points out, a hybrid between *C. dentata* and *C. pumila*.

[4] Hough, *loc. cit.*                                    [5] *Hortus Vilmorianus*, 55 (1906).

[6] Since this article was corrected for the press, a leaflet has been issued by the U.S. Forest Service, on Chestnut Bark disease, which is caused by a fungus, known as *Diaporthe parasitica* or *Valsonectria parasitica*. This has recently destroyed an immense number of trees in the north-eastern states, spreading with great rapidity. As the disease, if once introduced, may be equally destructive in Europe, we think it well to warn arboriculturists against importing American chestnuts at present.

# FRAXINUS

*Fraxinus*, Linnæus, *Gen. Pl.* 318 (1737); Bentham et Hooker, *Gen. Pl.* ii. 676 (1876); Wenzig, in Engler, *Bot. Jahrb.*, iv. 165 (1883); Lingelsheim, in Engler, *Bot. Jahrb.* xl. 185 (1907).

TREES or shrubs, belonging to the natural order Oleaceæ; leaves opposite, compound, unequally pinnate, rarely reduced to a single leaflet; stipules absent. Buds, large terminal and small axillary, the former usually with four scales visible externally, the latter with two outer scales; these scales are rudimentary leaf-stalks, often showing at their apex traces of the pinnate leaf, and increase in size after the bud opens, falling off eventually and leaving ring-like scars at the base of the shoots.

Flowers polygamous or diœcious, in panicles or fascicled racemes, terminal on leafy shoots of the year, or developed from separate buds either in the axils of the leaf-scars of the previous year, or at the base of the young branchlets. Calyx absent in some species; when present, campanulate and four-lobed. Corolla absent in many species; when present, of two to four (rarely five to six) petals, free or connate in pairs at the base. Stamens two, rarely three or four, affixed to the base of the petals or hypogynous. Ovary, with a style divided above into a two-lobed stigma, two-celled, each cell containing two pendulous ovules. Fruit, a samara, indehiscent, convex or compressed below, with a dry pericarp produced into an elongated terminal and more or less decurrent wing,[1] usually one-celled and one-seeded. Seed pendulous; embryo erect in a fleshy albumen; cotyledons flat.

The genus Fraxinus is widely distributed over the temperate regions of the northern hemisphere, three[2] species, however, occurring within the tropics in Cuba and the Philippines, and south of the equator in Java. The genus consists of nearly sixty species, many of which are imperfectly known and require further study in the field. Even in the case of the Mediterranean species, authorities are at variance. The present account deals only with the species which have been seen in the living state.

The genus is divided into five sections :—

I. *Ornus*, Persoon, *Syn. Pl.* ii. 605 (1807).

Calyx and corolla both present, the calyx persisting under the samara. Panicles terminal on leafy shoots or axillary on the branchlets of the current year. About eighteen species.

---

[1] Abnormal fruit with three wings, has been observed in several species, as *F. americana*, *F. caroliniana*, *F. Berlandieriana*.

[2] *F. caroliniana*, a native of the United States, is met with in Cuba. *F. Eedenii*, Boerl et Koord, occurs in Java; and *F. philippinensis*, Merrill, in the Philippine Islands.

II. *Ornaster*, Koehne and Lingelsheim, *Mitt. Deut. Dendr. Gesell.* 1906, p. 66.
Calyx present, persistent under the samara. Corolla absent. Flowers in terminal panicles, appearing with the leaves. Seven species.

III. *Sciadanthus*, Cosson et Durieu, *Bull. Soc. Bot. France*, ii. 367 (1855).
Calyx present, persistent under the samara. Corolla absent. Flowers in dense fascicled cymes, axillary on the preceding year's shoot. Two species.

IV. *Leptalix*, Rafinesque, *New Flora*, iii. 93 (1836).
Calyx present, persistent under the samara. Corolla absent. Flowers in panicles, axillary on the preceding year's shoot. About fifteen species.

V. *Fraxinaster*, De Candolle, *Prod.* viii. 276 (in part) (1844).
Calyx and corolla both absent. Flowers in panicles or racemes on the preceding year's shoot. About twelve species.

These sections, based on the characters of the flowers, are not available in practice in the determination of living trees, flowering specimens of which are often not obtainable ; and the following key groups the species according to the characters of the branchlets and foliage :—

### KEY TO THE SPECIES IN CULTIVATION

I. *Leaves simple or with two to three leaflets.*

*\* Branchlets four-angled.*

1. *Fraxinus anomala*, Watson. Colorada, Utah, Nevada. See p. 898.
Leaves usually simple, ovate or obovate, glabrous beneath.

*\*\* Branches terete.*

2. *Fraxinus angustifolia*, Vahl., var. *monophylla*. See p. 880.
Leaves opposite, simple or two- to three-foliolate, lanceolate, glabrous beneath.

3. *Fraxinus excelsior*, Linnæus, var. *monophylla*. See p. 866.
Leaves opposite, simple or two- to three-foliolate, ovate or oval, pubescent beneath at the base.

4. *Fraxinus syriaca*, Boissier. Western Asia. See p. 883.
Leaves in whorls. Leaflets usually three (occasionally five to seven occurring on the same branch), lanceolate, glabrous.

II. *Leaves with five or more leaflets.*[1]

A. *Branchlets, leaf-rachis, and leaflets quite glabrous.*

*\* Leaflets stalked.*

5. *Fraxinus potamophila*, Herder. Turkestan. See p. 885.
Leaflets seven to nine, ovate, serrate.

[1] Cf. *F. syriaca* (No. 4), which has occasionally five to seven leaflets.

*\* \* Leaflets sessile.*

6. **Fraxinus angustifolia**, Vahl.    S. France, Spain, Portugal, N. Africa.    See p. 879.

Leaflets seven to thirteen, lanceolate.    Leaf-rachis strongly winged, the wings meeting above; groove interrupted.

7. **Fraxinus Willldenowiana**, Koehne.    Origin unknown.    See p. 884.

Leaflets, seven to eleven, ovate or lanceolate, increasing markedly in size from the base to the apex of the leaf.    Leaf-rachis with a continuous open groove.

B. *Branchlets glabrous; leaflets pubescent on part of the lower surface.*

\* *Leaf-rachis strongly winged on the upper side, the wings meeting in part above, forming an interrupted open groove.*

8 to 11. *Leaf-rachis not conspicuously bearded at the nodes.*

8. **Fraxinus oxycarpa**, Willdenow.    Italy, S.E. Europe, Asia Minor, Caucasus. See p. 882.

Leaflets seven to thirteen, ovate or lanceolate; serrations few, ending in long incurved points.

9. **Fraxinus excelsior**, Linnæus.    Europe, Caucasus.    See p. 864.

Leaflets, five to eleven, oblong-lanceolate; serrations crenate, numerous, exceeding in number the lateral nerves.

10. **Fraxinus excelsior**, Linnæus, var. *rotundifolia*.    See p. 866.

Leaflets nine to thirteen, $1\frac{1}{2}$ to $2\frac{1}{2}$ inches long, ovate, oval, or orbicular, coarsely bi-serrate.

11. **Fraxinus Elonza**, Dippel.[1]    Origin unknown.    See p. 883.

Leaflets, eleven to thirteen, small, less than $2\frac{1}{2}$ inches long, irregularly serrate, oblong, lanceolate or oval; under surface with brown tomentum near the base.

12, 13. *Leaf-rachis with conspicuous tufts of brownish-red tomentum at the nodes.*

12. **Fraxinus nigra**, Marshall.    N. America.    See p. 898.

Leaflets, seven to eleven, oblong-lanceolate, rounded or broadly cuneate at the base, sessile.

13. **Fraxinus mandshurica**, Ruprecht.    Eastern Asia.    See p. 893.

Leaflets seven to thirteen, oblong-lanceolate, gradually tapering at the base, sub-sessile.

\* \* *Leaf-rachis with a continuous open groove on its upper side, which is sometimes almost obsolete.*

† *Some or all of the leaflets distinctly stalked.*

14, 15. *Leaflets white beneath.*

14. **Fraxinus americana**, Linnæus.    N. America.    See p. 901.

Leaflets seven to nine, 4 to 6 inches long, long-acuminate, dull light green above; rachis with an extremely slight groove.

---

[1] The groove on the leaf-rachis is variable in this species, sometimes being open its whole length.

15. *Fraxinus texensis*, Sargent.   Texas.   See p. 907.

Leaflets five to nine, $2\frac{1}{2}$ to $3\frac{1}{2}$ inches long, shortly acuminate, shining bluish-green above; rachis with a very slight groove.

16. *Leaflets green beneath; rachis glabrous.*[1]

16. *Fraxinus caroliniana*, Miller.[2]   S.E. United States, Cuba.   See p. 904.

Leaflets, five to seven, about 3 inches long, shortly acuminate; rachis with a well-defined but shallow groove.

17 to 23. *Leaflets green beneath; rachis slightly pubescent at the nodes.*

17. *Fraxinus rhynchophylla*, Hance.   N. China, Manchuria.   See p. 892.

Leaflets five to seven, 3 to 4 inches long, coriaceous, terminating in an obtuse-tipped acumen, entire or very obscurely serrate.

18. *Fraxinus chinensis*, Roxburgh.   Central and Southern China.   See p. 895.

Leaflets seven to nine, 3 to 4 inches long, coriaceous, shortly cuspidate at the apex, crenately serrate.

19. *Fraxinus obovata*, Blume.   Japan.   See p. 895.

Leaflets five to seven, 2 to 3 inches long, membranous, variable in shape, irregularly serrate, with minute curved bristles on the lower surface and petiolules, which are also present on the rachis of the leaf.

20. *Fraxinus longicuspis*, Blume.   Japan.   See p. 897.

Leaflets five, 3 to 4 inches long, membranous, very pale beneath, abruptly contracted into a long cuspidate apex, crenately serrate.

21. *Fraxinus Ornus*, Linnæus.   S. Europe, Asia Minor.   See p. 887.

Leaflets five to nine, 2 to 3 inches long, membranous, shortly acuminate, serrate.

22. *Fraxinus floribunda*, Wallich.   Himalayas, Upper Burma.   See p. 890.

Leaflets seven to nine, 4 to 6 inches long, membranous, apex long-acuminate, serrate; lateral nerves prominent and numerous.

23. *Fraxinus quadrangulata*, Michaux.   N. America.   See p. 900.

Branchlets quadrangular and four-winged.   Leaflets seven to nine, 3 to 5 inches long.

†† *Leaflets sessile or sub-sessile.*[3]

24. *Fraxinus Spaethiana*, Lingelsheim.   Japan.   See p. 897.

Leaflets seven to nine, 4 to 6 inches long, coriaceous, lanceolate, long-acuminate, irregularly and often crenately serrate.   Distinguished from all other species in cultivation by the dilated swollen base of the leaf-stalk.

25. *Fraxinus lanceolata*, Borkhausen.   N. America.   See p. 906.

Leaflets seven to nine, 3 to 6 inches long, lanceolate, long-acuminate.   Rachis grooved.

26. *Fraxinus Berlandieriana*, De Candolle.   Texas, Mexico.   See p. 907, note 1.

Leaflets, five to seven, 2 to 3 inches long, oval or obovate.   Rachis grooved.

---

[1] *Fraxinus lanceolata* (cf. No. 21), has occasionally the leaflets distinctly stalked, and might on that account be sought for here.

[2] Cf. No. 31A.   Two forms of this species occur in cultivation, differing in the absence or presence of pubescence on the branchlets and leaf-rachis.

[3] *Fraxinus Elonza* (cf. No. 11), sometimes having an open continuous groove on the rachis, might be sought for here.

27. *Fraxinus dimorpha*, Cosson et Durieu. N. Africa. See p. 884.
Leaflets, seven to nine, about $\frac{3}{4}$ inch long, ovate. Rachis with wide-spreading wings.

## C. *Branchlets minutely pubescent; leaflets glabrous.*

28. *Fraxinus Mariesii*, J. D. Hooker. Central China. See p. 892.
Leaflets five, coriaceous, about $2\frac{1}{2}$ inches long, oval, stalked, crenately serrate.
29. *Fraxinus Bungeana*, De Candolle. North China. See p. 891.
Leaflets five to seven, membranous, about $1\frac{1}{2}$ inch long, mostly stalked, oval or rhomboid, long-acuminate, irregularly serrate.
30. *Fraxinus raibocarpa*, Regel. Turkestan. See p. 886.
Leaflets five to seven, oval, entire, about $1\frac{1}{2}$ inch long; upper leaflets sub-sessile, lower leaflets stalked.

## D. *Branchlets, leaf-rachis, and leaflets pubescent.*
### * *Leaflets distinctly stalked.*

31. *Fraxinus Biltmoreana*, Beadle. United States. See p. 905.
Leaflets seven to nine, oval, about 4 inches long, white beneath; rachis very slightly grooved.
31A. *Fraxinus caroliniana*, Miller.[1] S.E. United States, Cuba. See p. 912.
Leaflets five to seven, oval, about 3 inches long, green beneath, rounded or broadly cuneate at the base; rachis with a well-defined but shallow groove.

### ** *Leaflets stalked, sessile, or sub-sessile.*

32. *Fraxinus pennsylvanica*, Marshall. N. America. See p. 908.
Leaflets, seven to nine, lanceolate, 4 to 5 inches long, green beneath, pubescent on both surfaces, long-acuminate, tapering at the base; rachis densely white-pubescent, and with a narrow, shallow groove. Buds reddish pubescent.
33. *Fraxinus pubinervis*, Blume. Japan. See p. 896.
Leaflets five to seven, lanceolate, 3 to 4 inches long, glabrous above, green and pubescent beneath, acuminate, tapering at the base; rachis grooved, with pubescence densest at the nodes. Buds greyish pubescent.

### *** *Leaflets sessile or sub-sessile.*

34. *Fraxinus oregona*, Nuttall. Western United States. See p. 910.
Leaflets seven to nine, oval, 3 to 4 inches long, green beneath, shortly acuminate, entire or obscurely crenate in margin.
35. *Fraxinus velutina*, Torrey. Texas to California. See p. 912.
Leaflets three to five, about $1\frac{1}{2}$ inch long; lateral leaflets variable in shape and serration, terminal leaflet obovate.
36. *Fraxinus xanthoxyloides*, Wallich. Baluchistan, Afghanistan, N.W. Himalayas. See p. 885.
Leaflets, five to nine, about $\frac{3}{4}$ inch long, ovate. Rachis with wide-spreading wings.

[1] Cf. No. 16.

37. *Fraxinus holotricha*, Koehne.   Origin unknown.   See p. 887.

Leaflets nine to thirteen, lanceolate or ovate-lanceolate, about 2 inches long, sharply serrate.

(A. H.)

## FRAXINUS EXCELSIOR, Common Ash

*Fraxinus excelsior*, Linnæus, *Sp. Pl.* 1057 (1753); Loudon, *Arb. et Frut. Brit.* ii. 1214 (1838); Willkomm, *Forstliche Flora*, 658 (1897); Mathieu, *Flore Forestière*, 241 (1897).

A large tree, attaining 140 feet in height.   Bark smooth and greyish when young, becoming rough and fissured in old trees.   Branchlets glabrous.   Leaflets (Plate 262, Fig. 4), 9 to 15, sessile and articulate, oval- or oblong-lanceolate, acuminate at the apex, tapering at the base, where the margin is entire, elsewhere crenately serrate, the serrations more numerous than the lateral nerves; upper surface glabrous and green; lower surface paler with pubescence on the midrib, extending over the basal part of the leaflet; venation pinnate, the lateral nerves forming loops near the margin.   Rachis glabrous or pubescent, strongly winged, the wings meeting above,[1] except opposite the insertion of the leaflets where there is an open channel, and below the leaflets where the rachis is flattened or broadly grooved.

Flowers,[2] opening before the leaves appear, fertilised by the wind, in dense axillary panicles, polygamous or occasionally diœcious, without calyx or corolla. Male flowers with two stamens more or less connate below.   Female flowers with a two-celled superior ovary, the style being dilated above into two thick stigmas. Perfect flowers with an ovary and two stamens.[3]

Fruit, of two carpels, joined together to form the body of the samara, which is compressed at right angles to the partition and is produced in front into a veined membranous wing.   The samaræ are very variable in shape, but are usually linear-oblong or elliptic, obtuse at both ends, and notched at the tip.   They hang in racemes on long stalks, and, ripening in autumn, generally remain on the tree till the following spring; and are ultimately carried by the wind a short distance away from the parent tree.

## Seedling [4]

The young plant on appearing raises the samara out of the soil, the two cotyledons being united together at first by a cap formed of the albumen.   The

---

[1] Rain collecting on the leaflets drains into the ducts thus formed, inside of which are hairs and peltate groups of cells that gradually absorb the water, which is retained for several days after a fall of rain.   See Kerner, *Nat. Hist. Plants*, Eng. Transl. i. 231, fig. 54 (1898).   [2] Section *Fraxinaster*.

[3] Schulz, in *Ber. Deutsch. Bot. Ges.* x. 401 (1892), has shown that trees of the common ash greatly vary in the kind of flowers which they produce.   Trees bearing only male flowers are common; while those with only female flowers or with only perfect flowers are rare.   In many cases two of the three kinds of flowers are borne on the same tree; and what is very remarkable, a tree is not necessarily of the same sex in successive years.   Ash trees do not flower, as a rule, regularly every year; and fruit is much more abundant in some years than in others.

[4] Figured in Lubbock, *Seedlings*, ii. 214, fig. 512 (1892).

cotyledons, when developed, are about ¾ to 1½ inch long, oblong, obtuse, entire, glabrous, pale beneath, tapering at the base into a very short winged petiole. Caulicle terete, 1 to 3 inches long, ending in a long yellow, fleshy, flexuose tap-root. Young stem, green and glabrous, terete below, angled above. First pair of leaves, arising ¼ to 1 inch above the cotyledons, simple, ovate, acuminate or acute, irregularly serrate and ciliate, minutely pubescent, on a winged petiole about ½ inch long. Second pair of leaves, three-foliolate, on a petiole about an inch long, the terminal leaflet the largest. Third pair with either three or five leaflets.

## IDENTIFICATION

The common ash is only liable to be confused with species like *F. angustifolia* and *F. oxycarpa*; but is readily distinguished by its black buds, and the crenate serrations more numerous than the lateral nerves in the leaflets.

In winter, the twigs are stout, shining-grey or olive green, compressed towards the tip, swollen at the nodes. Leaf-scars, opposite, obliquely set on projecting pulvini, semicircular or almost orbicular, often with lateral projecting horns, and showing an almost circular row of bundle-dots. Terminal buds black, conical, quadrate, with four scales visible externally, but consisting altogether of seven to eight pairs of scales. Lateral buds smaller, given off at a wide angle, with two or three external scales.

## VARIETIES

The common ash, though distributed over a wide area, varies little in the wild state; and such varieties, as have been based on the form of the fruit, cannot be considered as well established. Near Perpignan a form with small leaves has been collected, which is var. *australis*, Godron et Grenier, *Flore de France*, ii. 471. In the province of Talysch in the Caucasus, a remarkable form occurs with large leaflets, velvety pubescent underneath; and the shoots, buds, and leaf-rachis are densely pubescent. This variety, which was described by Scheele[1] as a distinct species (*F. coriaræfolia*), is said by Koch[2] to be met with occasionally in cultivation in gardens, where it is known as *Fraxinus expansa*.

A curious variation in the common ash was observed by A. D. Richardson[3] in the case of four young plants, found growing in a clump of several hundreds, on the banks of the Boyne near Navan in Co. Meath. The leader shoots had the leaves alternate in a 2/5 spiral arrangement, instead of the normal opposite and decussate one.

Numerous varieties have been obtained as seedlings in nurseries or as isolated specimens growing wild.

---

[1] *Linnæa*, xvii. 350 (1843).  [2] *Dendrologie*, ii. 1, 243 (1872).  [3] *Gard. Chron.* xxxvi. 133, fig. 55 (1904).

1. Var. *monophylla*, O. Kuntze, *Flora von Leipzig*, 82 (1867).

*Fraxinus monophylla*, Desfontaines, *Tab. de l'École de Bot.* 52 (1804).
*Fraxinus heterophylla*, Vahl. *Enum. Pl.* i. 53 (1804); Loudon, *Arb. et Frut. Brit.* ii. 1228 (1838).
*Fraxinus simplicifolia*, Willdenow, *Sp. Pl.* iv. 1098 (1805).
*Fraxinus excelsior*, var. *diversifolia*, Aiton, *Hort. Kew.* iii. 445 (1789).

This remarkable variety, which is usually known as the laurel-leaved or simple-leaved ash, is met with in the wild state in the forests near Nancy in France, and also rarely in England and Ireland.   It appears[1] occasionally when a quantity of ash seeds is sown, and intermediate forms are found with three, five, and seven leaflets. The shoots and buds are identical with those of the common ash.   In the ordinary form of the variety, the leaf (Plate 262, Fig. 2) is simple, not being divided into leaflets.   Occasionally there is a large leaflet, with one or two small leaflets at its base; and this form is known as var. *monophylla laciniata*.   The simple-leaved form or the terminal leaflet in the two- to three-leaflet form, has a stalk about half as long as the blade or a little longer, and is variable in shape, being usually oval in outline with an obtuse, acute, or acuminate apex; margin coarsely serrate; lower surface pubescent except towards the apex; petiole widely grooved on its upper side.   A form of the single-leaved ash with variegated leaves, was discovered, according to Loudon, in 1830 at Eglantine, near Hillsborough, Co. Down, Ireland.

The simple-leaved ash is very distinct in appearance and thrives well in towns. It is usually propagated by grafting.

At Beauport, Sussex, there is a tree of this kind, 70 feet by 5 feet 9 inches; and self-sown seedlings reproducing the variety have been observed by us there. Other large specimens occur: at Syon, a tree[2] 84 feet by 7 feet 6 inches; at Sidmouth, measured by Miss Woolward in 1904, two trees 9 feet 4 inches each in girth, the taller being 86 feet high; also three good trees in the grounds at Woburn growing beside the lake.   Lord Kesteven reports one at Stubton Hall, Newark, which was 67 feet high by 8 feet 1 inch in girth in 1906.   Elwes has seen others from 50 to 70 feet high at Scampston Hall, Yorkshire; at Sharpham, near Totnes; and at Dodington Park, Gloucestershire.   A tree at Oxford, near the east end of the broad walk in Christchurch meadow, mentioned by Walker[3] in 1833, is, according to Mr. Druce, about 65 feet high and 4 feet 7 inches in girth.   It is crowded by other trees and is not thriving.

On the Pitfour estate near Mintlaw in Aberdeenshire, a tree 55 feet high by 7 feet 9 inches is reported by Mr. Ainslie; and Elwes saw one at Gordon Castle, which in July 1907 was covered with fruit and measured about 60 feet by 9 feet 2 inches in girth.

There is a very good specimen in Stephen's Green, Dublin; one at Beauparc House in Co. Meath measured, in 1905, 40 feet high by 6 feet 2 inches in girth; and another at Curraghmore, Co. Waterford, was 50 feet by 5 feet in 1907.

2. Var. *rotundifolia*.   A tree growing in a wood at Strete Ralegh, near Exeter,

---

[1] Cf. Mathieu, *Flore Forestière*, 241 (1897), and *Allgem. Garten-Zeitung*, iii. 6 (cited by Loudon).
[2] This tree in 1849 was 50 feet high by 4 feet in girth; it is now beginning to decay at the top.
[3] *Flora of Oxfordshire*, 3 (1833).   Cf. also Dyer, in *Journ. Bot.* ix. 147 (1871).

the seat of H. Imbert Terry, Esq., who has sent specimens, is remarkably distinct in the shape of the foliage from any ash known to me; but is probably only a variety of the common ash, with which it agrees in bark and in buds. In the absence of flowers and fruit, this identification is not quite certain, and on that account a full description is now given: Leaflets (Plate 266, Fig. 32), nine to thirteen, $1\frac{1}{2}$ to $2\frac{1}{2}$ inches long; terminal leaflet stalked, with a long cuneate base; lateral leaflets, sessile, broadly oval or ovate, unequal at the usually cuneate, but occasionally broad and rounded base, acute or slightly acuminate at the apex, coarsely bi-serrate, slightly scabrous with scattered stiff hairs on the upper surface, pale beneath with dense woolly pubescence on the sides of the midrib and lateral nerves near the base. Leaf-rachis, strongly winged, the wings meeting above in its apical half, but forming a wide open groove towards the base; pubescent on the dorsal side with scattered stiff hairs, densest at the nodes.

This ash resembles in foliage the figure of *F. rotundifolia*, Aiton,[1] which is given by Willdenow.[2] The latter species, according to Aiton, Willdenow, and Loudon,[3] is a small tree of Italy, with flowers and buds like *F. Ornus*; and the Strete Ralegh tree cannot be identified with it, as in all essential characters[4] it resembles the common ash.

Nothing is known of the origin of the tree at Strete Ralegh, which Miss Woolward found in 1905 to be about 75 feet in height, the bole dividing near the ground into two stems, 3 feet 1 inch and 2 feet 7 inches in girth respectively.

3. Var. *angustifolia*, Schelle. A variety[5] with small narrow leaves (Plate 262, Fig. 5), which differs in no essential character from the common ash, of which it has the buds and the characteristic serrations and pubescence of the leaflets; and in this way can be readily distinguished from such species as *F. angustifolia*, Vahl, and *F. oxycarpa*, Willd.

4. Var. *crispa*, Loudon (also known in gardens as var. *atrovirens* and var. *cucullata*). Leaflets dark green, curled and twisted. Plant usually rigid and stunted, of very slow growth.

5. Var. *nana*, Loudon (also known in gardens as var. *polemoniifolia* and var. *globosa*). A compact slow-growing dwarf form, with very small leaves.

6. Var. *aurea*, Loudon. With yellow branches. A pendulous form of this is known.

7. Var. *asplenifolia*, Koch. Leaflets very narrow, almost linear.

8. Var. *fungosa*, Loddiges. Bark remarkably wrinkled, with corky ridges.

9. Var. *verticillata*, Loudon. Leaves whorled, not opposite as in the common form.

10. Var. *monstrosa*, Koch. Young branches fasciated.

11. Var. *erosa*, Persoon. Leaflets incised.

---

[1] *Hort. Kew.* iii. 445 (1789). Cf. our remarks on this species under *F. Ornus*, p. 888.

[2] *Berlin. Baumzucht*, 116, fig. vi. 1 (1796).      [3] *Arb. et Frut. Brit.* ii. 1244 (1838).

[4] Bark and buds especially. The pale under surface of the leaf, which is thin in texture, is seen in common ash seedlings and in some forms of var. *monophylla*. The strongly-winged rachis of the leaf is characteristic of *F. excelsior* and its near allies.

[5] Var. *elegantissima*, in cultivation at Aldenham, obtained from Simon-Louis, is scarcely to be distinguished from this variety.

12. Var. *verrucosa*, Desfontaines. Branchlets warty.

13. Certain variegated forms are known, as—var. *albo-marginata*, leaflets edged with white, and var. *albo-variegata*, in which the white colour appears as blotches on the leaflets.

Most of the foregoing varieties are of little or no beauty or interest, and do not, so far as we know, become large or shapely trees. (A. H.)

14. Var. *pendula*, Aiton. The weeping ash in some form or other is found in almost every garden, but rarely as a large tree. Loudon describes several forms of it, and says that the original tree was discovered near Wimpole in Cambridgeshire 150 years or more ago, and was decaying in 1835.

Another form, the Cowpen ash, which grew near Morpeth, is figured by Loudon;[1] and I have seen two trees which have naturally assumed a very similar habit. One stands by the road in the village of Ollerton, Notts. The other is in a field at Marsden, in the parish of Rendcombe, Gloucestershire.

A third form, called by Loudon the Kincairney Ash, grew in the parish of Caputh, near Dunkeld, Perthshire, and was distinguished by its alternately pendulous and upright branches. It was propagated at the Perth Nurseries, but I have not noticed any of this variety now in cultivation.

To make an effective tree, the weeping ash should be grafted on a very tall stock, and if the branches are attended to, may be trained into a shady arbour like a great umbrella. But if the stock is also allowed to grow as well as the graft, the effect will be more curious than beautiful; and the weeping ash is not so much admired or planted as it was formerly when trained and clipped trees were more in fashion. At Heanton Satchville, in North Devon, the seat of Lord Clinton, I saw it trained in combination with a trellis of living ashes which were planted all round the central weeping tree, and had their stems woven together when young so as to form the walls of the arbour; but in the course of time this had become ragged; and as the ash does not bear clipping like the beech or hornbeam, I should prefer either of those trees for such a purpose.

By far the finest grafted weeping ash that we know of is growing in the gardens of Elvaston Castle, near Derby, the seat of the Earl of Harrington. It was reported[2] in 1905 to be 98 feet high, with long weeping branches hanging vertically from the summit of the tree, one of them descending to about 20 feet from the ground; but when I saw it in 1906 I did not think it was more than about 90 feet high, the bole, 6 to 7 feet in girth, being straight and clean. (Plate 238.) This tree was grafted by Barron about 1848. Another larger tree also exists here, which has a bole 50 feet by $12\frac{1}{2}$ feet, and was apparently grafted with weeping ash at the same time, but in this case the branches of the stock have outgrown the grafts.

In Ireland there is a very handsome and well shaped weeping ash at Castle-wellan, 41 feet high, with a trunk 5 feet in girth with branches hanging to the ground all round it. (H. J. E.)

[1] *Arb. et Frut. Brit.* ii. 1216, f. 1045 (1838).      [2] *Garden*, 1905, lxviii. 400, with figure.

## Distribution

The common ash is spread through almost all Europe, and occurs also in the Caucasus.

The northern limit in Europe passes from the Trondhjem fjord in Norway, about lat. 63° 40', through Sweden at about lat. 61°, and in Finland extends to lat. 62°, descending from there through Russia in a S.E. direction to Riazan, whence it continues N.N.E. to Kazan, its extreme eastern point in Europe. From Kazan, the eastern limit descends in a S.W. direction through Penza, Saratof, and Voronej to Kharkof, and then passes by Ekaterinoslav to the Crimea. In the Caucasus [1] the ash does not occur north of the rivers Kuban and Terek. The southern limit extends from the Transcaucasian province of Talysch through Asia Minor and Turkey to Dalmatia, and across Italy and southern France to the Pyrenees. In the Iberian peninsula [2] it is met with, according to Willkomm, in the mountains of Catalonia, Aragon, Burgos, Santander, Leon, Asturias, Galicia, and northern Portugal. The western limit takes in the western coast of France and the British Isles. Outside the range mentioned above, it occurs as small scrub in rare situations, as in Norway at Tromsö (lat. 69° 40'), and in the government of St. Petersburg.

An ash occurs in the western Himalayan region which, according to most of the authorities, is *Fraxinus excelsior*. It has been distinguished as a distinct species by Wenzig,[3] and, so far as I can judge from dried specimens, is very different indeed from the common ash. Sir George Watt informs me that this ash is always an insignificant tree, never attaining more than 30 feet in height and a foot in girth.

The ash is generally met with growing wild as isolated trees or in small groups in the continental forests, but pure woods of some extent occur in moist situations, as in river valleys subject to flooding, in Hungary, Slavonia, Poland, and Russia. In northern regions it is rather a tree of the plains and valleys than of the mountains ; but in southern Europe it is only met with in the mountains. It ascends in south Tyrol to 4000 feet, and in Switzerland to about 4400 feet.

The ash is a true native of the British Isles, and has been found in a fossil state in the interglacial beds at Hitchin in Hertfordshire, and in neolithic deposits at Crossness in Essex.[4]

It may be said to occur wild in every part of the British Isles, except in the northern part of Scotland, where, however, it bears the climate in plantations. In Yorkshire[5] it ascends to 1250 feet elevation. In Braemar, H. B. Watt[6] observed it up to 1200 feet. In Ireland[7] it is frequent in woods, hedges, and rocky places ; and ascends in Donegal to 800 feet, in Down to 1000 feet, and in Wicklow to 1300 feet.

---

[1] It ascends in the Caucasus to 6000 feet, according to Radde, *Pflanzenverbreitung in den Kaukasusländern*, 181 (1899).

[2] Cf. Captain Widdrington's account of the distribution of the ashes in Spain, given under *Fraxinus angustifolia*, p. 880.

[3] *Fraxinus Hookeri*, Wenzig, in Engler, *Bot. Jahrb.* iv. 179 (1883). It differs from the common ash in having fewer leaflets, usually five, rarely seven. The bud is also very distinct, being dark yellow in colour, and covered with minute warts. The leaflets are sessile, oval, broadly cuneate at the base, acuminate at the apex, pale beneath, with pubescence on each side of the midrib, minutely crenulate-serrate. The rachis appears to have a wide open groove above ; and the fruit in its lower part is longitudinally and deeply grooved in the middle line.

[4] C. Reid, *Origin Brit. Flora*, 133 (1899).

[5] Lees, *Flora of W. Yorkshire*, 322 (1888).

[6] *Cairngorm Club Journal*, iv. 114 (1903).

[7] *Cybele Hibernica*, 236 (1898).

Ash woods, supposed to be wild, occur on limestone in hilly districts in Yorkshire, Derbyshire, and Somerset. In the latter county, they are very numerous in the Mendip Hills, and have been mapped by Mr. C. E. Moss,[1] who gives an interesting account of their distribution and peculiar features. The ash is often pure, with a dense undergrowth of hazel, or it is mixed with yew and whitebeam. Mr. Moss notices the prevalence of dog's mercury and wood garlic[2] in many of these ash woods.

The Gaelic name of the ash, according to Sir H. Maxwell,[3] is *uinnse* (inshy), and becomes Inshawhill in Wigtonshire, and the plural *uinnsean* (inshan) takes the peculiar form of Inshanks, the name of two places in that county, and Inshewan, near Kirriemuir; while the common alternative form *uinnseog* (inshog) remains as Inshock in Forfarshire, Inshaig in Argyllshire, Inshog near Nairn. Analogous forms,[4] with the initial letter *f*, appear in names of places in the south and west of Ireland, as the river Funcheon in Co. Cork. (A. H.)

## CULTIVATION

Though the oak will always be looked on as the premier tree of Great Britain, yet now that its most important use has passed away, the ash must be considered as the most economically valuable of all our native trees, and is perhaps the only hardwood from which a quick and certain return can be expected by the planter.

It is almost the only tree whose value has not fallen in consequence of foreign competition, and, though a good deal of American and some Hungarian and Japanese ash is now imported, yet the timber of these is not considered equal for toughness, strength, and elasticity to the best English ash, for which no foreign wood forms an efficient substitute. And as the tree can be grown over all parts of our islands, and attains a great size wherever suitable soil is found, it should be planted more largely in all favourable situations, where it produces timber of good quality.

In considering the requirements of the ash, one must always remember that it is a bad neighbour both to other trees and to crops, and that it is far more valuable as timber when grown in woods where it can be drawn up to a good height, than in hedgerows where it produces many branches. It likes a deep, rich soil, neither too wet nor too dry, and grows very well on limestone formations, even on a shallow soil, if the rock is sufficiently disintegrated for the roots to penetrate the crevices.

It is short-lived on wet or swampy soils, and the timber is inferior on sandy or peaty land. The finest trees are generally in sheltered situations, but though it is the latest of our native trees to come into leaf, none suffers more from late frosts; and therefore when planted in low situations it is often severely injured when young. It will grow up to a great elevation, and in the most exposed situations, though here it becomes stunted and branchy. No hardwood except that of the chestnut becomes valuable at so early an age, but the wood of old trees, even when sound, usually becomes discoloured or "black-hearted," and ash is never

---

[1] *Geog. Distribution of Vegetation in Somerset*, 41 (Roy. Geog. Soc., 1907).

[2] Mr. A. C. Forbes, in *Eng. Estate Forestry*, 72 (1904) says :—"We have always noticed that the existence of the wild garlic, *Allium ursinum*, is an almost certain indication of good ash ground."

[3] *Scottish Land Names*, 109 (1894).      [4] Cf. Joyce, *Irish Names of Places*, 488 (1870).

more valuable than when it is from 3 to 6 feet in girth, with a clean stem, a size often attained at fifty to sixty years of age.

In plantations ash is often mixed with other trees, and if allowed to take the lead will do them more harm than oak, but a few ash should be introduced in the best soil of larch or other plantations, because the seedlings, which spring up abundantly, will, when the conifers are cut, renew the plantation naturally, and the parent trees will throw up vigorous shoots from the stool after felling.

In the midland counties, ash is the commonest, and by far the most profitable underwood, being cut at intervals of twelve to twenty years; when the poles are much in demand for many purposes, especially for sheep hurdles. But in most places during the last twenty years ash poles have fallen in value, though larger timber has increased in price; and so much damage has been done to the stools by rabbits that large areas are now becoming very thin, and the crop inferior. No tree except beech suffers more from rabbits than ash, and where they are allowed to increase, and are not killed before winter, the bark of old trees as well as of underwood is sure to be peeled, and the natural reproduction from seed checked. I believe that where the soil is stiff, young ash will pay for some cultivation when young, as their shade is not dense enough to keep down grass and weeds, and if they become stunted, as they often do after planting, it is better, and, indeed, necessary, to cut them down to the ground two to five years later.

Self-sown ash seem to grow more vigorously than planted ones, if not too crowded, and their rate of growth is sometimes extraordinary. An ash self-sown in my nursery, at three years old was 7 to 8 feet high, whilst the transplanted seedlings on the same ground were only 3 feet high at the same age. I have seen shoots 6 to 7 feet long the first year from strong healthy stools, and poles worth £15 to £20 per acre at sixteen years old, on land which for agricultural purposes was not worth 8s. an acre. The stools, however, often become worn out and hollow at the base after five or six cuttings, and these should be replaced with seedling plants every time the crop is cut.

Some years ago, when ash coppice began to fall in price, I left the strongest and straightest pole on every stool at the rate of about 160 per acre, with the object of converting the coppice into timber trees. But though, where the soil is good and the stools young, these poles are likely to make useful trees at fifty to sixty years from the last cutting over, yet where the land is poor they have increased but little, and have a hidebound appearance, owing, no· doubt, to the want of shelter and the exposure to the sun and wind. I should advise all intending planters of ash to examine carefully the best local ash plantations, and inquire into the probable demand for poles before adopting this course.

The ash is always raised from seed, except in the case of varieties which are grafted on stocks of the common ash. The keys are ripe in late autumn, and often hang on the tree till the following summer, especially when they are not mature. The best-ripened seed, I believe, usually falls first, and should be gathered before winter, and put in a shallow layer mixed and covered with earth or sand, and kept fairly dry until the following winter, when it should be sown. It is advised in books that they should be turned over several times during the year before sowing, but I have not found this necessary; and with regard to the time of sowing, it should be put off as long as possible, because the natural germination of the seed takes place six to eight weeks before the tree comes into leaf, and the tender seedlings are thus often injured and killed by late frost. Therefore I advise storing them in a cold place, and not sowing until they begin to germinate. If they come up too thickly and survive the first spring, they may be transplanted in the following March or April, which will tend to check their early leafing, but if thin on the ground they may be allowed to stand two years before transplanting into rows. At three, or at most four, years old, they will be fit to go out permanently, the stronger side branches and double leads, if any, being first pruned. If intended for copsewood, they must be cut over in the month of April, two or three years after planting, and any pruning necessary to older trees should be done in summer or early autumn, so that the wounds may heal as soon as possible. The tree makes an abundance of fibrous roots, and unless these are allowed to become dry, the proportion of loss from transplanting should be very small, and transplanting may be done later than in the case of most hard woods.

For ash coppice, 4 or 5 feet apart is the right distance; for timber trees, they may be alternated with spruce or larch, which will keep them from becoming branchy. The cutting of the stools must be done with a sharp knife or axe as near the ground as possible, and with an upward cut, and the poles removed at latest by the middle of May, as much harm is caused by getting the poles away after the stools have begun to push new growth.

One of the best examples of copse-grown ash that I have seen in England is the Walk Copse near Buckhold, Berks, where a number of tall, slender, clean poles, believed by Dr. Watney to be about sixty years old, have originally sprung up from seed, in a plantation once largely composed of silver firs. Though the soil is a flinty clay, now of little agricultural value, the majority of these trees are 90 to 100 feet high, by 3 to 4 feet in girth, and quite clean to 50 or 60 feet. One of the best was quite straight and clean to 65 feet high, but only about 3 feet in girth. Such poles as these are much sought for by agricultural implement and coach makers, and are worth from 2s. 6d. to 3s. per foot.

## REMARKABLE TREES

I do not know of any ash at present alive in England which equals in size a tree mentioned by Loudon as growing near Moccas Court, Herefordshire, on the edge of a dingle. This had immensely large roots, running on the surface for 50 feet or more down the steep hillside, and a clear trunk of 30 feet long, 7 feet in diameter at 15 feet from the ground. This, including three large limbs, was estimated to contain 1003 cubic feet of timber. This ash is remembered as a marvellous tree, though quite decayed, by the Rev. Sir George Cornewall, who told me that not a vestige of it now remains.

The tallest living ash trees I have seen or heard of are in a grove near the old heronry at Cobham Hall, Kent, the seat of the Earl of Darnley, a place which contains taller and finer ash trees and hornbeams, and more of them, than any that I have seen in England. Strange to say, Strutt, who figured several trees at Cobham, overlooked these; but in Francis Thynne's continuation of Holinshed's *Chronicles*, p. 1512, I find the following, which shows that Cobham was renowned for its trees more than three centuries ago. Speaking of William, the last Lord Cobham but one, he says :—

"Besides which, owerpassing his goodlie buildings at the Blackfriers in London, in the year of Christ 1582, and since that the statelie augmenting of his house at Cobham Hall, with the rare garden there; in which no varietie of strange flowers and trees do want, which praize or price maie obtaine from the farthest part of Europe, or from other strange countries, whereby it is not inferior to the garden of Semiramis."

The largest ash here, described by Loudon, was a tree 120 feet high, with a trunk 6 feet 8 inches in diameter, straight, and without a branch for a great height. This was perhaps the same whose trunk I saw in July 1905 lying on the ground, where it had fallen several years ago. But those which remain are not only the tallest ash trees, but the tallest trees of any sort with one exception that I have measured in England, and there are so many of them that I can well believe that I did not measure the tallest. The tallest tree, measured in April 1907, by Lord Darnley, is 146 feet high by 12 feet in girth. Another growing by the side of a drive, which he christened Queen Elizabeth's ash, I measured 143 feet high by 12 feet 7 inches in girth. In the grove near it are several, very nearly if not quite, as tall, one of which I made 141 feet by 13 feet 1 inch, with a bole 50 feet high, and a roughly estimated contents of 700 to 800 feet. (Plate 239.) Another, 140 feet by 12 feet 9 inches, with a bole of 48 feet, which, judging from the large mass of fungus growing on its root, is probably decaying. There are many other trees in this grove which are 125 to 130 feet high, and stand pretty close together, growing in a sheltered situation,[1] on what appeared to be a deep but rather sandy loam.

Other remarkable ash trees at Cobham are the Twisted Ash, whose trunk is spiral, and measures 116 feet by 17 feet 9 inches. (Plate 240.) The View Ash, a tree nearer the house, is only about 80 feet high by 17 feet 9 inches at 5 feet, but is 29 feet in girth at the base, and has its trunk and most of its branches covered with green and healthy twigs.

Next to Cobham in respect of its great ash trees is Knole Park, also in Kent, where, in a sheltered valley near the gate from the Sevenoaks Road, called "The Hole in the Wall," are a number of very fine sound trees, from 125 to 130 feet high or more, and from $13\frac{1}{2}$ to 16 feet in girth, one of which has a bole 35 feet long, and probably contains over 700 feet of timber. Here again the soil is a deep sandy loam, which grows splendid beech, oak, and chestnut, but I cannot guess the age of the ash, though they are probably over 200 years.

One of the most perfect examples, from a timber point of view, is a tree growing in a wood called Poultridge, just outside Ashridge Park (Plate 241), which

---

[1] This is about 330 feet above sea-level, and is situated between the Medway and the Thames.

is about 125, perhaps 130 feet high, but difficult to measure on account of the surrounding trees. It had, in 1906, an absolutely straight, clean stem, about 75 feet in height by 11 feet 10 inches in girth, and looked as if it would square 27 to 28 inches halfway up, in which case it contains about 400 feet of faultless timber in one length. It is surrounded by other trees and underwood, and a photograph could not have been taken if Mr. Liberty, forester to Earl Brownlow, had not been obliging enough to clear away the intervening brushwood.

At Chilham Castle, in Kent, the seat of C. S. Hardy, Esq., there are some splendid ash trees. The best of these was recently cut down in the heronry, where I saw its stump in 1907, and counted about 185 rings, the diameter being about 4 feet. The soil here does not appear to be deep, and is on a chalk subsoil, as at Ashridge, but the tree grew in a very sheltered position, was drawn up to a height of 132 feet, and nourished by the beech which surrounded it. The first length contained 236 cubic feet. A tree of very similar character was growing in the park not far off, and measured about 115 feet by 13 feet 6 inches, with a straight, clean bole about 45 feet long. I estimated its contents at 280 feet in the first length.

At Godinton, Kent, there are many very fine tall ash trees in what is called the "Tole," a splendid clump mainly composed of chestnut, which contains as large a quantity of clean, fine timber as I ever saw on the same area in England. I measured an ash here about 110 feet high, and only 6 feet 5 inches in girth, which hardly fell off at all in thickness up to 50 or 60 feet.

At Woburn there was in Strutt's time an immense ash which he figures on Plate 22, and gives the height as 90 feet, the girth at 3 feet, 15 feet 3 inches, the bole 28 feet high, and the diameter of the branches 113 feet. The estimated contents were 872 feet. This tree was still healthy in Loudon's time but I can find no trace of it at present.

At Arley Castle there is a fine ash tree 120 feet by 14 feet as measured in 1904 by Mr. R. Woodward; and at Althorp there was a very fine ash which is now much damaged by wind and old age. When measured in 1890 by Mr. Mitchell, now forester at Woburn Abbey,[1] it had a bole 36 feet high with immense limbs spreading on all sides, was 17 feet 3 inches in girth at 3 feet, and had a cubic content of 800 feet. At Hatfield there is also a very fine ash tree growing near the big elm in a hollow, which I made 104 feet by 15 feet 8 inches. An immense ash butt was bought by Mr. Miles of Stamford in February 1894, and hauled into his yard by eleven horses. I am informed by Mr. C. Richardson that it measured 20 feet long by 50 inches quarter-girth, equal to 435 cubic feet of timber, and that it was the largest he had ever seen or heard of. Mr. A. B. Jackson in 1908 saw some tall trees at Kedleston Park, Derby, one measuring 120 feet by 10 feet 2 inches, and another 125 feet by 8 feet 10 inches. He measured also a tree at Elvaston Hall, 110 feet by 12 feet 7 inches.

At Studley Royal, Yorkshire, there are some tall, straight, and clean ash, the best that I measured being 119 feet by 10 feet 6 inches; and at Castle Howard and other places in the same county, and in parts of Lincolnshire, the ash is a more

[1] *Trans. Roy. Scott. Arb. Soc.* xiii. 88 (1891).

profitable timber tree than any other. The largest that I have seen in the north is one at Lowther Castle, known as Adam's Ash, which is 21 feet 10 inches in girth.

Lees in *Gardeners' Chronicle*, November 7, 1874, figures several curiously split and distorted relics of ash trees which existed in the district he knew so well round Worcester; but does not mention any of extraordinary size. He suggests that such wrecks may often have escaped being converted into firewood, owing to the superstition which formerly prevailed, that rickety children might be cured by passing them through a fissured ash tree; and relates an instance known to him of the curious superstition, that a similar passage through a cleft ash would induce fertility in barren women.

White, in his *Natural History of Selborne* (Ed. Allen, 1900, p. 266), speaks of a row of pollard ashes at Selborne which had been in former times cleft and held open by wedges while ruptured children were pushed through the aperture, in the belief that they would be cured of their infirmity. He also states that there were then several people living in his parish, who in childhood were supposed to be so cured. He describes an old pollard ash which for ages had been looked on with no small veneration as a "shrew-ash," and whose twigs were, when applied to the limbs of cattle, supposed to cure the pains caused by a shrew-mouse having run over the affected part. A "shrew-ash" was made by boring a hole in the trunk and putting a live shrew into the hole, where it was plugged up with several quaint incantations now long forgotten.

In Wales I have not seen or heard of any larger than a tree by the slaughter-house in Dynevor Park. Though not very well shaped and somewhat past its prime, I found it in 1908 to be 104 feet by 22 feet 9 inches.

In Scotland there are also many great ash trees, of which perhaps the largest recorded anywhere has long ago completely decayed. It was described in Walker's *Essays*, p. 17, as growing in a deep, rich soil, in the churchyard of Kilmalie, and was considered to be the largest and most remarkable tree in Scotland; and said to measure in 1764 no less than 58 feet in girth at the ground. Another celebrated old tree is in the hotel garden at Logierait, Perthshire, which Hunter[1] described in 1883 as measuring 47 feet 7 inches at 1 foot from the ground, and 32 feet 5 inches at 5 feet. It was then completely hollow and covered with ivy, with an opening 5 feet 9 inches wide on one side, in which a summer-house had been made. But the late Sir R. Menzies informed me in 1903 that this tree is now very much decayed.

Strutt, *Sylva Scotica*, Plate 8, figures a very fine ash tree, at Carnock, Stirlingshire, then in perfect vigour, and said to have been planted about 1596 by Sir T. Nicholson of Carnock. He gives its measurement as 90 feet high by 19 feet 3 inches at 5 feet. This tree, however, died and was broken up about 40 years ago. The tallest ash that I have myself measured in Scotland is at Gordon Castle, which is a fine healthy tree in the home park, 101 feet high with a bole of 30 feet, which in 1904 girthed 12 feet 6 inches. J. Webster records one at the same place

---

[1] *Woods, Forests, and Estates of Perthshire*, 545 (1883).

70 feet by 15 feet 8 inches in 1881. Sir Archibald Buchan-Hepburn reports a tree at Smeaton-Hepburn 124 feet by 11 feet in 1908.

But these are much exceeded in height by a tree west of the Beech Walk at Mountstuart in the island of Bute, which James Kay[1] states to have been in 1879 no less than 134 feet high by 9 feet 4 inches, with a bole of 35 feet 6 inches, and if this measurement was correct it must have been the tallest hardwood tree in Scotland. It was estimated to contain 273 feet of timber. I could not, however, identify this tree when I visited Mountstuart in 1906, and fear that, like some of the splendid beech trees which grew there, it has fallen.

At Ochtertyre, Perthshire, Hunter[2] records an ash supposed to be about 400 years old, which measured in 1881 34 feet 10 inches at 1 foot and 20 feet 8 inches at 5 feet, and I am informed by the widow of the late gardener, Mr. Conacher, that the tree still remains in very good condition. At Keir, near the Bridge of Allan, there is a remarkable ash stool, from which four stems proceed, averaging 6 feet 4 inches in girth and 103 feet in height. A tape 16 feet 10 inches long, girths the four stems at 5 feet from the ground. At Dupplin Castle I measured a fine tree about 100 feet high with a stem clean to 45 feet and 10 feet 7 inches in girth. At Dalswinton, Dumfriesshire, Henry measured in 1904, a tree 110 feet by 8 feet 3 inches, with a fine clean stem ; and another 93 feet by 13 feet 3 inches.

Near Cawdor Castle, Nairnshire, the property of the Earl of Cawdor, and one of the most beautifully situated of the really old inhabited castles in Scotland, there is a very large, though branchy and ill-shaped ash no less than 21 feet in girth. At Brodie Castle, Morayshire, there is a very large tree, 18 feet 8 inches in girth, of which the owner has kindly sent me a photograph ; and at Darnaway, in the same county, an immense tree of great age, much damaged by storms, existed in 1881. Even as far north as Conon House, Ross-shire, the seat of Sir Kenneth MacKenzie of Gairloch, the ash grows extremely well in a low-lying flat. Here I saw a lot of beautifully grown, though not very large trees, which would have been a credit to any woodland in the south.

On the shore of Loch Fyne, a mile north of Minard Castle, a curious ash grows on the beach at high-water mark, which is known as the " Nine Sisters," because nine stems sprang from the same root, the largest of which when I saw them in 1907 were 7 to 9 feet in girth.

In Ireland the ash thrives exceedingly well ; and often attains an immense size. In Co. Meath, where the soil is remarkably fertile, it has in many parts expelled all the other trees from the hedgerows ; and one may drive long distances on the roads between lines of flourishing ash trees, without seeing a single oak or beech.

In the latter part of the eighteenth century several ash trees of enormous size were still living, of which Hayes gives an account.[3] He relates, on the authority of an official of the Dublin Society, that a tree was then standing at Donirey near Clare Castle in Co. Galway, which measured in girth 42 feet at 4 feet and 33 feet at 6 feet. The trunk had long been hollow, having been used 25 years before as a

[1] *Trans. Scot. Arb. Soc.* ix. 75 (1879).          [2] *Op. cit.* 454.
[3] *Practical Treatise on Planting,* 137, 142, 148 (1794).

school. In 1794, it had only a few branches remaining, which were, however, still vigorous. At Curraghmore, the seat of the Marquess of Waterford in the county of that name, there were many enormous ash trees, one of which was 22½ feet in girth at a height of 13 feet 9 inches, the girth of the butt being 33 feet 9 inches. In 1792 Hayes measured the famous ash tree at Leix in Queen's County, which was 40½ feet round at 1 foot, and 25 feet at 6 feet up, where the girth was least. Marsham in a letter to Gilbert White, dated 12th February 1792, says that a print of this tree was then being engraved in London, but we have never been able to see a copy of this, if it was ever published.

An ash at Castledurrow, which Hayes says was the finest he ever saw, in October 1793 measured 18 feet in girth and nearly as large at 14 feet high, the branches extending 45 feet from the stem in every direction. Another at Kennity Church, King's County, was 21 feet 10 inches in girth with a bole of 17 feet. The funeral parties used to stop and say prayers under this tree, after which they threw a stone at its foot.

None of these relics of antiquity now remain; and most of the fine ashes have been cut down. Probably the tallest now left in Ireland is one at Woodstock in Co. Kilkenny, in an alluvial flat beside the river Nore and close to the village of Inistioge (Plate 242). Henry, in 1904, made it 127 feet in height by 16 feet 9 inches in girth; but the forester's records give it as 136½ feet high in 1901. This tree was 11 feet in girth in 1825, 11 feet 8 inches in 1834, 13 feet 2 inches in 1846, and 16 feet 9 inches in 1901, according to the same records. Another tree beside it, which was 9 feet 5 inches in girth in 1825, was 11 feet 9 inches in 1901; and a third tree, 10 feet 5 inches in 1825, had attained 15 feet 3 inches in 1901.

At Mitchellstown, in Co. Cork, there was in 1903 a remarkably fine ash, with a tall clean stem, which was 111 feet high by 9 feet in girth. Another tree, 91 feet high, was 27 feet in girth at 2 feet above the ground, but only 15 feet 2 inches at 6 feet up. At Kilmacurragh in Co. Wicklow there is a good ash, the height of which exceeds 100 feet, but is impossible to measure on account of the situation of the tree; the girth in 1903 was 13 feet at 5 feet above the ground. There are fine trees of great girth (17 to 18 feet), but not remarkable for height, at Doneraile Court, Co. Cork. Henry measured in 1904, on Lord Oranmore's property at Castle Macgarrett, Co. Mayo, a tree 119 feet by 14 feet 2 inches, containing 216 cubic feet of timber. At Castlewellan, Co. Down, there are some very fine ash in the park, one of which near the garden gate is figured on Plate 243. It measured in 1908 about 70 feet high, and 18 feet in girth.

The finest ash that I have seen on the continent is in the Royal park of Jægersborg near Copenhagen, and measured in 1908, 125 feet by 11 feet 8 inches.

## ASH DISEASES

The commonest disease in the ash in my own district is a canker which affects trees during most of their life without killing them, though the timber is worthless except for firewood. This disease is described and figured by the late Mr. Wilson

Saunders, F.R.S.;[1] and Mr. Sidney Webb fully explained the manner in which extensive canker resulted from minute wounds, at a meeting of the Scientific Committee of the Royal Horticultural Society, February 11, 1879. A full account of the disease is given by Dr. Masters in the *Gardeners' Chronicle*, 1879, p. 208, where it is stated that the injury is originated by the larva of a minute moth called *Prays curtisellus*.[2] Plate 244 shows a bad case of this canker in an ash at Staple, near Colesborne; and there is a tree at least 70 feet high by the roadside, close to the sixth milestone from Cirencester to Cheltenham, which has had this disease from its base to near the summit as long as I can remember.

In *Trans. Scot. Arb.* x. 235, there is a useful paper on the Ash Bark Beetle, *Hylesinus Fraxini*, a pest which seems to be dangerous only where the ash is already unhealthy. As the eggs of this insect are laid in spring only under the bark of felled, dead, or sickly trees, wherever this pest is troublesome, all such should be removed from the neighbourhood of the healthy trees by April, and ash loppings should not be left on the ground. A curious malformation occurs in a tree growing close to Cirencester, on the east side of the Tetbury road nearly opposite the Kennels. Plate 244 shows the remarkable growths on its branches, specimens of which were sent to Kew and found to contain numerous examples of *Hylesinus Fraxini*.

## TIMBER

For coach, waggon, and agricultural implement making, and for all purposes in which strength, toughness, and durability are required, ash timber has no equal, and no substitute has been found among foreign trees which can be relied on as well. In consequence, it is now the easiest to sell, if not quite the highest priced, of all English timbers; and its growing scarcity seems to point to a great future for it.

It varies much, however, in strength, toughness, and elasticity, according to the soil on which it is grown, and the age at which it is cut. I am informed by Mr. Clutterbuck of the Gloucester Waggon Works, who has had long experience with English and foreign timber, that there is no better ash in England than that grown on the Cotswold hills; but if left standing too long, it becomes discoloured at the heart, and is probably never worth more per foot than when 60 or 70 years old.

It is now perhaps the only wood worth growing as copsewood, and, when established on good land and cut every ten to fifteen years, still makes as much as a pound per acre per annum. The poles are used for making hurdles, hoops, crates, and many other purposes, and as hop-poles are only second to chestnut. I have found that for the rails of light field-gates in a hunting country, on account of their elasticity nothing is better; and when well made they last thirty years or more. Ash wood takes creosote well, which very much increases its durability; and some sheep-hurdles which I had creosoted thirteen years ago are still sound, though when not so treated, they do not last more than three to five years. Ash, however, soon decays in contact with the soil and is unfit for building purposes, though it was

---

[1] *Journ. Roy. Hort. Soc.* v. 135 (1879).
[2] Also known as *Tinea curtisella*, Don. Cf. Schlich's *Man. Forestry*, iv. 344 (1907), where the moth is figured and described.

formerly used for staircases. Loudon says that the staircase at Wroxton Abbey, near Banbury, was made of this wood, and in Hatfield House some of the inside work is made of ash. Slabs from flat-sided trees often show a very ornamental curly grain which makes very handsome panelling, and might be used for door panels with good effect.

The burrs are also cut into beautiful veneers which, when polished, are used in cabinetmaking, and which sometimes in Hungary and south Russia are of great size and perfectly sound, though in England usually small and faulty. I purchased in Manchester, under the name of Circassian ash burr, some splendid veneers of this wood, which measure about 5 feet by 3 feet, and are made up of small, closely crowded knots, which take a fine polish and are of a greyish white or pinkish grey colour.[1] English ash, however, seldom or never assumes the wavy grain which is found in Hungary and Russia, and is one of the most beautiful woods I know. This is known as fiddle-back ash, because wood of this character, usually maple, is selected for the backs of violins. It varies very much in colour, the most valuable being the whitest; and also in the size and character of the figure; but when a combination of small waves with eye-like patches is combined, it is superior to the best American maple. Such wood was formerly much used for decorating railway carriages, and for furniture, but from some reason which I cannot explain is now out of fashion. I believe that the waving rarely extends throughout the tree, the best figure being always near the outside, and the causes which produce it are, so far as I know, as yet without any scientific explanation.          (H. J. E.)

## FRAXINUS ANGUSTIFOLIA, NARROW-LEAVED ASH

*Fraxinus angustifolia*, Vahl, *Enum.* i. 52 (1804); Loudon, *Arb. et Frut. Brit.* ii. 1229 (1838).
*Fraxinus australis*, Gay, ex Koch, *Dendrologie*, ii. 1. 247 (1872).

A tree attaining 70 to 90 feet in height. Shoots glabrous, green, slender. Leaflets (Plate 262, Fig. 6), seven to thirteen, $1\frac{1}{2}$ to 3 inches long, smooth and slightly coriaceous, shining above, usually pretty uniform in size, subsessile, lanceolate, base cuneate, apex acuminate, glabrous on both surfaces; coarsely and sharply serrate except near the base; serrations few, spreading, often with incurved points (occasionally deeply serrate with long bristle points). Rachis of leaf glabrous, strongly winged, the wings meeting above and only showing a groove opposite the insertions of the leaflets. Flowers (section *Fraxinaster*) without calyx or corolla, few in erect racemes, arising from the axils of the leaf-scars of the preceding year's shoot. Fruit lanceolate, obliquely truncate and entire at the apex; but apparently variable.

This species is distinguished from all forms of the common ash by its absolutely glabrous leaflets, which have fewer, sharper, and more spreading serrations than in that species. The terminal buds are also different, being small, dark brown, quadrate,

---

[1] I have seen in London a fine old cabinet, supposed to be veneered with Amboyna wood, so like the Circassian ash in pattern, though the colour was yellower, that I much doubt whether the two could be distinguished when made up.

conical, usually glabrous, with four outer scales—six external scales, however, occurring in individual trees with leaves in whorls of threes.

## VARIETIES

1. Var. *monophylla*[1] (*F. Veltheimi*, Dieck). A form in which the leaves (Plate 262, Fig. 3) are simple, unequally two-foliolate, or three-foliolate; terminal leaflet or single leaf lanceolate, acuminate, coarsely serrate or dentate; lateral leaflets, when present, much smaller but similar in shape; petiole with a wide open groove on the upper side. Shoots green, glabrous, with pink lenticels.

This variety can only be confused with *Fraxinus excelsior*, var. *monophylla*, from which it is readily distinguishable by the glabrous leaflets, different in texture and usually narrower, being lanceolate and not ovate. This tree appears to do well in cultivation, but will probably not attain a large size. We know of no trees of this variety in England except in the collection at Kew.

2. Var. *lentiscifolia* (*F. lentiscifolia*, Desfontaines[2]?). In the typical form of *F. angustifolia*, both as it occurs wild and under cultivation, the leaflets are set close upon a short rachis and point forwards towards its apex. In this variety, the leaflets are set wide apart upon an elongated rachis, from which they spread out at right angles and are not directed forwards; they also differ slightly in colour and texture from the type.

Willdenow[3] considered *F. parvifolia*, Lamarck,[4] *F. tamariscifolia*, Vahl,[5] and *F. lentiscifolia*, Desfontaines,[6] to be identical. I have not been able to follow Koehne or Dippel in their treatment of these forms as distinct. It is not in the least certain what species was intended by Lamarck's *parvifolia*, a name which has been given by some writers to a small-leaved variety of *F. oxycarpa*. In the Kew Herbarium, a garden specimen, collected by Bentham in 1854 and labelled "*F. lentiscifolia*," agrees with our variety of *F. angustifolia*; and a wild specimen from Italy is indistinguishable from it.

Both pendulous and dwarf forms of var. *lentiscifolia* are known in cultivation.

## DISTRIBUTION

The narrow-leaved ash, with glabrous leaves, is common in the south of France, and occurs also in Spain, Portugal, Morocco,[7] and Algeria.

Captain Widdrington[8] says of this tree: " The ash is extraordinarily rare in the south and central regions of Spain. It is not now cultivated, and the only specimens I saw growing wild were in the wilder parts of Estremadura and the Sierra Morena where they were generally by the side of water-courses. The only species that I have seen in these regions is the *lentiscifolia*. The first year that I was in the

---

[1] Stated in the *Kew Hand List of Trees*, 545 (1902), to be a cross between *F. parvifolia* and *F. excelsior monophylla*. It shows no evidence of hybrid origin, and is evidently a variety of *F. angustifolia*.

[2] *Table de l'École de Bot.* 52 (1804).     [3] *Sp. Pl.* iv. 2, 1101 (1805).     [4] *Encycl.* ii. 546 (1786).

[5] *Enum.* i. 52 (1804).     [6] *Table de l'École de Bot.* 52 (1804).

[7] This ash, referred to *F. oxyphylla* by Ball, was seen by him in Morocco, between Tangiers and Tetuan, and also at the base of the Atlas mountains. Cf. *Journ. Linn. Soc.* (*Bot.*) xvi. 564 (1878).

[8] *Spain and the Spaniards*, i. 390 (1844).

interior of Spain, I picked up the seed of an ash near the Escorial; but the leaves having fallen, I did not ascertain the species, but sending them to England they vegetated, and are now growing in Northumberland. This is the same tree, and I have never seen it farther north than New Castile; at the same time I think it probably may exist as far as Leon, where, the instant you cross the chain, the *Fraxinus excelsior*, our common species, supplies its place; at least, I could make out no difference. The timber of *F. lentiscifolia* is heavy and less elastic than that of our species, but the elegance of the tree, and its perfect hardiness in a dry soil, should make it more common than it is in our ornamental collections."

This species[1] replaces the common ash in Algeria, where it is only found wild in quantity in the forests of the plains and along the banks of streams and rivers; but it ascends as isolated trees occasionally to 6000 feet in the mountains, and is reported to be common in the Djurdjura range. I saw it growing in fields near the forest of Akfadou, inland from Bougie, where the trees have a mutilated appearance, owing to the annual lopping of their branches by the natives, who feed their cattle with the leaves.[2] In Algeria[3] the tree attains a large size, and grows in good soil with great rapidity, reaching a height of 90 feet by 3 feet in diameter at seventy years old. The wood is similar to that of the common ash, though slightly inferior in quality. Dr. Trabut informed me that he had sent seed to Australia, where the tree is said to thrive well, succeeding better than the common ash.

A tree at Chiswick House measured, in 1903, 75 feet by 7 feet 5 inches. Another at Whitton, near Hounslow, in 1905 was 56 feet by 6 feet 2 inches. At Williamstrip Park, Gloucestershire, in 1904 Elwes measured a tree 60 feet by 6 feet 9 inches.

The variety *lentiscifolia* has been identified by us at Syon, 55 feet by 5 feet 6 inches in 1905; at Hardwick, Bury St. Edmunds, a grafted tree, 72 feet by 6 feet 3 inches below the graft and 7 feet 10 inches above it; at Bicton, another grafted tree, over 50 feet by 6 feet; and at Stowe, also a grafted tree, 68 feet by 7 feet 8 inches. From the similarity in appearance of the trees at Hardwick, Stowe, and Bicton, there is little doubt that they were all propagated and planted at the same time. Elwes has also seen at the Hendre, Monmouthshire, a tree of similar appearance, which was 71 feet high by 5 feet 4 inches below, and 6½ feet above the graft. There is also a healthy grafted tree at Ware Park, Herts, growing on sandy soil, which Mr. H. Clinton Baker measured as 78 feet by 6 feet 3 inches in 1908. Another (Plate 245) on the lawn at Rougham Hall, the seat of F. K. North, Esq., is 76 feet high by 8½ feet above the graft, and 7½ feet below it. A third at Ribston Park, Yorkshire, was 68 feet by 6 feet 7 inches.

It is hardy as far north as Denmark, where Elwes measured in the park of Count Friis, in 1908, at Boller near Horsens, a grafted tree about 60 feet by 4 feet 4 inches, which was bearing immature fruit.                    (A. H.)

---

[1] The Algerian tree has been distinguished as var. *numidica* (*F. numidica*, Dippel), with broader and larger leaflets; but specimens gathered by me at Akfadou are typical *angustifolia*.

[2] M. Maurice L. de Vilmorin, in *Bull. Soc. Amis des Arbres*, 1895, states that this ash is much planted around villages in Kabylia, where its leaves, which are stripped off the tree in September, are an indispensable fodder for cattle, sheep, and goats at this season when no grass is available. The foliage of a single tree is usually worth 50 francs; and he was shown a very old wide-spreading tree, the owner of which sold its leaves annually for 300 francs.

[3] Cf. Lefebvre, *Forêts de l'Algérie*, 348 (1900).

## FRAXINUS OXYCARPA

*Fraxinus oxycarpa*, Willdenow, *Sp. Pl.* iv. 2, 1100 (1805); Loudon, *Arb. et Frut. Brit.* ii. 1230 (1838).

*Fraxinus oxyphylla*, M. Bieberstein, *Fl. Taur. Cauc.* ii. 450 (1808); Boissier, *Flora Orientalis*, iv. 40 (1879).

*Fraxinus rostrata*, Gussone, *Pl. Rar.* 374, t. 63 (1826).

A tree of moderate size. Shoots green, glabrous; lenticels pink. Leaflets (Plate 263, Fig. 11), nine to thirteen, usually small, $1\frac{1}{2}$ to 3 inches long, sessile or subsessile; lanceolate, oval or ovate; base tapering, apex acuminate; serrations few, sharp, spreading and often ending in incurved points; lower surface pubescent on the midrib and veins towards the base. Rachis of the leaf glabrous, winged, the wings meeting above and only forming an open channel opposite the nodes. Flowers (section *Fraxinaster*) without calyx or corolla, in short racemes in the axils of the leaf-scars of the previous year's shoot. Fruit broad, oblanceolate, acute or acuminate at the apex.

This species closely resembles *F. angustifolia*, differing in the leaflets being always pubescent beneath. The terminal buds are conical, quadrate, long and slender, with four outer narrow scales of equal length, dark brown and pubescent.

It is doubtful if the differences in the samaræ relied upon for the separation of this species from *F. angustifolia* are really constant. The fruits in this group of *Fraxinus* are extremely variable, and do not appear to give specific characters.

While *F. angustifolia* seems to be confined to the western part of the Mediterranean region, *Fraxinus oxycarpa* is widely distributed in Italy, Asia Minor, Persia, and the Caucasus.

Var. *parvifolia*, Wenzig.[1] Leaflets small, oval-oblong. Boissier considers this to be rather a bushy or sterile juvenile form of the species than a distinct variety, and records it from various localities in Asia Minor. An ash identified by Mathieu,[2] with *F. parvifolia*, Lamarck, grows in the neighbourhood of Montpellier in France, as a shrub about 5 to 10 feet high, and belongs to this variety.

At Kew, the ash, cultivated as *F. parvifolia*, Lamarck, is a variety of *F. oxycarpa*, distinguished by having leaflets (Plate 263, Fig. 9) shorter than in the type, broader in proportion to their length, and more closely set upon the rachis.

*Fraxinus oxycarpa* is much rarer in cultivation in England than *F. angustifolia*; but small trees are growing at Kew, Woburn, Eastnor Castle, Oxford Botanic Garden, and at Grayswood, Haslemere, where it has produced seed, from which plants have been raised by Elwes, and are growing at Colesborne. It does not seem in this country to be so vigorous a tree as *F. angustifolia*.          (A. H.)

---

[1] In Engler, *Bot. Jahrb.* iv. 175 (1883).          [2] *Flore Forestière*, 245 (1897).

## FRAXINUS SYRIACA

*Fraxinus syriaca*,[1] Boissier, *Diag.* Ser. I. ii. p. 77 (1849).
*Fraxinus oxyphylla*, M. Bieberstein, var. *oligophylla*, Boissier, *Fl. Orient.* iv. 40 (1879).

A tree attaining 60 feet in height. Shoots glabrous, green, stout, conspicuously marked by very prominent leaf-bases; lenticels white. Leaves (Plate 263, Fig. 10), small, always in whorls of threes or fours. Leaflets usually three, occasionally five to seven on some of the branchlets, sessile, lanceolate to ovate, base cuneate, apex acuminate, sharply and coarsely serrate, the serrations with incurved points, glabrous on both surfaces. Rachis of the leaf narrowly winged, the wings not meeting above, but forming an open groove.

Flowers (section *Fraxinaster*) in short racemes in the axils of the leaf-scars of the preceding year's shoot; without calyx or corolla. Fruit ovate-oblong; apex rounded, truncate or acuminate, ending in a mucro.

This species occurs in Syria, Kurdistan, Persia, Baluchistan, and Afghanistan.

The occurrence always of the leaves in whorls, a phenomenon met with in individual instances in other species, appears to be constant in this species. On strong shoots leaves with five to seven leaflets exceptionally appear; those with three leaflets, however, being by far the most common. Small specimens of this tree are growing at Kew, but it does not seem likely to be worth growing in England.                                                    (A. H.)

## FRAXINUS ELONZA

*Fraxinus Elonza*, Dippel, *Laubholzkunde*, i. 87, fig. 46 (1889); Koehne, *Deutsche Dendrologie*, 513 (1893).

A small tree. Branchlets green, glabrous; lenticels few, oval, white. Buds laterally compressed and not quadrangular, narrowed and rounded at the apex; external scales four, densely brown pubescent, inner pair longer than the outer pair. Leaflets (Plate 266, Fig. 27), eleven to thirteen, 1 to $2\frac{1}{2}$ inches long, sessile, oval or lanceolate, with unequal base and acuminate apex, sharply and irregularly serrate, some of the serrations being often triangular and spreading; under surface pubescent near the base with brown tomentum, often occurring only on the inner side of the midrib. Leaf-rachis, with scattered pubescence, densest at the nodes; strongly winged, the wings meeting above in part of their length. Fruit described as broadly linear, with almost parallel sides, truncate and emarginate at the apex.

The native country of this species is unknown; and it is possibly a hybrid, as

---

[1] *Fraxinus sogdiana*, Bunge, *Mém. Sav. Etrang. Acad. Pétersbourg*, vii. 390 (1851), occurring in Turkestan, formerly supposed to be identical with this species, is considered distinct by Koehne in *Gartenflora*, 1899, p. 288, and by Lingelsheim, in Engler, *Bot. Jahrb.* xl. 222 (1907).

Koehne suggests. It occurs in England in cultivation under the erroneous name of *F. chinensis*,[1] the plants at Kew having been obtained from Sir C. W. Strickland, who tells us that it was sent out by the Royal Horticultural Society some years ago. Plants cultivated as *F. Elonza* are usually *F. oxycarpa*.                     (A. H.)


## FRAXINUS WILLDENOWIANA

*Fraxinus Willdenowiana*, Koehne, *Deutsche Dendrologie*, 515 (1893).
*Fraxinus parvifolia*, Willdenow, *Berlinische Baumzucht*, 124, t. 6, f. 2 (1796) (*non* Lamarck); and
   *Sp. Pl.* iv. 1101 (1805).

A small tree. Shoots glabrous, lenticels white. Leaflets (Plate 265, Fig. 24), seven to eleven, 2 to 3½ inches long, subsessile (except the terminal one, which is much the largest and stalked), ovate, base broad and unequally cuneate, apex acuminate; serrations coarse and sharp with minute incurved points; both surfaces glabrous. The leaflets increase in size from the base to the apex of the leaf, the rachis of which is winged, the wings usually not meeting on the upper side, but forming an open groove. Fruit unknown.

This species was considered by Willdenow to be different from *F. parvifolia*, with which he had first identified it, yet he left it with this name. Koehne has accordingly given it a new name. It is sometimes met with in cultivation under the name of *F. rotundifolia*. It is readily distinguished by the large terminal leaflet and the open-grooved rachis from the other glabrous species. Its native home is uncertain. It is perfectly hardy at Kew and has very distinct foliage.                     (A. H.)


## FRAXINUS DIMORPHA

*Fraxinus dimorpha*, Cosson et Durieu, *Bull. Soc. Bot. France*, ii. 367 (1855); Mathieu, *Flore
   Forestière*, 245 (1897).
*Fraxinus xanthoxyloides*, Wallich, *var. dimorpha*, Wenzig, in Engler, *Bot. Jahrb.* iv. 188 (1883).

A small tree attaining 40 feet in height. Young shoots purple, slender, glabrous, obscurely quadrangular. Leaves (Plate 262, Fig. 1) on barren branchlets with seven to nine small leaflets, which are ½ to ¾ inch long, sessile or subsessile, ovate or oval, crenulate-serrate, and glabrous except for pubescence on the midrib towards the base on the under surface; leaf-rachis usually glabrous, strongly winged, the wings spreading and forming a very open channel. The leaves on flowering shoots are larger, with seven to eleven leaflets, oblong-lanceolate, acute and serrate. Flowers (section *Sciadanthus*) perfect, without a corolla, but with a calyx which persists under the fruit, grouped in fascicled cymes on the previous year's shoot

---

[1] *F. chinensis*, Roxburgh, a native of China, is entirely distinct from *F. Elonza*, which is closely allied to *F. oxycarpa*, a Mediterranean species. Koch, in *Dendrologie*, ii. pt. 1, 247 (1872), mentions *F. Elonza* as having been in cultivation some years, and considered it to be probably a variety of *F. angustifolia*.

in the axils of the leaf-scars. Fruit oblong; body compressed with a longitudinal furrow on each surface, and many-rayed; wing long and obliquely truncate at the apex. Buds very small, with two outer scales pinnately lobed.

This remarkable ash was observed by Sir Joseph Hooker in South Morocco.[1] It also occurs in the mountainous regions of Algeria in the valleys at 4000 to 6000 feet altitude. In dry situations it remains a bush with very rigid and almost spiny branches, and rarely flowers even when very old. On the banks of streams it grows to be a small tree and produces flowers and fruit. The wood is very hard and heavy, with a satin-like lustre.

This species is rarely seen[2] except in botanical gardens, where, as at Kew, it grows to be a small tree, remarkable for its diminutive foliage. It has smooth grey bark.
(A. H.)

## FRAXINUS XANTHOXYLOIDES

*Fraxinus xanthoxyloides*, Wallich, *List*, 2833; C. B. Clarke, in Hooker, *Fl. Brit. India*, iii. 606 (1882); Brandis, *Indian Trees*, 444 (1906).

This species is probably only a pubescent geographical form of *Fraxinus dimorpha*, which it resembles exactly in habit. The young shoots are covered with a minute dense pubescence. The leaflets only differ from those of *F. dimorpha* in having a scattered pubescence all over the lower surface; the rachis of the leaf is also pubescent.

This species occurs in Baluchistan, Afghanistan, and the north-west Himalayan region, at altitudes of 3000 to 9000 feet, growing mainly in dry valleys, where it is often gregarious. It is reported to attain a height of 25 feet.

It is rare in cultivation, and forms a small tree, scarcely distinguishable in appearance from *F. dimorpha*.
(A. H.)

## FRAXINUS POTAMOPHILA

*Fraxinus potamophila*, Herder, in *Bull. Soc. Imp. Mosc.*, xli. 65 (1868); Dippel, *Laubholzkunde*, i. 98, fig. 54 (1889).

A small tree; branchlets glabrous. Leaflets (Plate 262, Fig. 8) small, seven to nine, about $1\frac{1}{2}$ to 3 inches long, stalked (petiolules glabrous, $\frac{1}{2}$ inch or more in length), ovate, tapering unequally at the base, acute at the apex, coarsely serrate, the serrations often ending in long points; glabrous on both surfaces. Rachis of the leaf with angled edges on its upper side, enclosing a shallow groove. Flowers

---

[1] Hooker and Ball, *Tour in Morocco*, 176 (1878); Ball, in *Journ. Linn. Soc.* (*Bot.*), xvi. 564 (1878).

[2] A tree at Coombe Wood, which had attained almost 25 feet in height, was destroyed in 1907. Mr. A. B. Jackson has seen small specimens at Barron's nursery, Elvaston, where they were erroneously named *E. lentiscifolia*.

unknown. Fruit described as linear-oblong, nearly two inches long, acute, blunt or obliquely truncate.

This ash grows along the banks of rivers in Turkestan and Songaria, occurring in the Ili region at 1000 to 2500 feet elevation. It was introduced into cultivation by the Botanic Garden at St. Petersburg; and small trees are doing fairly well at Kew.

*Fraxinus Regelii*, Dippel,[1] of which I have seen no authenticated specimen, is said to be also a native of Turkestan, and was considered by Koehne[2] to be probably identical with *F. potamophila*, Herder. There are young plants in the Kew collection, raised from seed sent in 1900 by M. Scharrer, Director of the Botanic Garden at Tiflis, and named *F. Regelii* on his authority, which are remarkably distinct from any ash known to me, and differ from Dippel's description of *F. Regelii* in the larger size of the leaflets, which are crenate and not dentate in serration. The Tiflis plants have the young branchlets glabrous, purplish; leaflets (Plate 265, Fig. 25), five or seven, about 3 inches long, stalked, the base of the leaflet often decurrent on one side of the petiolule to its insertion; terminal leaflet obovate or rhomboid; lateral leaflets ovate or oval; all shortly acuminate or cuspidate at the apex, unequal at the base, crenately serrate; bluish green and glabrous on the upper surface; pale green and slightly pubescent on the sides of the base of the midrib on the lower surface; rachis elongated, terete, glabrous, with a shallow groove on its upper side. The identification of these plants with *F. Regelii* must be left uncertain.

(A. H.)

## FRAXINUS RAIBOCARPA

*Fraxinus raibocarpa*, Regel, in *Act. Hort. Petrop.* viii. 685 (1884).

A small tree. Branchlets brown, minutely pubescent, glandular. Leaflets (Plate 266, Fig. 29), five, upper subsessile, lower stalked, about 1½ inch long, oval, unequal and rounded at the base, acute or obtuse at the apex, usually entire in margin without cilia; under surface glabrous, with a few minute brown glands. Leaf-rachis slightly glandular, with a wide open groove on its upper side. Fruit in leafy panicles, arising on the current year's shoot; samara surrounded at the base by the persistent calyx, curved, falcate; body terete and rayed; wing terminal, very broad, spathulate-obovate, obtuse.

This species, of which the flowers are unknown, belongs apparently to the section *Ornus*. It was discovered in 1882 by Regel at elevations of 6000 to 7000 feet in the mountain valleys of eastern Bokhara and Turkestan; and was introduced into cultivation shortly afterwards by the St. Petersburg Botanic Garden. Small plants at Kew have grown very slowly, and this species does not seem likely to be worth cultivating in this country.

(A. H.)

---

[1] *Laubholzkunde*, i. 97, fig. 53 (1889), described from plants sent out by the St. Petersburg Botanic Garden as *F. sogdiana*, an entirely different species, referred to on p. 883, note 1.

[2] *Deutsche Dendrologie*, 515 (1893).

## FRAXINUS HOLOTRICHA

*Fraxinus holotricha*, Koehne, in *Mitt. Deut. Dendrol. Ges.* 1906, p. 67.

A small tree. Branchlets grey, densely covered with short stiff erect pubescence. Leaflets, nine to thirteen, subsessile, lanceolate or ovate-lanceolate, about 2 inches long, ½ to ⅝ inch wide; apex prolonged into a sharp-pointed and often curved acumen; base unequal and tapering; margin ciliate, unequally and sharply serrate; both surfaces covered with a scattered grey pubescence. Rachis of the leaf, with pubescence like that of the branchlets, densest at the nodes; narrowly grooved on the upper side.

Flowers (section *Fraxinaster*), without calyx or corolla, in short racemes, perfect in the single specimen seen; ovary pubescent. Fruit unknown.

This species resembles *F. angustifolia* in the shape and size of the leaflets; but differs in the copious pubescence on the branchlets and leaves. The buds are quadrate, with four dark brown scales, very pubescent at the tips.

*F. holotricha* was discovered by Koehne in Späth's nursery near Berlin, and in the botanic gardens at Berlin and Dresden, where it had received the erroneous name of *F. potamophila*. Its native country is unknown. The specimen, from which I have drawn up the above description, was sent me from Metz by Messrs. Simon-Louis. Three young plants were sent by Späth to Kew in 1908.　　(A. H.)

## FRAXINUS ORNUS, Flowering Ash, Manna Ash

*Fraxinus Ornus*, Linnæus, *Sp. Pl.* 1057 (1753); Bentley and Trimen, *Medicinal Plants*, iii. 170, fig. 170 (1880); Hanbury, *Science Papers*, 362-368 (1876).

*Fraxinus paniculata*, Miller, *Dict.* No. 4 (1759).

*Fraxinus florifera*, Scopoli, *Fl. Carn.* ii. 282 (1772).

*Ornus europæa*, Persoon, *Syn. Pl.* ii. 605 (1807); Loudon, *Arb. et Frut. Brit.* ii. 1241 (1838).

A tree attaining about 60 feet in height and 6 feet in girth, with smooth ashy-grey bark. Young branchlets slender with white lenticels, usually glabrous, occasionally glandular pubescent, marked with a ring of brown hairs at the base of the shoot. Leaflets (Plate 265, Fig. 26), five to nine, 2 to 3 inches long, the terminal one obovate and stalked, the lateral ones with distinct stalklets, which are about ¼ inch long and brownish pubescent; ovate to oblong, base rounded or broadly cuneate, apex shortly acuminate, finely and irregularly serrate, under surface glabrous except for brown woolly pubescence on the midrib. Rachis of the leaf grooved, the groove deepest in the upper part of its length, usually glandular pubescent, with tufts of brown hairs opposite the insertions of the leaflets.

Flowering branches developed from terminal buds, which contain leaves as well as flowers, the inflorescences being usually accompanied by two pairs of leaves and consisting of five panicles, the largest one terminating the branch, four smaller ones arising in the axils of the leaves, the whole forming a drooping compound panicle. Pedicels long and slender. Calyx divided into four triangular acute persistent sepals.

Corolla deeply divided into four strap-shaped wavy wide-spreading petals. Stamens two, hypogynous, the filaments twice as long as the petals. Flowers apparently perfect, but functionally behaving as if distinctly staminate and pistillate. Fruit compressed, with a terminal flat obovate-linear wing, blunt or emarginate at the apex.

## IDENTIFICATION

In summer the tree is readily distinguished by its smooth bark and stalked leaflets showing the characters just enumerated. In winter the twigs show a slight pubescence towards the apex and a ring of hairs at the base of the shoot. Leaf-scars parallel to the twig on projecting leaf-cushions, semi-orbicular to crescentic, the ends of the horns truncate, marked on the surface with a curved row of separate bundle scars. Terminal bud large, greyish to greyish brown, ovoid, four-sided, rounded (rarely acute) at the apex, the two outer scales gaping above and densely pubescent. Lateral buds smaller, densely pubescent, arising from the twigs at a wide angle.

## VARIETIES

*Fraxinus Ornus*, occurring over a wide area, both as a wild tree in forests, and cultivated in sunny arid regions, as in Sicily, shows considerable variation in the size, shape, and texture of the leaflets; and several varieties have been established.[1] The only one of those which is truly distinct is var. *argentea*, Grenier et Godron,[2] a remarkable form, growing wild in the forests of Corsica and Sardinia, distinguished by the leaflets being silvery white beneath, firm in texture, crenulate-serrate, usually smaller than the type, often subsessile, though stalked leaflets also occur on the same branch, ovate or oval in outline, occasionally approximating to an orbicular shape. This singular variety has been made a distinct species;[3] but modern French and Italian botanists regard it as only a peculiar geographical variety, which seems to be sporadic in forests where the type is also met with.

Aiton[4] and Willdenow[5] described, as the true manna ash, *F. rotundifolia*, with broad ovate or almost orbicular, deeply serrate leaflets. Willdenow's figure of the foliage corresponds with the ash which I have described (p. 866) as *F. excelsior*, var. *rotundifolia*, and it is possible that both he and Aiton were in error in considering their plant to have flowers like those of *F. Ornus*.

Lamarck's[6] *Fraxinus rotundifolia*, differs, according to the description, from Aiton and Willdenow's species of the same name, and is considered by Wenzig[7] and Lingelsheim[8] to be *F. Ornus*, var. *rotundifolia*; but is kept up by Koehne[9] as a distinct species. Hanbury[10] states that the manna ash cultivated in Sicily shows

---

[1] Fiori et Paoletti, *Flora Analitica d'Italia*, ii. 341, mention, besides the typical form, var. *rotundifolia* with broad elliptical leaflets, and var. *lanceolata* with lanceolate leaflets.

[2] *Flore de France*, ii. 473 (1850).  [3] *Fraxinus argentea*, Loiseleur, *Flora Gallica*, ii. 697 (1806).

[4] *Hort. Kew*, iii. 445 (1789).

[5] *Berlin. Baumzucht*, 116, fig. vi. 1 (1796). The figure was copied by Loudon, *Arb. et Frut. Brit.* ii. 1244 (1838), who merely repeats Willdenow's description, and was probably unacquainted with the tree.

[6] *Encycl.* ii. 546 (1786).   [7] In Engler, *Bot. Jahrb.* iv. 169 (1883).   [8] *Ibid.* xl. 212 (1907).

[9] *Deutsche Dendrologie*, 508 (1893).                    [10] *Science Papers*, 368 (1876).

great variation, but that no special form can be singled out as deserving the name *rotundifolia* ; and I have seen no specimens of *F. Ornus*, which could be separated as a var. *rotundifolia*, much less any which could be separated as a distinct species.[1]

A variegated form of *F. Ornus*, of which I have seen no specimens, is said to occur in cultivation. A simple-leaved form, *var. diversifolia*, Roch., is of rare occurrence in the wild state, and has been noticed in the canton of Tessin in Switzerland.[2]

## Distribution

The manna ash is widely distributed in southern Europe and Asia Minor. In France it only occurs wild in the department of the Maritime Alps ; and in Switzerland it is met with in a few places in mountain woods about Lake Lugano in the canton of Tessin.[2] It grows in the southern Tyrol, where it ascends to 2000 feet elevation, in Carniola, Istria, Dalmatia, Croatia, Slavonia, and Banat, reaching its northern limit in Hungary on the south side of the Carpathian chain. It is common in eastern Spain, Corsica, Sardinia, Italy, Servia, Bosnia, Greece, and Asia Minor.

The wood of the flowering ash is excellent, and the foliage is used as fodder in the southern countries of Europe ; but its chief economic importance is due to its being the source of manna. The manna of commerce, according to Hanbury, is exclusively collected in Sicily, where the plantations are known as Frassinetti. However, of late years, attempts have been made to cultivate it on a large scale for manna in the southern parts of the Austrian empire. In Sicily trees begin to produce manna when they are about eight years old ; and they are tapped in subsequent years annually until they are about twenty years old, when they are cut down and their place taken by coppice shoots from the stools. During July and August transverse incisions are made in the bark, so as just to reach the wood ; and the manna exudes as a clear liquid, which solidifies on the stem of the tree or on pieces of straw or wood that are inserted in the incisions. Manna consists mainly of a peculiar sugar called mannite, which is a mild laxative and is employed as an officinal drug in many countries.  (A. H.)

## Remarkable Trees

The manna ash is said to have been introduced into England by Dr. Uvedale of Enfield about 1710. It is commonly cultivated as an ornamental tree, on account of its beautiful appearance when in flower ; and it thrives and attains a large size, especially in the southern parts of England. The largest I have measured is an old

---

[1] Tenore, in *Syll. Pl. Fl. Neap.* 10 (1831) considers Lamarck's plant to be a variety of *F. Ornus*. Both Tenore and Bertolini, *Fl. Italica*, i. 54 (1833), were of opinion that *F. rotundifolia*, Willdenow, was a distinct species. Lingelsheim, in Engler, *Bot. Jahrb.* xl. 212, 213 (1907), retains *F. rotundifolia*, Tenore, *loc. cit.*, as a distinct species, confined to a small area in south Tyrol, Bosnia, and Dalmatia ; and creates a new species, *F. cilicica*, occurring in Cilicia in Asia Minor. These supposed species appear to be glabrous forms of *F. Ornus*.

[2] Cf. Bettelini, *Flora Legnosa del Sottoceneri*, 145 (1904).

tree in the park at Godinton, Kent, which is about 60 feet high, 10 feet 4 inches in girth below the graft, and 8 feet 8 inches above it.

There is a fine tree growing at Escot, Ottery St. Mary, Devonshire, the residence of Sir John Kennaway, which was in 1905, according to Miss F. H. Woolward, 61 feet in height, with a girth of stem at 5 feet up of 6 feet 9 inches, the circumference at the base being 11 feet 9 inches. This tree flowers abundantly every year. Another at Carclew is 50 feet by $9\frac{1}{2}$ feet.

A tree, about 60 feet high, grows near the stables at Tottenham House, Savernake, the seat of the Marquess of Ailesbury; a photograph of this tree, which was taken in June 1908, from the roof of the stables, shows it in full flower.

There is a good tree at Kew, close to the North Gallery, which measured, in 1907, 42 feet by 5 feet; and another, on the mound, near the Cumberland gate, is 60 feet high and 7 feet 6 inches in girth at three feet from the ground, dividing above into three main stems.

At Syon there is a tree 62 feet by 6 feet 6 inches, and good trees occur at Beauport, Sussex, and Whiteaway, Devon. A very fine one is reported by the Hon. Vicary Gibbs to be growing at Rook's Nest, near Oxted, Surrey. At Brocklesby Park, Lincolnshire, a tree was 50 feet by 7 feet 1 inch in 1904.

In Scotland there is a tree at Gordon Castle about 50 feet high.

In Ireland, a tree at Fota measures 65 feet by 5 feet 6 inches; and trees about 40 feet high are growing at Narrow Water, near Warrenpoint, in Down, and at Glenstal, Limerick.

Henry saw specimens in the botanic gardens at Copenhagen and Christiania, about 40 feet high and 3 feet in girth, in 1908.　　　　　　　　　　　(H. J. E.)

## FRAXINUS FLORIBUNDA

*Fraxinus floribunda*, Wallich, *List*, 2836, and in Roxburgh, *Fl. Ind*. i. 150 (1820); Hooker, *Fl. Br. India*, iii. 605 (1882); Gamble, *Man. Indian Timbers*, 471 (1902); Brandis, *Indian Trees*, 443 (1906).
*Ornus floribunda*, A. Dietrich, *Sp. Pl*. i. 1, 249 (1831).

A large tree, attaining in the Himalayas over 100 feet in height. Shoots compressed, purple, glabrous, with scurfy glands; lenticels white and prominent. Leaflets (Plate 264, Fig. 17), seven to nine (rarely five), the upper pair subsessile, the others on glabrous petiolules, 4 to 6 inches long, oblong, except the terminal one which is obovate, base unequal and rounded, apex long-acuminate, midrib and principal veins prominent and pubescent beneath, regularly and sharply serrate. Rachis of the leaf winged, the wings enclosing a broad open groove on the upper side, pubescent in the groove and at the insertions of the leaflets.

Flowers (section *Ornus*) in large panicles; corolla lobes, $\frac{1}{8}$ to $\frac{1}{4}$ inch, linear-oblong. Samaras very narrow, obtuse or emarginate.

This is the largest and finest of the flowering ashes, and attains a great size in its native home, Sir D. Brandis mentioning trees in the Chenab Valley planted near villages and temples, which reach 120 feet in height and 15 feet in girth. It is the only valuable species of ash in the Himalayas, where it grows on rich moist soils, generally on limestone. It is distributed throughout the Himalayas from the Indus to Sikkim, between 5000 and 9000 feet, but is only common locally. It is also met with in Baluchistan, Afghanistan, and in the Shan Hills of Upper Burma.

We have seen no large trees of this species in England; but it appears to do well at Kew, and probably would succeed as an ornamental tree in at any rate the warmer parts of the British Isles. (A. H.)

## FRAXINUS BUNGEANA

*Fraxinus Bungeana*, De Candolle,[1] *Prod.* viii. 275 (1844); Franchet, *Pl. David.* i. 203 (1884); Hemsley, *Journ. Linn. Soc.* (*Bot.*), xxvi. 84 (1889); Sargent, *Garden and Forest*, vii. 4, fig. 1 (1894).

*Fraxinus parvifolia*, Lingelsheim, in Engler, *Bot. Jahrb.* xl. 214 (1907) (not Lamarck).

A shrub about 5 feet high. Branchlets grey, minutely pubescent, with a dense ring of hairs at the base of the shoot. Buds ovoid, with dark puberulous scales. Leaflets (Plate 266, Fig. 31), five to seven, 1 to 1½ inch long, thin, membranous; usually on pubescent stalklets, ¼-inch long, upper pair sometimes subsessile; oval or rhomboid, broadly cuneate or rounded at the base, abruptly contracted into a long acuminate apex, crenately serrate, pale beneath; both surfaces quite glabrous. Leaf-rachis, grooved on the upper side, minutely pubescent, pubescence densest opposite the nodes.

Flowers, in terminal panicles, polygamous; petals 4, linear-obovate; calyx minute, 4-lobed. Fruit with a short slightly flattened many-nerved body, margined to about the middle by the decurrent base of the wing, which is oblong with a rounded, often emarginate apex.

This species is common on the hills near Peking and in the adjacent parts of Mongolia, and is the representative there of the European *F. Ornus*. Dr. Bretschneider sent seeds in 1881 to the Arnold Arboretum, Massachusetts, where plants were raised which are perfectly hardy in New England. They are pretty shrubs with abundant clusters of white flowers. This species does not seem to be in cultivation[2] in England. (A. H.)

---

[1] This species was founded by De Candolle on specimens (one of which is at Kew) of a shrub, collected by Bunge in 1831, near Peking, and named by the latter *F. floribunda* in *En. Pl. Chin. Bor.*, 61 (1832). Maximowicz, followed by Koehne and Lingelsheim, have erroneously applied De Candolle's name to *F. rhynchophylla*.

[2] Plants, sent to Kew by Sargent in 1903, cannot be found, and are supposed to have died. Since this article was corrected for the press, four plants from the Arnold Arboretum have arrived at Kew.

## FRAXINUS MARIESII

*Fraxinus Mariesii*, J. D. Hooker, *Bot. Mag.*, 6678 (1883); Hemsley, *Journ. Linn. Soc.* (*Bot.*), xxvi. 86 (1889).

A small tree. Branchlets slender, terete, purple, minutely pubescent, especially towards the tip; lenticels white. Leaflets (Plate 264, Fig. 18) five, $1\frac{1}{2}$ to 3 inches long, coriaceous, oval, acute or acuminate at the apex, base rounded or slightly tapering; regularly and crenately serrate, occasionally entire in margin; glabrous beneath. Petiolules $\frac{2}{4}$ to $\frac{3}{8}$ inch, scurfy pubescent. Rachis of the leaf purple, finely pubescent, grooved on its upper side.

Flowers very showy in erect panicles from the uppermost axils, about as long as the leaves. Calyx minute, four-cleft. Petals five or six, $\frac{1}{4}$ inch long, linear-oblong. Stamens two to four, as long as the petals.

This was discovered by Maries in the Lushan Mountains near Kiukiang in central China; and it has not been found elsewhere by subsequent collectors. Maries sent home seeds in 1879 from which the plant was raised by Messrs. Veitch. It flowered[1] for the first time, at Coombe Wood, as early as 1882.

It appears to be perfectly hardy in England, and is an ornamental small tree or shrub of considerable value, on account of the creamy-white large panicles of flowers, which appear about the end of June, and the bronze tint of the foliage.

(A. H.)

## FRAXINUS RHYNCHOPHYLLA

*Fraxinus rhynchophylla*, Hance, *Journ. Bot.*, vii. 164 (1869); Franchet, *Pl. Davidianæ*, i. 203, t. 17 (1884); Sargent, *Garden and Forest*, vi. 484, fig. 70 (1893); Komarov, *Fl. Manshuriæ*, iii. 248 (1907).
*Fraxinus Bungeana*, Maximowicz, *Mél. Biol.* ix. 396 (1874) (not De Candolle).
*Fraxinus chinensis*, Roxburgh, var. *rhynchophylla*, Hemsley, *Journ. Linn. Soc.* (*Bot.*), xxvi. 86 (1889).

A large tree. Shoots glabrous, lenticels few and scattered. Terminal buds remarkable, obtuse, conical, somewhat four-sided, with four or six outer scales, which are ovate, strongly keeled, with acute points directed outwards, pubescent and grey on the dorsal surface, and densely ferruginous woolly pubescent on their inner surface and edges; lateral buds, ovoid to rounded, small, with two to four outer scales, directed outwards at an open angle.

Leaflets (Plate 265, Fig. 23) five to seven, coriaceous, stalked or subsessile, ovate to ovate-lanceolate; base cuneate; apex with a long acumen, which is blunt or rounded and tipped with a short mucro; margin remotely and crenulately serrate

[1] *Hortus Veitchii*, 367 (1906).

in its upper half, sometimes almost entire; upper surface glabrous; lower surface glandular and glabrous except for some pubescence along the midrib towards the base. Terminal leaflet largest, basal leaflets smallest. Rachis of the leaf with a wide and open shallow groove on its upper side, pubescent in its whole length and bearded at the insertions of the leaflets, the pubescence being continued on the upper side of the petiolules.

Flowers (section *Ornaster*) in compact terminal panicles, polygamous, without a corolla; calyx persistent at the base of the samaræ, which are long, narrow, and erect on filiform pedicels.

This is the large ash tree[1] which is common in the mountains about Peking; and it also occurs in the adjacent parts of Mongolia, and in Manchuria and in northern Korea. It was discovered[1] by Père D'Incarville in the eighteenth century. Dr. Bretschneider sent seeds to the Arnold Arboretum in 1881, and plants were raised there, which are growing vigorously and promise to become large trees. They had already in 1893 produced flowers and fertile seed.

The species is remarkably distinct, and is very different from *F. chinensis*, Roxburgh, of which it has been supposed to be a variety. It appears to be scarcely known in Europe, the only specimens which I have seen being from Tortworth, where small plants are reported to be growing badly; and from Aldenham, where the foliage of a young tree, growing freely, is remarkable for its large size and glossy appearance. (A. H.)

## FRAXINUS MANDSHURICA

*Fraxinus mandshurica*, Ruprecht, *Bull. Phys. Math. Acad. Sc. Pétersb.*, xv. 371 (1857); Maximowicz, *Prim. Fl. Amurensis*, 194, 390 (1859), and *Mél. Biol.* ix. 395 (1874); Hemsley, *Journ. Linn. Soc. (Bot.)*, xxvi. 86 (1889); Komarov, *Fl. Manshuriæ*, iii. 248 (1907).

A large tree, attaining 100 feet in height and 12 feet in girth, with bark like that of the European ash. Branchlets glabrous. Leaflets (Plate 266, Fig. 28), seven to thirteen, 3 to 5 inches long, oblong-lanceolate, sessile, or with very short pubescent stalklets, tapering and unequal at the base, long-acuminate at the apex, sharply and irregularly serrate; glabrescent above; under surface with scattered coarse hairs on the sides of the midrib and lateral nerves. Leaf-rachis, with dense tufts of rusty-brown tomentum at the nodes, winged, the wings meeting above in part of its length, elsewhere deeply grooved.[2]

Flowers (section *Fraxinaster*) diœcious, in panicles in the axils of the leaf-scars of the preceding year, without calyx or corolla. Fruit, in loose clusters, oblong-lanceolate, apiculate or emarginate.

[1] See Bretschneider, *European Bot. Discoveries in China*, 53, 336, 1058 (1898). There are two ashes in the Peking mountains, one a large tree, *F. rhynchophylla*; the other a small shrub, *F. Bungeana*.

[2] The rachis of the uppermost two leaves is usually fringed at its insertion, close to the terminal bud of the branchlet, with rusty-brown pubescence.

The Manchurian ash is the representative of the common ash in eastern Asia, and is very similar in appearance to *F. nigra*,[1] being mainly distinguishable by the longer points to the leaflets, which are more tapering at the base, often shortly stalked, and usually more sharply serrate. (A. H.)

*F. mandshurica* is widely spread throughout Manchuria, Amurland, Korea, Saghalien, and Japan, and is a large tree, Ruprecht having measured specimens at the mouth of the Ussuri 12 feet in girth.

In Japan this fine ash is known as *yachidamo*, and is one of the commonest trees in Hokkaido, but is only known to occur in the north of the main island. I did not see it in Aomori or Akita. It seems to grow best in Hokkaido in the deep rich flats of black alluvial soil which are now being rapidly brought under cultivation by the Japanese settlers, but even here seldom attains the dimensions of the European ash ; the average size of the trees in the virgin forests being from 80 to 100 feet high by 6 to 8 feet in girth, though no doubt, if they had room to spread they would grow much thicker than this. The general habit of the tree and of its leaves and seed is very similar to that of *F. excelsior*, and the wood also seems similar, but it is not apparently used for the same purposes as in Europe. A great quantity of it is made into railway sleepers, which are now being exported largely to Korea and China, and which, if I can judge from what I saw on the Hokkaido railroads, will not last very long unless creosoted. There is, however, a particularly handsome variety of this wood, which seems to be found only near the butt and on the outside of old trees growing in damp places, which goes by the name of Tamu, and which, if known in Europe would certainly command a high price for veneers. A large wardrobe, which has been made for me from Japanese woods, is fronted with veneer cut from a billet of this wood, which I brought home in 1904, and is extremely handsome ; showing a figure like that of the best Hungarian ash but of a pale pinkish-brown colour. This wood is much used in Japan for veneering railway carriages, for doors, and for the posts used for supporting and for fitting the sliding screens of Japanese houses. It is liable to warp, however, and requires very careful seasoning to prevent cracking.

*F. mandshurica* was introduced[2] at Kew in 1891 from the St. Petersburg Botanic Garden ; but only one plant now survives, which has a stunted and unhealthy appearance. Like many plants from Manchuria and Amurland, the foliage appears early in the spring, and is badly injured by late frosts every year. I have raised seedlings from seed sent me from Japan by Professor Shirasawa, but they are too young to judge of the probable success of the tree in England.

Sargent, however, says[3] that this tree has proved perfectly hardy in the Arnold Arboretum, where it has been introduced for some years. He considers it one of the noblest of all the ashes, and one of the most valuable timber trees of Eastern Asia. None of the other ashes of Japan, so far as I saw, attain any great size, or are likely to have any economic value in Europe. (H. J. E.)

---

[1] *Lingelsheim*, in Engler, *Bot. Jahrb.* xl. 223 (1907), unites this species with *F. nigra*, of which he considers the Manchurian tree to be only a geographical variety.

[2] The St. Petersburg plants were probably raised from seed from Amurland. Cf. *Gartenflora*, xxvii. 13 (1879).

[3] *Forest Flora of Japan*, 52 (1894).

## FRAXINUS CHINENSIS

*Fraxinus chinensis*, Roxburgh, *Fl. Ind.* i. 150, Carey's edition (1820); Hanbury, *Science Papers*, 271, fig. 17 (1876); Hance, *Journ. Bot.* xxi. 323 (1883); Hemsley, *Journ. Linn. Soc. (Bot.)* xxvi. 85 (1889); Lingelsheim, in Engler, *Bot. Jahrb.* xl. 216 (1907).

A small tree. Branchlets glabrous. Leaflets seven to nine, 2 to 4 inches long, coriaceous, shortly cuspidate at the apex, crenately serrate; terminal leaflet the largest, obovate-oval, and long-stalked; lateral leaflets oval or elliptic, subsessile or with short pubescent winged stalklets, unequal and broadly rounded or tapering at the base; dark green and glabrous above; under surface pale green and pubescent on the sides of the midrib and lateral nerves. Leaf-rachis deeply channelled throughout, with brownish tufts of tomentum on the upper side of the nodes; base of the rachis of the uppermost two leaves fringed with brown hairs.

Flowers (section *Ornaster*) in terminal and lateral glabrous panicles; calyx 4-toothed; corolla absent. Fruit, about $1\frac{1}{2}$ inch long, $\frac{1}{4}$ inch wide, oblanceolate, acute or rounded at the apex.

This species in the wild state is very variable as regards the shape of the leaves, five varieties being distinguished by Lingelsheim. The above description is drawn up from small specimens cultivated at Kew, which were received in 1891 from St. Petersburg under the name *F. Bungeana*. A distinct variety has been in cultivation at Kew for some years under the erroneous name *F. longicuspis*, which has five or seven leaflets, obovate-lanceolate or narrow-oblong, 2 to $3\frac{1}{2}$ inches long, about 1 inch wide, cuspidate at the apex, cuneate at the base, indistinctly serrate. It agrees exactly with a dried specimen at Kew, gathered in the Ningpo mountains, where the broad leaflet form also occurs.

This species is widely spread throughout the central and southern provinces of China, and is noteworthy as being one of the trees on which the wax insect lives. It is very rare in cultivation, the only specimens we have seen being those at Kew and a small shrub at the Edinburgh Botanic Garden. (A. H.)

## FRAXINUS OBOVATA

*Fraxinus obovata*, Blume, *Mus. Bot. Lugd. Bat.* 311 (1850); Franchet et Savatier, *Enum. Pl. Jap.* i. 310 (1875).

*Fraxinus Bungeana*, Maximowicz, *Mél. Biol.* ix. 396 (1873) (not De Candolle); Franchet et Savatier, *Enum. Pl. Jap.* ii. 434 (1879); Lingelsheim, in Engler, *Bot. Jahrb.* xl. 214 (1907) (in part).

A small tree. Branchlets glabrous. Leaflets five or seven, 2 to 3 inches long, membranous, terminal one largest and long-stalked; lower pairs shortly-stalked, obovate or oval; cuneate and unequal at the base; acute, acuminate or rounded at the apex; remotely and irregularly serrate; upper surface shining, smooth, glabrous; lower surface light green, glabrous, but roughened on the midrib and nerves by minute, curved, stiff bristles, which are also present on the petiolules and on the margin of the blade. Rachis of the leaf, grooved throughout, pubescent on the upper side of the nodes, and armed near the nodes and at the base with minute curved bristles, which are also occasionally present towards the apex of the branchlet. Flowers (section *Ornus*), in terminal glabrous panicles, on filiform pedicels; calyx with five long acuminate teeth. Fruit linear, $1\frac{1}{2}$ inch long, $\frac{1}{6}$ inch wide.

This species, of which we have seen the type specimen, preserved at Leyden, and a specimen at Kew, lately received from Tokyo, differs from all the ashes known to me, in the occurrence of characteristic minute curved prickles on the under surface, margin, and petiolules of the leaflets and on the rachis of the leaf. Blume considered it to be only cultivated in Japan, and possibly an introduction from China. It appears to be unknown to Japanese botanists. However the only living specimen which we have seen is a small plant at Aldenham, which was raised from seed obtained from the Imperial Garden at Tokyo. (A. H.)


## FRAXINUS PUBINERVIS

*Fraxinus pubinervis*, Blume, *Mus. Bot. Lugd. Bat.* i. 311 (1850); Franchet et Savatier, *Enum. Pl. Jap.* i. 311 (1875), and ii. 435 (1879); Lingelsheim, in Engler, *Bot. Jahrb.* xl. 214 (1907).
*Fraxinus Bungeana*, De Candolle, var. *pubinervis*, Wenzig, in Engler, *Bot. Jahrb.* iv. 170 (1883).

A small tree, with smooth bark. Young branchlets densely covered with greyish pubescence, disappearing in the second year. Buds conical, densely covered with a brownish-grey pubescence. Leaflets, five or seven, coriaceous, 3 to 4 inches long, 1 to $1\frac{1}{2}$ inch wide; terminal largest, long-stalked and often broadest in its upper half; lateral, upper pair subsessile, lower pairs shortly stalked; lanceolate, cuneate at the base, acuminate at the apex, crenately serrate; upper surface glabrous; lower surface pale green, with dense whitish pubescence on the side of the midrib, spreading to the lateral nerves, and continued on the petiolules. Rachis of the leaf, with a continuous open groove, pubescent throughout, the pubescence densest at the nodes. Flowers (section *Ornus*), in large, terminal glabrous panicles, with early deciduous petals and long pedicels; calyx with long acuminate teeth. Fruit linear-spatulate, acute, $1\frac{1}{2}$ inch long, $\frac{1}{6}$ inch wide.

This species is known to the Japanese as *toneriko*; and in a native book on forest trees, is said to attain 30 feet in height and 3 feet in girth, but no accurate account of its distribution or habitat is given. There is a dried specimen at Kew, lately received from Tokyo. The only tree in cultivation known to me, is one at Aldenham, about 15 feet high, which was received six years ago from the Yokohama Nursery Company, and is growing vigorously.

## FRAXINUS SPAETHIANA

*Fraxinus Spaethiana*, Lingelsheim, in Engler, *Bot. Jahrb.* xl. 215 (1907).
*Fraxinus Sieboldiana*, Dippel, *Laubholzkunde*, i. 63, t. 27 (1889); Koehne, *Deutsche Dendrologie*, (1893) (not Blume).

A small tree. Branchlets glabrous, grey. Leaflets seven to nine, coriaceous, 4 to 6 inches long, 1½ to 1¾ inch wide, sessile or subsessile (terminal leaflet sessile or with a stalk up to ½ inch long); lanceolate-oblong; unequal and tapering at the base; abruptly contracted at the apex into a long, often curved acumen; margin non-ciliate, irregularly and often crenately serrate; lateral nerves fifteen to twenty pairs; glabrous except for slight pubescence along the midrib on the lower surface at the base. Rachis of the leaf, glabrous, with a continuous open groove on its upper side, dilated at its base into a swollen, dark-brown, shining sheath, which partly embraces the branchlet and conceals the glabrous dark-brown buds. Flowers, section *Ornus*. Fruit in large terminal glabrous leafless panicles. Samaræ linear-spatulate, 1½ inch long, ⅓ inch wide in the broadest part, rounded and entire at the apex; calyx with five short teeth.

This species, which has been a considerable time in cultivation, under the garden name of *F. serratifolia*, is readily distinguishable from all the other species[1] of ash which have been introduced, by the swollen base of the petiole, which somewhat resembles that of the plane tree. It is a native of Japan, where it appears to have been confused with *F. Sieboldiana*, which we consider to be a form of *F. longicuspis*. Specimens lately received at Kew, through the Hon. Vicary Gibbs, from Dr. Fukuba, Director of the Imperial Gardens at Tokyo, enable us to describe the fruit, which has hitherto been unknown.

It is a handsome and striking species, represented at Kew by a tree about 15 feet high, and at Aldenham by small plants. (A. H.)

## FRAXINUS LONGICUSPIS

*Fraxinus longicuspis*, Siebold et Zuccarini, in *Abhand. Baier. Acad. Wissen.* iv. 3, p. 169 (1846); Franchet et Savatier, *Enum. Pl. Jap.* i. 310 (1875); Shirasawa, *Icon. Ess. Forest. Japon*, text 126, t. 81 (1900); Lingelsheim, in Engler, *Bot. Jahrb.* xl. 214 (1907).
*Fraxinus Sieboldiana*, Blume, in *Mus. Lugd. Bat.* i. 311 (1850).

A tree, attaining, according to Shirasawa, 50 feet in height and 7 feet in girth. Young branchlets grey, glabrous. Leaflets five, occasionally seven; terminal largest, stalked; lateral, upper pair sessile, lower pair stalked; oblong, oblong-lanceolate, or

---

[1] *F. platypoda*, Oliver, a species, discovered by me in Central China, has a similar swollen base to the petiole; but it has never been introduced.

ovate, cuneate at the base, abruptly contracted above into a long cuspidate apex, crenately serrate; upper surface glabrous; lower surface pale green, slightly pubescent on the side of the midrib, elsewhere glabrous. Rachis of the leaf, grooved on its upper side, pubescent at the nodes, glabrous elsewhere. Buds purplish brown, minutely pubescent. Flowers (section *Ornus*) in terminal and lateral panicles; petals four, narrowly-linear. Fruit about an inch in length, subtended by a 4-toothed calyx, oblanceolate, with a long cuneate base and an obtuse emarginate apex.

This species is variable as regards the shape of the leaflets; and two varieties have been noticed, regarded as distinct species by Blume; one characterised by broad ovate leaflets (terminal leaflet 3 to 3½ inches long and 1½ inch broad), with slight pubescence along the midrib beneath; the other with narrow oblong leaflets (terminal leaflet 3½ to 4 inches long and 1 to 1⅛ inch wide), glabrous on the under surface; but, as Franchet points out, there are numerous specimens with inter-mediate characters.

*Fraxinus longicuspis*[1] is a native of the mountainous districts of Japan, attaining, according to Shirasawa, an elevation of 5000 feet in the central chain of Hondo. It is abundant in Nikko, Chichibu, and Kiso; and has been collected near Hakodate by Maximowicz, in Akita by Elwes, in the mountains of Yamagata by Faurie, and in the island of Tsu-sima by Wilford. On account of its small size, it is of no economic importance in Japan. It appears to be extremely rare in cultivation in this country, the only specimen which we have seen being a small plant at Kew, about 2 feet high, which was raised from seed sent by Sargent in 1894. (A. H.)

## FRAXINUS NIGRA, BLACK ASH

*Fraxinus nigra*, Marshall, *Arb. Am.* 51 (1785); Sargent, *Silva N. Amer.* vi. 37, tt. 264, 265 (1894), and *Trees N. Amer.* 764 (1905).
*Fraxinus sambucifolia*, Lamarck, *Dict.* ii. 549 (1786); Loudon, *Arb. et Frut. Brit.* ii. 1234 (1838).

A tree, attaining in America 90 feet in height and 5 feet in girth of stem, with bark divided into large irregular scaly plates. Shoots glabrous. Leaflets (Plate 264, Fig. 19), seven to eleven, 3 to 5 inches long, terminal one petiolulate, lateral leaflets sessile; oblong to oblong-lanceolate, rounded and unequal' at the base, long acuminate at the apex, remotely and finely crenulate-serrate; under surface glabrous, except for long reddish hairs along the nerves and midrib, densest towards the base, where they spread over the surface of the leaflet. Leaf-rachis winged on the upper side, the wings meeting above and not forming a continuous open groove; glabrous except opposite the insertions of the leaflets, where it is bearded all round with dense rufous hairs.

The terminal buds are blackish, broadly ovate and acute, with six scales visible externally, of which the outer pair, slightly puberulous, almost enclose the others.

[1] The Japanese botanists recognise two species :—*F. longicuspis*, called in Japanese *oshida* or *aotago*; and *F. Sieboldiana*, known as *shioji*. The latter name may possibly refer to *F. Spaethiana*. Cf. p. 897.

Flowers (section *Fraxinaster*) polygamous, without calyx or corolla, arising in the axils of the leaf-scars of the preceding year's shoot. Fruit, linear-oblong, with a short flattened faintly nerved body, surrounded by a thin wing very emarginate at the apex.

This is the American representative[1] of the common European ash, and it is easily distinguished from nearly all the other species by the dense ring of rufous pubescence on the leaf-rachis at the nodes. *Fraxinus mandshurica*, which has the rachis similarly bearded, but more deeply grooved, differs in having leaflets with a sharp serrate margin and a tapering base. (A. H.)

*Fraxinus nigra* is found, according to Sargent, in deep, cold swamps and on the low banks of streams and lakes, from southern Newfoundland and the north shores of the Gulf of St. Lawrence to Lake Winnipeg, and southward to the mountains of Virginia, southern Illinois, and north-west Arkansas. Macoun says that in Canada it is more widely distributed than the white ash, and more abundant than the latter throughout its range, from Anticosti to eastern Manitoba, in swamps and river bottoms. It grows on peat mosses, but remains small in such situations. I saw it in the woods about Ottawa, but of no great size, and Sargent gives 80 to 90 feet as its extreme height. Ridgway found it abundant in Knox County, Indiana, where a tree 83 feet high, of which the bole was 57 feet, only measured $1\frac{1}{2}$ foot in diameter at 5 feet from the ground. He says that it presents so very close a resemblance to the young Pecan tree (*Carya olivæformis*) as not to be readily distinguished except by experts.[2] In Carya, however, the leaves are alternate. Its wood, according to Emerson, is remarkable for toughness, and on this account was preferred to every other by the Indians for making baskets, and is still used for that purpose in preference to every kind of wood except that of a young white oak.

We have seen no specimens of this tree except small plants[3] at Kew and Colesborne, although it was probably planted in many places early in the nineteenth century; having been introduced, according to Loudon, in 1800. Like many other trees of the Atlantic side of North America, it is short-lived and does not thrive in our climate. Prof. Sargent informs me in a letter that at the Arnold Arboretum, near Boston, it is one of the most difficult of all trees to grow. At Angers in France, it does badly on its own roots, but succeeds when well grafted; a specimen there having attained 15 feet high in eight years. (H. J. E.)

[1] According to Cobbett, *Woodlands*, Art. 136 (1825), the seeds of this species, like the English ash, do not come up until the second year.

[2] Mr. G. B. Sudworth, however, tells me that the bark of the two is so distinct that they can readily be distinguished. He adds that other species of ash, hickory, oak, Liquidambar, and Nyssa, are now used for basket-making.

[3] These plants were raised from seed sent from Michigan in 1895, and are not thriving, having been repeatedly injured by frost. They are now only 2 feet high.

## FRAXINUS ANOMALA, Utah Ash

*Fraxinus anomala*, Watson, King's *Rep.* v. 283 (1871); Sargent, *Silva N. Amer.* vi. 39, t. 266 (1894), and *Trees N. America*, 765 (1905).

A small tree, attaining about 20 feet in height, with bark shallowly fissured by narrow ridges. Branchlets quadrangular, slightly four-winged, glabrous, with inconspicuous reddish lenticels. Leaves (Plate 262, Fig. 7) simple (occasionally two- to three-foliolate), ovate or obovate, acute or rounded at the apex, base cuneate or cordate, slightly crenulate or entire in margin, glabrous beneath. Petiole flattened and grooved above, about half the length of the blade of the leaf.

Flowers (section *Leptalix*) in panicles from the axils of the leaf-scars of the preceding year's shoot, with a calyx, but corolla absent. Fruit oblong, with a rounded wing surrounding the long, flattened, striately nerved body.

This curious tree, so remarkable amongst the ashes, in its usually simple leaves and quadrangular stems, occurs in Colorado, Utah, and Nevada.

It is in cultivation at Kew, where it is perfectly hardy, and is worthy of a place in collections on account of its peculiarities. (A. H.)

## FRAXINUS QUADRANGULATA, Blue Ash

*Fraxinus quadrangulata*, Michaux, *Fl. Bor. Am.* ii. 255 (1803); Loudon, *Arb. et Frut. Brit.* ii. 1235 (1838); Sargent, *Silva N. Amer.* vi. 35, t. 263 (1894), and *Trees N. Amer.* 761 (1905).

A large tree, attaining 120 feet in height and 9 feet in girth. Bark separating irregularly into large thin plates. Branchlets glabrous, stout, quadrangular; with four wings between the nodes, persisting and becoming corky in the second year; lenticels white. Leaflets (Plate 265, Fig. 22), five to nine, 3 to 5 inches long; on pubescent stalklets, $\frac{1}{8}$ to $\frac{1}{4}$ inch long; ovate or oval, rounded or broadly cuneate at the base, acuminate at the apex, regularly serrate, glabrous above; under surface covered with scattered whitish tomentum, densest towards the base. Leaf-rachis pubescent, with a shallow open channel on its upper side.

Flowers (section *Leptalix*) in panicles from the axils of the leaf-scars of the previous year's shoot, perfect, calyx obsolete, corolla absent. Fruit oblong; body long, flat, with numerous faint rays, surrounded by the base of the broad wing, which is emarginate at the apex.

This species is readily distinguished from all the other species with numerous leaflets by the conspicuous wings on the branchlets. (A. H.)

The blue ash though little known outside of botanic gardens in Europe is, next to the white ash, the largest of its genus found in the United States. It is unknown in New England or Canada, though hardy at Ottawa, where I saw it in the arboretum of the Experimental Farm. Sargent says that it grows usually on rich limestone hills from southern Michigan to Iowa, central Missouri, and north-eastern Arkansas, and southward to northern Alabama and east Tennessee on the Big Smoky mountains, where it attains a great size. Usually it is from 60 to 70 feet high, with a trunk 2 to 3 feet in diameter; but Ridgway says that four freshly cut trees, in the Wabash valley in Illinois, were 116 to 124 feet high, with clear trunks 51 to 76 feet long, and 2 to 2½ feet diameter on the stump. Here it was common in rich hilly woods, but I saw none standing of anything like these dimensions.

The tree was discovered by the elder Michaux in 1795 and introduced[1] by him into France, and his son speaks of the beautiful stocks that were growing in Europe; but I have seen none of considerable size, the best perhaps in England being a tree at Tortworth, which was 34 feet by 1 foot 10 inches in November 1905, and reported by Lord Ducie to be growing freely. Michaux says that the wood of this species in the western states is extensively used for waggon-building and wheels, and also for flooring houses; but it does not now seem to be known to English importers. Mr. G. B. Sudworth informs me, however, that this timber is still found in considerable quantity in the Ohio valley, and can be obtained in logs as large as 24 to 30 inches in diameter. He adds that for farm tool handles it is preferred to any other ash on account of its superior strength and elasticity. (H. J. E.)

## FRAXINUS AMERICANA, WHITE ASH

*Fraxinus americana*, Linnæus, *Sp. Pl.* 1057 (1753); Loudon, *Arb. et Frut. Brit.* 1232 (1838); Sargent, *Silva N. America*, vi. 43, tt. 268, 269 (1894), and *Trees N. America*, 767 (1905).
*Fraxinus alba*, Marshall, *Arbust. Am.* 51 (1785).
*Fraxinus acuminata*, Lamarck, *Dict.* ii. 547 (1786).
*Fraxinus juglandifolia*, Lamarck, *Dict.* ii. 548 (1786).
*Fraxinus epiptera*, Michaux, *Fl. Bor. Am.* ii. 256 (1803).

A large tree, attaining in America 120 feet in height and 15 to 20 feet in girth. Bark deeply divided by narrow fissures into broad flattened scaly ridges. Shoots stout, green, glabrous, with white lenticels. Leaflets (Plate 265, Fig. 21), seven to nine, 4 to 6 inches long, distinctly stalked (the petiolules glabrous and ¼ to ½ inch long), lanceolate or oval, rounded or broadly cuneate at the base, acuminate at the apex; entire, crenulate, or coarsely serrate; under surface whitish and pubescent along the midrib and nerves, or in some cases throughout. Rachis of the leaf terete and apparently not grooved on the upper side; but usually a slight groove can be made out on close examination.

Flowers (section *Leptalix*) diœcious, in glabrous panicles in the axils of the leaf-scars of the preceding year's shoot; corolla absent. Fruit in crowded clusters, surrounded by the persistent calyx at the base, lanceolate or oblong, with a terete rayed oblong body, much shorter than the terminal wing, which is pointed or emarginate at the apex.

[1] It was introduced, according to Loudon, into England in 1823.

## Varieties

A form is known with very small fruit, var. *microcarpa*, Gray.[1] In regard to the leaf, two varieties have been described,[2] viz. :—

Var. *acuminata*, Wesmael. Leaflets dark green above, very white and almost glabrous beneath, nearly entire in margin. This form is more common in the southern states.

Var. *juglandifolia*, Rehder. Leaflets usually broader than in the preceding variety, more or less pubescent beneath, coarsely serrate at least above the middle. This is the northern form of the species.

These two varieties occur in cultivation in England, the leaves of both remaining unchanged in colour until they fall in autumn. A form of var. *acuminata* occurs, in which the leaflets are narrow at the base, and turn reddish brown before they fall.

There is also said to exist a horticultural variety, var. *albo-marginata*, in which the leaflets are edged with white.

## Identification

As *Fraxinus americana* may be a valuable tree for economic planting in England, its correct identification is important. Reputed trees of *F. americana* growing slowly usually turn out on examination to be *F. viridis* or some other species. In summer, the leaflets white beneath and distinctly stalked, the rachis terete and practically not grooved, and the glabrous branchlets, will readily distinguish *F. americana*. The only species which closely agrees with it, *F. texensis*, has, when adult, fewer, usually five (rarely seven) leaflets, which are smaller and quite different in shape, being broadly oval with a rounded or acute apex.　(A. H.)

## Distribution

The white ash is one of the best known and most highly valued trees in New England and Canada; where it occurs according to Sargent, from Nova Scotia and New Brunswick through Ontario to northern Minnesota, southward to northern Florida, and westward to Kansas, Indian Territory, and eastern Nebraska, and the valley of the Trinity River, Texas; being less common and smaller west of the Mississippi. It attains in the forest a height of 120 feet with a diameter of 4 feet and upwards, and thrives best in a deep loamy soil near the banks of streams, just as the common ash does in England. When standing alone it assumes a spreading habit with large branches.

Ridgway measured a sound tree in Wabash County, Illinois, which was 144 feet high with a clean stem 83 feet long and 9 feet in girth at the top, and 13 feet at the base, which, according to English measure, would have contained over 500 cubic feet; and Dr. C. Schneck measured a tree in the same county 144 feet high with a stem 90 feet long and 17½ in girth above the swell at the base; and this tree, if it carried its girth up, might have contained 1000 feet of timber.

---

[1] This variety is common in the Gulf States, the fruit being less than ½ inch long; whereas it usually attains 1½ to 2 inches. Sargent, however, states that both large and small fruit may occur on the same individual, and even on the same branch.　　[2] Cf. Rehder in *Cycl. Amer. Horticulture*, 607 (1900).

I saw no large specimens of this species in New England or Canada where most of the best have been felled. I noticed at Ottawa that in the autumn its leaves assume a rich purplish colour, which the black and red ashes did not show.

Michaux says[1] that on large trees the bark is deeply furrowed, and divided into small squares from 1 to 3 inches in diameter; and that it grows in Maine in company with the white elm, yellow birch, white maple, hemlock, and black spruce, and in New Jersey with red maple, shellbark hickory, and button-wood (*Platanus occidentalis*), in places that are constantly wet or occasionally flooded. Pinchot and Ashe[2] figure a splendid tall straight forest tree with small head and rough bark resembling that of the common ash, and say that its average height in North Carolina is 50 to 80 feet with a diameter of 2 feet to 3 feet.

Professor Sargent says[3] that the white ash when planted with the common ash in the high regions of central Europe comes into leaf still later than that species, and thus escapes the spring frosts; it is less able to resist drought than the green ash, and is usually found on moist soil, though it does not like wet swamps, like the black ash. In the forest it sends up a perfectly straight and slender stem to a great height, ash poles 100 feet high and not over a foot in diameter being often seen. A photograph taken in Chester County, Pennsylvania, shows a tree in the open with a rather spreading head and a trunk 15 feet 10 inches in girth.

## CULTIVATION

*Fraxinus americana* is stated by Aiton[4] to have been introduced into England in 1724 by Catesby. Sargent, however, points out that the ash described by Catesby is another species.

Though very seldom seen, I believe that the American ash will grow in this country, in some places at least, almost as fast as the native ash. My attention was first called to this fact by the very straight, clean, and rapid growth of a young tree at Kew which stands by the walk not far from the Old Conservatory, now the Museum of Timbers. The date of planting is unknown. It measured in 1907 63 feet by 2 feet 5 inches, while *F. lanceolata*, growing near it is 43 feet by 2 feet 1 inch, and *F. oregona*, 44 feet by 2 feet 8 inches.

Knowing that the wood is considered better for oars than that of the native ash, and used exclusively for the heavy oars of our navy, I thought it worth trying as a timber tree, and raised a large number of plants from seed sent me by Messrs. Meehan of Philadelphia as *F. americana*, but which I found out three years later to be *F. lanceolata*.[5] Later on I raised seedlings of the true white ash and found that at first they do not grow nearly so fast as *F. lanceolata*. They do not ripen the young

[1] *N. Amer. Sylva*, iii. 49.
[2] *Trees of North Carolina*, p. 71, plate 6.
[3] *Garden and Forest*, vii. 402.
[4] *Hort. Kew*. iii. 445 (1789).
[5] Mr. G. B. Sudworth informs me that ten or fifteen years ago Messrs. Meehan became aware that they had been selling two or three species of ash seed as *F. americana*, and submitted samples to him for identification. To his great surprise he found that none of it was pure, but contained a mixture of *F. americana, F. lanceolata,* and *F. pennsylvanica*. Since then they have been more careful. He adds that continental tree planters have, to his knowledge, been planting green and red ash in mistake for white ash; and it is probable that most of the young trees grown under that name on the continent are incorrectly named.

wood well until they attain a certain age, but are perfectly hardy and will probably grow best in the east and south-east of England.

The only large trees I know of in England are two at Kew which grow on the mound near the Cumberland gate and which measure 85 feet by 8 feet 6 inches and 85 feet by 8 feet respectively (Plate 246). There is a tall slender white ash at Croome Court, Worcestershire, crowded by other trees, which measures about 80 feet by 4 feet 6 inches, and another[1] at Arley Castle 60 feet by 5 feet 4 inches. There is also a smaller one at Syon. Another at Tortworth was 46 feet by 3 feet in 1907. Loudon says that at St. Anne's Hill in Surrey there was one 33 feet high which had been planted thirty-six years, and that near London young plants are generally injured by spring frost, which I have not found to be the case at Colesborne.

Sir Charles Strickland tells me that he has planted the tree at Hildenley at Yorkshire, but does not find it succeed so well as the Oregon ash.

A tree at Fota, in the south of Ireland, is 50 feet high and 6 feet in girth.

Cobbett says:[2] "This tree grows much faster than ours. I have abundant proof, for the American white ash plants which I have at Kensington, which were not sown till last April, are now (1825) full as tall again as any of the English ash of the same age that I ever saw. This, therefore, is above all others the ash which I recommend to be put into plantations in England, whether for ornament, for timber, or for underwood." But Cobbett in this case, as in many others, was rather apt to jump to conclusions after too short experience; for if the tree had continued to grow as it did at first, there must by this time have been many good-sized ones in England.

## TIMBER

The timber of the white ash is as highly valued in America as ours is in England for the same purposes, and is largely imported to England where it is used as a substitute for English ash. Laslett in writing of American white ash says: "It is tough, elastic, clean, and straight in the grain, and stands well after seasoning, hence we get from this tree the best material for oars for boats that can be produced. They are much and eagerly sought after by foreign governments as well as our own, and also by the great private steamship companies and mercantile marine. The best quality wood has a clean, bright uniform whitish colour; the second is slightly stained with red and yellow shades alternating; the third and least valuable quality is that in which the red and yellow colours predominate. It is much slower in growth than the English, and is probably not so durable." On visiting the principal importers of this wood in Liverpool, I found large quantities of American oars imported ready-made; and was told that the timber had now become so scarce in the east, that it came from the west side of the Mississippi.[3] I have not been able to procure a sample grown in this country for comparison; but I am indebted to

---

[1] *Catalogue of Hardy Trees, Arley Castle*, No. 43 (1907).  [2] *Woodlands*, art. 135.

[3] A lot of oars which I saw in the Portsmouth Dockyard were stamped on the blade "De Valls Bluff. Ark. U.S.A." I am informed by experienced naval carpenters, and officers, that they believe that English ash, if it can be procured of sufficient straightness and length, would be at least as good, if not better.

Mr. A. Howard for a plank believed to be of this species, which was imported from Canada, and was cut from a log measuring 56 feet long by 24 inches quarter-girth at the top. It cost £66 and was cut into twenty-four boards containing 2763 feet board measure. (H. J. E.)

## FRAXINUS TEXENSIS, TEXAN ASH

*Fraxinus texensis*, Sargent, *Silva N. Amer*. vi. 47, t. 270 (1894), and *Trees N. Amer*. 768 (1905).
*Fraxinus albicans*, Buckley, *Proc. Philad. Acad*. 1862, p. 4 (in part).
*Fraxinus americana*, Linnæus, var. *texensis*, Gray, *Syn. Fl. N. Amer*. ii. part i. 75 (1878).
*Fraxinus americana*, Linnæus, var. *albicans*, Lingelsheim, in Engler, *Bot. Jahrb*. xl. 219 (1907).

A small tree, with a short stem, rarely attaining 50 feet in height and 9 feet in girth. Bark dark grey, deeply divided by narrow fissures into broad scaly ridges. Branchlets glabrous. Leaflets, five to seven, often nine in young plants, 2 to $2\frac{1}{2}$ inches long, distinctly stalked, with glabrous petiolules about $\frac{1}{8}$ to $\frac{1}{4}$ inch long, ovate or oval (the terminal leaflet often obovate), base rounded and unequal, apex acute or rounded (acuminate in young plants), crenate-serrate and non-ciliate in margin; upper surface shining bluish green, glabrous; lower surface whitish, usually pubescent on the sides of the midrib and lateral nerves, the latter being forked near the margin; rachis slender, glabrous, terete, slightly grooved on the upper side.

Flowers (section *Leptalix*) diœcious in glabrous panicles in the axils of the leaf-scars of the preceding year; corolla absent. Fruit, spatulate, with persistent calyx at the base; body short, terete; wing terminal, rounded or emarginate at the apex.

This species, while very close to *F. americana* in technical characters, is distinct in appearance, and differs in the smaller leaflets, which are shining bluish green, and not dull on the upper surface.

It was discovered in 1852 by Dr. J. M. Bigelow, and grows on high dry limestone bluffs and ridges in northern, central, and western Texas, from near Dallas City to the valley of the Devil's River.

Young plants are growing in the nursery at Kew, which are thriving, and about 6 to 8 feet in height, after six years' growth. They were raised from seed sent by Mr. Bush in 1901. These plants have usually 7 to 9 leaflets, while specimens in the Kew Herbarium from Texas have only five leaflets; but this is probably a juvenile character. It is also growing well at Aldenham.

(A. H.)

## FRAXINUS BILTMOREANA, Biltmore Ash

*Fraxinus Biltmoreana*, Beadle, *Bot. Gazette*, xxv. 358 (1898); Sargent, *Trees N. Amer.* 773, fig. 618 (1905).
*Fraxinus catawbiensis*, Ashe, *Bot. Gazette*, xxxiii. 230 (1902).

A tree attaining in America, according to Ashe, over 100 feet in height, with a girth of about 7 feet. Young shoots covered with a dense white pubescence, retained in the second year; lenticels few, conspicuous, narrow, long, white. Leaflets (Plate 266, Fig. 30), seven to nine, about 4 inches long, oval or oblong (the terminal one on a long stalk, broadly oval or obovate), abruptly tapering and unequal at the base, acuminate at the apex, remotely serrate (the serrations often obsolete, so that the margin is nearly entire), with occasional scattered cilia; distinctly stalked with pubescent petiolules, $\frac{3}{8}$ to $\frac{1}{2}$ inch long; upper surface dark green, glabrous except for a little pubescence towards the base of the leaflet; lower surface white in colour, with a thin fine short pubescence, densest on the sides of the midrib and nerves. Rachis of the leaf slender, terete, finely pubescent, not grooved or only slightly grooved towards the apex.

Flowers (section *Leptalix*) dioecious in pubescent panicles in the axils of the leaf-scars of the previous year; corolla absent. Fruit girt at the base by the persistent calyx; body, short, elliptical, many-nerved; wing not decurrent, only slightly narrowed at the ends, emarginate at the apex.

Buds shortly ovoid, with four outer visible scales, equal in length, the external pair overlapping the inner pair; scales carinate, obtuse at the apex, orange-coloured, and covered with a scaly pubescence.

This species, as regards leaf-characters, looks like a pubescent *Fraxinus americana*;[1] and is readily distinguished from *F. pennsylvanica* by the leaflet being white in colour beneath, with a finer pubescence, and being more abruptly tapering and unequal at the base, with shallower and remoter serrations, which often become obsolete. The rachis of the leaf is like that of *F. americana*.

According to Sargent, it occurs on the banks of streams from northern West Virginia through the foothills of the Appalachian Mountains to northern Georgia and Alabama, and to middle Tennessee.

This species in the United States has apparently been considered to be a form of *F. pennsylvanica*, and has been in cultivation probably as long. At Fawley Court, Oxfordshire, the residence of W. D. Mackenzie, Esq., there are two fine trees. The largest, in a shrubbery, rather crowded by other trees, is about 80 feet in height by 7 feet 3 inches in girth. The other, standing in the open, is a very well-shaped vigorous tree (Plate 247), measuring 68 feet by 6 feet 6 inches. Both are bearing mistletoe and grow in good alluvial soil. The bark is grey in colour, and fissured like that of the white ash. (A. H.)

---

[1] Beadle says that this species bears the same relation to *F. americana* as *F. pennsylvanica* bears to *F. lanceolata*. Lingelsheim, *op. cit.* 191, 222, considers this species to be a hybrid, between *F. americana* and *F. pennsylvanica*; but its wide distribution and abundance in the forest are not favourable to his view.

## FRAXINUS LANCEOLATA, GREEN ASH

*Fraxinus lanceolata*, Borkhausen, *Handb. Forst. Bot.* i. 826 (1800); Sudworth, *Check List Forest Trees U.S.* 107 (1898).

*Fraxinus viridis*, Michaux, *Hist. Arb. Amer.* iii. 115, t. 10 (excl. fruit), (1813).

*Fraxinus juglandifolia*, Willdenow, *Sp. Pl.* iv. 1104 (1805) (not Lamarck); Loudon, *Arb. et Frut. Brit.* ii. 1236 (1838).

*Fraxinus pennsylvanica*, Marshall, var. *lanceolata*, Sargent, *Silva N. America*, vi. 51, t. 272 (1894), and *Trees N. America*, 771 (1905).

A tree rarely attaining more than 60 feet in height, with a girth of stem of 6 feet. Shoots dark green, glabrous, with conspicuous white lenticels. Leaflets (Plate 264, Fig. 16), seven to nine, usually subsessile, 3 to 6 inches long, pale green beneath, ovate-lanceolate, tapering and unequal at the base, long-acuminate at the apex, under surface glabrous except for slight pubescence along the midrib, variable in serration. Rachis of the leaf glabrous, distinctly grooved on its upper side.

Flowers and fruit similar to those of *F. pennsylvanica*.

This species[1] is considered by Sargent to be a variety of *F. pennsylvanica* because west of the Mississippi trees occur, which are intermediate in character and can be as readily referred to one species as to the other. As seen in cultivation in England, it is very distinct, and on account of its glabrous shoots, it is very often mistaken for *F. americana*. The green ash is, however, readily distinguished from that species by the usually subsessile leaflets, which are pale green and not white beneath.[2] The rachis is, moreover, more deeply grooved than is ever the case in *F. americana*, and the buds in the two species are different. (A. H.)

This tree was described and figured by the younger Michaux, who says, "The green ash is easily recognised by the brilliant colour of its young shoots and leaves, of which the two surfaces are nearly alike." He found it more common in the western districts of Pennsylvania, Maryland, and Virginia, than anywhere else, and speaks of it as a tree of moderate dimensions, laden with seed when only 25 to 30 feet high.

Sargent, who treats it as a variety of *F. pennsylvanica*, and distinguishes it by the leaves being rather narrower, shorter, and usually with more sharply serrate leaflets, bright green on both surfaces, says that it rarely exceeds 60 feet high, and occurs from the shores of Lake Champlain through the Alleghany mountains to western Florida and west to the valley of the Saskatchewan, the valley of the Colorado river, Texas, the eastern ranges of the Rocky Mountains, the Wasatch range, Utah, and Arizona. It is comparatively rare east of the Alleghany mountains, and most abundant in the Mississippi basin. East of the Mississippi it seems distinct, but westward is connected with the red ash by intermediate forms.

[1] A small tree growing in Kew Gardens and labelled *F. coriacea*, resembles in many respects the green ash. It is apparently the Texan form of the Mexican *F. Berlandieriana*, DC., which was formerly considered to be a variety of *F. lanceolata*. The Kew tree has glabrous shoots like those of the green ash; but the leaflets (Plate 263, Fig. 14) are smaller, more coriaceous and very reticulate, usually three to five in number, the terminal one largest and obovate, the lateral ones oval and acuminate at the apex, all sparingly and irregularly serrate in the upper two-thirds, with the under surface pale green and glabrous except for some pubescence along the midrib. Rachis of the leaf glabrous and slightly grooved. *F. coriacea*, Watson, a tree occurring in desert regions from Utah to California is probably a variety of *F. Berlandieriana*, differing in the pubescent branchlets and leaf-rachis. It is in cultivation at Aldenham, where a young tree, about 10 feet high, obtained from Barbier's nursery at Orleans, is thriving.

[2] Mr. G. B. Sudworth, adds, that the leaflets are usually sharply serrate, while those of *F. americana* are commonly undulate, entire, or with only a few teeth.

Ridgway, in his *Additional Notes on the Trees of the Wabash*,[1] says that Dr. J. Schneck of Mount Carmel measured a tree 92 feet high and 5 feet in girth; but when I visited the remains of this wonderful forest, in September 1904, I saw no ash trees of considerable size.

Pinchot and Ashe[2] say that the wood is inferior in quality to that of the white ash, but in North Carolina is not distinguished from it commercially.

Having raised a large quantity of plants from seed sent me as that of *Fraxinus americana*, which I did not identify as the green ash until Mr. F. V. Coville of Washington saw them growing in my nursery in 1904, I have distributed them to many friends as *F. americana*, and it is probable that the tree will thus become common in England under a wrong name, as has happened in so many cases before. For this mistake, which was unavoidable, I now apologise; but as the tree grows faster than any other American ash in a young state, and is likely to make useful poles, if not large trees, I have planted out some thousands of them at Colesborne.

Like all the American ashes which I have raised, the seed germinates quickly after sowing, and though liable to be injured by late frosts is at least as hardy as the common ash. When young the shoots continue to grow late in autumn and do not ripen their young wood, which for three to four years at least is liable to be killed back by winter frosts. Some of these seedlings are now, at four years old, 6 to 7 feet high and growing very vigorously. Michaux says that this species was introduced by his father to France in 1785, but I cannot hear of any surviving under this name.

Loudon says that at Stackpole Court, Pembrokeshire, it had in forty years attained a height of 60 feet, and had ripened seeds from which many plants had been raised and distributed in the plantations, but the Earl of Cawdor tells me that his gardener can find no trees in the woods which resemble the American ash, and that none of the men on the place can remember any peculiar ash trees there. In 1906 I also searched the woods at Stackpole without finding any trace of these trees. Loudon also mentions a tree in the garden of Pope's villa at Twickenham, which no longer exists. There are several young trees in Kew Gardens; and the tallest, about 40 feet high, is widely branching in habit, differing remarkably from a white ash of the same height beside it, which has narrow branches and a straighter stem. Their foliage is also very different in colour.                    (H. J. E.)


## FRAXINUS PENNSYLVANICA, Red Ash

*Fraxinus pennsylvanica*, Marshall, *Arb. Amer.* 51 (1785); Sargent, *Silva N. Amer.* vi. 49, t. 271 (1894), and *Trees N. Amer.* 770 (1905).
*Fraxinus pubescens*, Lamarck, *Dict.* ii. 548 (1786); Loudon, *Arb. et Frut. Brit.* ii. 1233 (1838).

A tree, attaining 60 feet in height and 5 feet in girth of stem. Bark brownish red and slightly furrowed, with scaly ridges. Young shoots stout, covered with

---

[1] *Proc. U.S. Nat. Mus.* xvii. 411 (1894).          [2] *Timber Trees of North Carolina*, 73.

dense white pubescence, which is retained in the second year; lenticels white, inconspicuous. Leaflets (Plate 263, Fig. 13), seven to nine, occasionally five, 3 to 5 inches long, three times as long as broad, ovate-lanceolate to oblong-lanceolate, tapering at the base, acuminate at the apex, finely serrate and ciliate in margin; upper surface with scattered fine pubescence; lower surface densely pubescent and green in colour. The leaflets are usually distinctly stalked, with pubescent petiolules; but forms occur in which they are subsessile, the substance of the leaflet being prolonged to its insertion. Rachis of the leaf densely white pubescent, with a distinct shallow groove on its upper side.

Flowers (section *Leptalix*) dioecious, in tomentose panicles in the axils of the leaf-scars of the preceding year's shoot; corolla absent. Fruit linear-spatulate, surrounded at the base by the persistent calyx; body slender, terete, many-rayed; wing slightly decurrent, narrow, and rounded or acute at the apex.

For the distinctions between this species and the Oregon ash, see under the latter. *Fraxinus Biltmoreana* differs conspicuously in having the leaflets white beneath. *Fraxinus profunda*, Bush,[1] which is remarkably distinct in its fruit, differs also in having the leaflets entire or undulate in margin, their base being usually very asymmetrical.

There are several forms of *F. pennsylvanica* in cultivation, some having the leaflets very firm in texture and set close on the rachis, others having thin leaflets wider apart on the rachis. The leaflets also vary in the length of their stalklets, in the size of the serrations, and in the shape of the base, which may be gradually tapering or abruptly tapering and almost rounded. *F. Richardi*, *F. Boscii*, and *F. glabra*, names given to certain horticultural varieties, are all probably referable to this species.

Var. *aucubæfolia* (*F. aucubæfolia*, Kirchner, *Arb. Musc.* 507 (1864)), in which the leaves are variegated with yellow, is considered by Lingelsheim[2] to be a hybrid between *F. pennsylvanica* and *F. lanceolata*. At Aldenham[3] this forms a handsome tree about 30 feet high. (A. H.)

It is neither so large nor so common a tree as the white ash in the United States where, according to Sargent, it has nearly the same distribution as the latter; being most common and largest in the north Atlantic States, smaller and less abundant west of the Alleghanies. Macoun says[4] that in Canada it ranges farther west than the white and black ashes, growing along the Assiniboine river and the tributaries of Lake Manitoba. It is usually 40 to 60 feet high, with a diameter rarely exceeding 18 inches to 20 inches; and is here of no value for timber, but makes good firewood, even when green. Emerson measured a tree at Springfield in September 1840 which was 9 feet in girth at 3 feet from the ground; and Ridgway says[5] that Dr. Schneck measured a tree in the Wabash forests 138 feet high by 16 feet in girth.

---

[1] See Sargent, *Trees N. America*, 772. This ash, which is probably not yet introduced, grows to a great size in river swamps in Missouri, Arkansas, and Florida. It is considered by Lingelsheim, in Engler, *Bot. Jahrb.* xl. 220 (1907), to be a variety of *F. pennsylvanica*.　　　　[2] *Op. cit.* 222.

[3] It is cultivated here under the erroneous name, *F. americana*, var. *aucubæfolia*, which is given in *Kew Handlist of Trees*, 533 (1902).　　　[4] *Forest Wealth of Canada*, p. 23.　　　[5] *Proc. U.S. Nat. Mus.* xvii. 411 (1894).

*Fraxinus pennsylvanica* was introduced into England in 1783[1]; and it is often met with in cultivation in public parks and botanic gardens, where it grows well as a small tree. One in the Botanic Garden at Oxford measures about 50 feet by 3 feet, and I have seen smaller trees at Stowe and elsewhere. I have raised it from American seed, and it seems to grow as fast as the white ash, but not so fast as the green ash. It ripens its wood better, and when young loses its leaves earlier than either of these.

Macoun says[2] that the red ash and the green ash are not separated commercially from the other species, the wood of the latter resembling that of the white ash, while that of the former is more like the black ash. Therefore there is some doubt whether Laslett, who writes of the Canadian ash, whose timber is often confounded with that of the white ash, is speaking of this tree or of the black ash which he does not mention. He says that it was, until recently, imported in considerable quantity in the form of oars, and that it is reddish brown in colour, considerably darker than the wood of the English ash. (H. J. E.)

## FRAXINUS OREGONA, Oregon Ash

*Fraxinus oregona,* Nuttall, *Sylva,* iii. 59, t. 99 (1849); Sargent, *Silva N. Amer.* vi. 57, t. 276 (1894), and *Trees N. Amer.* 776 (1905).

A tree attaining 80 feet in height and a girth of stem of 12 feet. Bark deeply divided by interrupted fissures into broad flat scaly ridges. Young shoots stout, covered with dense white tomentum, which persists in the second year; lenticels white, inconspicuous. Leaflets (Plate 263, Fig. 15), 3 to 4 inches long, subsessile, usually seven, sometimes five or nine, oval, about twice as long as broad, base rounded or abruptly tapering, apex acute or shortly acuminate; margin entire or minutely and remotely crenate, ciliate; upper surface with scattered fine pubescence; lower surface covered with dense white tomentum. Rachis white tomentose, with a distinct shallow groove on its upper side, basal part wide and flattened.

Flowers (section *Leptalix*) diœcious in glabrous panicles rising out of the axils of the preceding year's shoot; calyx present, persisting under the fruit, corolla absent. Fruit obovate-oblong; body slightly compressed; wing long, decurrent, many-nerved, and rounded, apiculate or emarginate at the apex.

*Fraxinus oregona* can only be confused with sessile forms of *F. pennsylvanica,* which has longer serrate leaflets, with more nerves, tapering gradually to the base. In *F. oregona* the leaflets are shorter in proportion to their breadth, and are usually entire in margin; but this last character is not absolutely distinctive, as the leaflets of the two species vary in regard to the presence or absence and size of the serrations.

[1] Aiton, *Hort. Kew.* v. 476 (1813).  [2] *Forest Wealth of Canada,* p. 23.

This species was discovered by David Douglas in 1825 on the banks of the Lower Columbia river. It is mentioned by Koch[1] as having been in cultivation in the Botanic Garden of Berlin prior to 1872. (A. H.)

This tree is the common ash of north-west America, extending from Puget Sound through western Washington and Oregon and the coast region of California to San Francisco, and along the western foothills of the Sierra Nevada to San Bernardino and San Diego counties. It is an important timber tree in Oregon and Washington, but if it occurs in British Columbia,[2] is too scarce to be noticed by Macoun. I saw no trees of great size during either of my visits to the Pacific coast, but Sheldon[3] says that it attains as much as 100 feet in height and 1 to 4 feet in diameter, and describes the wood as hard, tough, firm, straight-grained, and taking a high polish.

In California, according to Jepson,[4] *F. oregona* only attains a length of 15 to 30 feet, growing along the Sacramento river, and on the banks of the streams of the coast ranges. The leaflets become glabrous with age; and on this account Lingelsheim[5] has distinguished the Californian form as var. *glabra*.

In England the tree grows well in a young state; but I know of none of any size, except one at Nuneham Court, Oxfordshire, which in 1907 was 63 feet high and 3 feet 8 inches in girth. A tree of *F. americana* growing near it, and believed to have been planted at the same time, was only 38 feet by 1 foot 8 inches. Sir Charles Strickland has a plantation at Hildenley, Yorkshire, of this species mixed with larch; and as these trees, when I saw them in 1901, were 20 to 30 feet high, and had ripened seed from which I have raised plants, I think their hardiness in this country is abundantly proved; and that the tree is more likely to succeed in the west and north of England than the eastern American ashes. At Tortworth Court Lord Ducie has made a mixed plantation of *F. americana* and this species, which are now 20 to 30 feet high at about fifteen years old; but the soil and situation are not very favourable. In 1908 the Oregon ash were bearing seed freely.

(H. J. E.)

[1] *Dendrologie*, ii. 1, p. 260 (1872).

[2] Piper, *Flora of the State of Washington*, 449 (1906), says its range is " British Columbia to California, in the coast region "; but gives no localities for British Columbia. It is not mentioned as occurring in Vancouver Island by the authors of *Postelsia*, the year-book of the Minnesota Seaside Station for 1906.

[3] *Forest Wealth of Oregon*, 32 (1904).  [4] *Flora W. Mid. California*, 385 (1901).

[5] In Engler, *Bot. Jahrb.* xl. 220 (1907).

## FRAXINUS CAROLINIANA, SWAMP ASH

*Fraxinus caroliniana*, Miller, *Dict.* No. 6 (1768); Loudon, *Arb. et Frut. Brit.* ii. 1237 (1838);
Sargent, *Silva N. Amer.* vi. 55, tt. 274, 275 (1894), and *Trees N. Amer.* 762 (1905).
*Fraxinus platycarpa*, Michaux, *Fl. Bor. Amer.* ii. 256 (1803).
*Fraxinus triptera*, Nuttall, *Gen.* ii. 232 (1818).
*Fraxinus cubensis*, Grisebach, *Cat. Pl. Cub.* 170 (1866).

A tree attaining 40 feet in height, with a stem 3 feet in girth; bark marked by irregularly-shaped brown patches, separating on the surface into thin scales. Branchlets glabrous or pubescent, with white minute scattered lenticels. Leaflets (Plate 263, Fig. 12) seven, occasionally five, stalked (petiolule ¼ to ⅜ inch), about 3 inches long, oval; unequal, rounded, or broadly cuneate at the base; apex shortly acuminate; finely and irregularly serrate; green and glabrous on the under surface except for some white pubescence along the sides of the midrib and nerves, or in some forms pubescent throughout. Leaf rachis, glabrous or pubescent, with two slight wings on the upper side, forming a groove.

Flowers (section *Leptalix*) diœcious in panicles arising in the axils of leaf-scars of the preceding year's shoot; calyx present, persisting under the fruit; corolla absent. Fruit broad, elliptic or spatulate; body short and compressed, surrounded by a pinnately-veined broad thin wing.

This species grows in river swamps in the coast regions of the Atlantic and Gulf States from southern Virginia to the valley of the Sabine river in Texas, extending through western Louisiana northwards to south-western Arkansas. It also occurs in Cuba.

This species was introduced into England in 1783, according to Loudon, who, however, mentions no trees of any size as growing in England in 1838. We have seen no specimens, except small trees at Kew, which are thriving.        (A. H.)


## FRAXINUS VELUTINA

*Fraxinus velutina*, Torrey, Emory's *Report* 149 (1848); Sargent, *Silva N. Amer.* vi. 41, t. 267
(1894), and *Trees N. Amer.* 774 (1905).
*Fraxinus pistaciæfolia*, Torrey, *Pacific R. Report*, iv. 128 (1856).

A tree, 40 feet high, with a girth of stem of 2 feet. Bark deeply divided into broad flat broken scaly ridges. Shoots purple, covered with dense white pubescence; lenticels white. Leaflets (Plate 265, Fig. 20), small, about 1½ inch long, three or five, occasionally seven or nine, or rarely only one, and variable in shape, margin, and insertion; usually sessile, occasionally stalked, the terminal leaflet often obovate, the lateral leaflets commonly lanceolate with cuneate base and acuminate apex; coarsely serrate in the upper half or two-thirds, but sometimes entire, ciliate; upper surface pubescent; lower surface densely white pubescent. Rachis of the leaf white pubescent and deeply grooved on its upper side.

Flowers (section *Leptalix*) diœcious in short panicles in the axils of the leaf-scars of the preceding year's shoot; calyx present and persisting under the fruit; corolla absent. Fruit spatulate, with terete body and terminal wing.

This species is readily distinguished by the dense white pubescence over the shoot, leaf-rachis, and leaflets, the latter being variable in number and smaller than those of the other pubescent ashes, except *F. xanthoxyloides*, which has still smaller leaflets with a broadly winged rachis and is much less strongly pubescent.

*Fraxinus velutina* occurs usually in elevated cañons beside streams in Texas, New Mexico, Arizona, southern Nevada, and south-east California.

Young trees are doing well at Kew, where they are of considerable interest from their peculiar foliage, which gives them a neat and elegant appearance. The oldest, planted in 1891, are now about 15 feet high. (A. H.)

# ZELKOVA

Zelkova,[1] Spach, in *Ann. Sc. Nat.* sér. 2, xv. 356 (1841); Bentham et Hooker, *Gen. Pl.* iii. 353 (1880); Nicholson, in *Woods and Forests*, 1884, p. 176.

*Abelicea*, Reichenbach, *Consp. Veg.* 84 (1828); Schneider, *Laubholzkunde*, i. 224 (1904).

*Planera*, Gmelin, subgenus *Abelicea*, Planchon, in *Ann. Sc. Nat.* sér. 3, x. 261 (1848).

DECIDUOUS trees or shrubs, belonging to the order Ulmaceæ. Branchlets slender, distichous. Leaves alternate, distichous, simple, shortly stalked, penni-nerved, crenately serrate. Stipules in pairs, membranous, lanceolate, caducous.

Flowers monœcious; corolla absent; calyx, four- or five-lobed. Staminate flowers, clustered, two to five together, on the branchlets below the leaves or in the axils of the lowermost leaves; disc absent; stamens, four or five, with short, erect filaments and exserted anthers; ovary rudimentary or absent. Pistillate flowers, solitary in the axils of the uppermost leaves; disc cupular, fleshy; staminodes present or absent; ovary sessile; styles, two, stigmatiferous on the inner side; ovule solitary, pendulous. Fruit, a small drupe, sessile, subtended by the persistent calyx, subglobose, oblique, veined or rugose on the surface, crowned by the remains of the styles, persisting as two minute beaks; with a membranous or slightly fleshy outer covering, and a thin, hard endocarp or stone, containing a compressed, concave, horizontal seed, without albumen. The fruit ripens late in autumn, and persists on the branchlets till the following spring.

In Zelkova no true terminal bud is formed, and the tip of the branchlet falls off in early summer, leaving a small circular scar at the apex of the twig. The base of the shoot is ringed with the scars of the inner scales of the previous season's bud, and shows, as a rule, a few of the outer scales persisting dry and membranous. The buds, all axillary, and composed of numerous imbricated scales, are often multiple, two being then developed, side by side, in a single axil. The leaf-scars are narrow, crescentic, and three-dotted; with a linear stipule-scar on each side. In Ulmus the buds are single in the axils, and none of the scales persist at the base of the shoot.

Three species of Zelkova[2] are known to exist in the wild state, two of which are large trees well known in cultivation, *Z. acuminata*, Planchon, a native of China and

---

[1] The name *Zelkova* is sanctioned, and *Abelicea* rejected in *Actes Congrès Internat. Bot. Vienne*, 77 (1906).

[2] *Hemiptelea Davidii*, Planchon, *Compt. Rend. Acad. Paris*, lxxiv. 1496 (1872), a thorny tree, occurring in northern and central China and Korea, is united with *Zelkova* by Bentham and Hooker, in *Gen. Pl.* iii. p. 353. It differs from that genus in having winged fruit. Cf. Schneider, *Laubholzkunde*, i. 224. This species does not appear to be in cultivation in Europe.

Japan, and *Z. crenata*, Spach, inhabiting the Caucasus and North Persia. The third species, *Z. cretica*, Spach, is a shrub growing in Crete and Cyprus, which has not yet been introduced, and does not come within the scope of our work. The following species, also a shrub, is only known in cultivation :—

*Zelkova Verschaffeltii*, Nicholson, *Kew Handlist of Trees*, 145 (1896).

*Zelkova japonica*, Dippel, var. *Verschaffeltii*, Dippel, *Laubholzkunde*, ii. 39, fig. 14 (1892).

A shrub or small tree. Branchlets slender, pubescent, with white hairs. Leaves (Plate 267, Fig. 8), coriaceous, variable in size, from $1\frac{1}{4}$ inch long by $\frac{3}{4}$ inch wide to $2\frac{1}{2}$ inches long by $1\frac{1}{4}$ inch wide, oval, acuminate at the apex, cuneate and unequal at the base, divided by the midrib into unequal halves, the larger half with six to eight nerves, the smaller half with four to seven nerves, each nerve ending in a long triangular tooth, tipped with a short cartilaginous point ; margin ciliate ; upper surface dark green, with scattered white pubescence ; lower surface light green, with downy white pubescence, densest on the midrib and nerves ; petiole, $\frac{1}{8}$ inch to $\frac{1}{4}$ inch, pubescent. Buds, often two together in an axil, small, globose, pubescent. Fruit similar to that of *Z. crenata*, but slightly smaller in size.

This species, which resembles an elm in having asymmetrical oblique leaves, was considered by Schneider[1] to be a peculiar variety of *Ulmus glabra*, and is occasionally met with in cultivation, as *Ulmus Verschaffeltii*, and *Ulmus pendula laciniata Pittcursii*.

A tree, 15 feet high, in the nursery of the Paris Municipality, specimens of which have been sent us by M. Vacherot, produced flowers and fruit this year ; and the fruit, hitherto unknown, proves to be that of a *Zelkova*. *Z. Verschaffeltii* is not known in the wild state, though Koehne[2] states that O. Kuntze collected specimens of *Z. crenata* in the Caucasus, which strongly resembled it. It is possibly a hybrid between *Z. crenata* and *Z. cretica*, and was first noticed by Dippel in 1892.

(A. H.)

## ZELKOVA CRENATA

*Zelkova crenata*, Spach, *Ann. Sc. Nat.* xv. 358 (1841) ; Boissier, *Fl. Orientalis*, iv. 1159 (1879).
*Zelkova carpinifolia*, Dippel, *Laubholzkunde*, ii. 38 (1892).
*Zelkova ulmoides*, Schneider, *Laubholzkunde*, i. 806 (1906).
*Rhamnus ulmoides*, Güldenstadt, *Itin.* i. 313 (1787).
*Rhamnus carpinifolius*, Pallas, *Fl. Rossica*, i. 2, 24 (1788).
*Planera Richardi*, Michaux, *Fl. Bor. Am.* ii. 248 (1803) ; Loudon, *Arb. et Frut. Brit.* iii. 1409 (1838).
*Planera carpinifolia*, Watson, *Dendrol. Brit.* 106, t. 106 (1825) ; Koch, *Dendrol.* ii. 1. 425 (1872).
*Planera crenata*, Desfontaines, *Cat. Hort. Paris* (1829).
*Abelicea ulmoides*, Kuntze, *Rev. Gen.* ii. 621 (1892) ; Schneider, *Laubholzkunde*, i. 224 (1904).

A tree attaining about 100 feet in height, and 15 feet in girth. Bark thin, smooth, greyish-brown, marked with persistent lenticels ; on older trees, scaling off in small

---

[1] *Laubholzkunde*, i. 226 (1904).          [2] *Deutsche Dendrologie*, 137 (1893).

irregular plates.    Branchlets covered with dense, white pubescence.    Leaves (Plate 267, Fig. 6), slightly coriaceous, about 3 inches long and $1\frac{1}{2}$ inch wide, oval-lanceolate, acute at the apex, cordate and unequal at the base, with nine to eleven pairs of nerves, each ending in a crenate serration, the apex of which is minutely pointed; margin ciliate; upper surface dark green, with scattered minute pubescence; lower surface more or less covered with white pubescence, densest on the midrib and nerves; petiole, $\frac{1}{8}$ to $\frac{3}{16}$ inch, pubescent.

Fruit, about $\frac{1}{5}$ inch long, pubescent, with a very slight depression on either side of the prominent ridge on the upper surface.

In winter the twigs are pubescent, and bear elongated conical buds, which are brownish, tinged with white in colour, owing to the scales, which are glabrous on the surface, being fringed with long, white cilia.

This species is readily distinguished by the pubescent oval leaves, acute and not acuminate at the apex, the serrations only showing minute points.    *Z. acuminata* has glabrous ovate leaves, with a long acuminate apex, the serrations ending in long, sharp, often recurved points.

## DISTRIBUTION

This species occurs in the Russian provinces, lying south of the main range of the Caucasus, and in northern Persia, in the territory bordering on the Caspian Sea, extending as far eastward as Asterabad.

The best account of its distribution in Transcaucasia is given by Scharrer,[1] who states that it grows wild in two distinct areas, one in the government of Kutais and the other in Talysch, while there are a few scattered trees at Araxes, in the Karabagh district.    In Kutais it grows in the Mingrelian plain, east of Sennakh, and ascends in the lower mountains of Imeritia to about 1000 feet, occurring at low levels in small groups in the oak forests, and at higher elevations mixed with ash, maple, and beech, and never forming pure woods.    In Talysch it is not found on the marshy plains, but is common in river valleys, ascending on the mountains to 5000 feet, and often forming pure and dense woods.    Scharrer measured a tree, 100 feet in height, with a stem 8 feet in diameter and free of branches to sixty feet. The climate in which it thrives is humid, with a rainfall of 50 or 60 inches; and it requires a moist, permeable, rich soil to come to perfection.    At Tiflis, however, where the rainfall is only $19\frac{1}{2}$ inches, the tree is met with growing on good loamy soil and on rocky mountain slopes, but is slower in growth than in Mingrelia.    It has borne without injury a temperature of $-24°$ C.

The elder Michaux,[2] who travelled in Persia in 1782, and saw the tree growing in the forests of Ghilan, states that it commonly attains a height of 80 feet, with a girth of 9 to 12 feet, with a straight trunk, branching at about thirty feet up, and resembling the hornbeam in its bark, fluted trunk, and mode of branching.    Scharrer, however, states that in the forest it produces clean stems, very uniform in thickness

---

[1] *Gartenflora*, xxxvi. 187 (1887).
[2] Cf. André Michaux, *Mémoire sur le Zelkoua* (Paris, 1831).

and free from branches, except at the summit, excelling in this respect most broad-leaved trees. It grows fast in youth, continuing its growth in height to sixty or eighty years old, afterwards mainly increasing in girth. It is moderate in its demands for light, and gives good coppice shoots.

## CULTIVATION

This species was introduced into cultivation in 1760, the oldest known tree in Europe being one[1] in the garden of M. Lemmonier at Petit Montreuil, near Paris, which was cut down in 1820, when it was 72 feet in height and 6 feet 8 inches in girth. It is probable that the elder Michaux, who saw this tree growing in northern Persia in 1782, also introduced seed.[1] Further consignments[1] were sent to France in 1831, by Chevalier Gamba, French Consul at Tiflis. Seeds from this source germinated after lying eighteen months in the ground; but Gay, in a note in the Kew Herbarium, states that this tardy germination was probably accidental, as seeds from Karabagh, which he sowed in the last days of March, produced seedlings, which were peeping out of the ground at the end of May. (A. H.)

This tree is now rarely seen in nurseries, though it is easily propagated by suckers, and seed could be procured without difficulty from its native country. In consequence it is hardly known to modern gardeners, though both from its ornamental habit and valuable timber it would be much better worth planting than many trees of more recent introduction.

The principal point to be attended to is to protect it from frost, and prune it carefully until the main stem has attained the desired height; and to plant it in a deep, rich alluvial soil, and warm, sheltered situation. So far as I have been able to learn, no tree has produced fertile seed in this country.

## REMARKABLE TREES

A remarkable tree at Wardour Castle, Tisbury, Wiltshire, is reputed by tradition to have been sent, when quite young, from North America by the second Lord Baltimore, about 1632, and has been supposed[2] to be *Planera aquatica*, Gmelin,[3] a native of swamps in the south-eastern United States. There must be some error in the tradition, as the tree is undoubtedly *Zelkova crenata*. It is known as the Iron tree, and the late Lord Arundell of Wardour assured me that this name, used in America for the Hornbeam and Hop Hornbeam, was a proof of its American origin. He also believed that the tree had been cut down during the siege of Wardour Castle in Cromwell's time, and had afterwards produced from the stool the seventeen

---

[1] André Michaux, *Mémoire sur le Zelkoua* (Paris, 1831).

[2] See a lengthened correspondence concerning this tree in *Garden*, xxiv. 370 (1883), xxvi. 38 (1884), and xxxii. 92 (1887).

[3] This species was introduced into England in 1816, according to Loudon, *Arb. et Frut. Brit.* iii. 1413 (1838); but it appears to be unsuitable for our climate, and no specimens are known to us to exist, except two plants in the Elm collection at Kew, about 8 feet high, which were introduced in 1897, and are thriving so far.

tall stems which it now shows (Plate 248). These measure from 5 to 8 feet in girth, the whole forming a group about 12 feet wide, and some of them reaching nearly 100 feet in height. There are many small suckers, as usual, and some of these have been transplanted to the front of the present mansion, where they were, when I saw them in 1903, about 25 feet high. Lord Arundell was good enough to give me two rooted suckers from the old tree, one of which is now planted at Tortworth Court, and the other at Colesborne.

At Holme Lacy, Hertfordshire, the seat of the Earl of Chesterfield, there is a very fine tree standing on a walled depression near the house. Its bole was until recently surrounded by laurels, which have now been cleared away, so that I was able in October 1908 to measure it carefully. I found it to be about 95 feet high by 19 feet in girth at the smallest point, about 3 feet from the ground. A photograph of this tree is at Kew, and it was reported in 1884 to be 70 feet high.

There are several trees in Kew Gardens, the largest growing in front of the Herbarium, and measuring 60 feet in height and 9 feet 3 inches in girth. A larger specimen was cut down a good many years ago, and a section of the trunk is exhibited in the Timbers Museum.

There are several fine trees at Syon,[1] the largest near the lake being, in 1905, 98 feet in height and 12 feet 7 inches in girth, while another is 89 feet by 13 feet. Both these trees are remarkable for their buttressed stems. Near the bridge there is a slender specimen, crowded by other trees, which is 92 feet high, with a stem 7 feet 2 inches in girth, and free of branches to about 50 feet.

At Albury there is a remarkable specimen, with a bole of only 4 feet in height, but 16 feet in girth, dividing into numerous stems.

In the Wilderness, at Croome Court, Lady Coventry showed me a tree of this species, which was supposed to be a species of hornbeam. It measured, in 1906, 65 feet by 14½ feet, and grew in an angle between two hedges, into which its suckers had spread profusely, and being clipped with the hedge, may eventually form part of it when the original tree dies. Another tree grows in the Temple Shrubberies at the same place, about 70 feet by 7 feet, with a clean trunk about 15 feet long, and was of a better shape, but is now partially decayed.

At Pitt House, near Chudleigh, Devonshire, the seat of Captain Morrison Bell, there are several Zelkovas which seem to thrive well in this climate. The largest is a very well-shaped and healthy tree measuring 80 to 85 feet by 13 feet. Its leaves were beginning to unfold on 15th April 1908. Some trees have thrown up suckers in the hedge by the high road here, but the gardener has not observed any flowers.

At Oxford, in the University Park, on the banks of the Cherwell, there are two, the larger of which, not a well-shaped tree, is about 80 feet by 12½ feet.

At Kyre Park there are two trees on the banks of a pond near the house, both of which have begun to decay, and large pieces of the smooth bark were dropping off them when I visited Kyre in 1904. The largest measured about 75 feet by 16 feet

---

[1] According to Loudon the largest tree at Syon was in 1834, 54 feet high and 2 feet 3 inches in diameter.

7 inches. Mrs. Childe tells me that in the hot summer of 1905 flowers were produced by this tree.

At Belshill, near Belford, Northumberland, the property of Sir W. Church, Bart., there is a fine tree in a sheltered situation which measures 70 feet by 9 feet 10 inches, and looks healthy, though it is believed to be over 100 years old.

In Scotland we have not seen or heard of any trees, though there is no doubt it would grow well in the south and west, where the climate is much better than at Belshill.

There are two trees at Glasnevin, one 50 feet and the other, a remarkably fine one, 61 feet in girth (Plate 249). Both are 9 feet in girth, and divide at 10 feet up into numerous branches.

A tree[1] at Verrières, near Paris, is 70 feet in height and 8 feet in girth.

There is a large tree in the grounds of the Petit Trianon at Versailles, about 90 feet by 10 to 12 feet, which appears to be grafted on the roots of an elm, and Mr. Hickel informed me that most of the older trees in France were so grown.

This species[2] is represented in the United States at Woodlands, Philadelphia, where there are growing in a cemetery a few low bushy trees, with short trunks, 4 feet in diameter, and numerous erect branches.


## TIMBER

According to Scharrer the wood is homogeneous, prettily veined, very tough and flexible, does not crack and warp, takes a fine polish, and is very durable even when placed in wet situations. It is very suitable for cabinet-work and carriage-building. The native name of the tree, *dzelkwa*, signifies "stone-wood," so-called on account of the hardness of the timber, into which nails are driven with difficulty.

The younger Michaux, who examined a tree cut down at Paris in 1820, states that the sapwood is white, and the heartwood reddish in colour, the latter being heavier and stronger than that of elm, while even the sapwood equalled the ash in strength and elasticity.

A plank of this wood cut from a tree which grew at Boynton, in Yorkshire, was given me by Sir Charles Strickland, and resembles the wood of the Japanese species in texture and colour. Mrs. Baldwyn Childe has also sent me a specimen of it from a branch of her tree. Though unknown in the trade, and, as far as I can learn, never cut for export, I believe that this wood would prove valuable for making furniture if it could be obtained at a reasonable price. (H. J. E.)

[1] *Hortus Vilmorinianus*, 52 (1906).      [2] *Garden and Forest*, x. 488 (1897).

## ZELKOVA ACUMINATA

*Zelkova acuminata*, Planchon, in *Compt. Rend. Acad. Paris*, lxxiv. 1496 (1872), and DC. *Prod.* xvii. 166 (1873).

*Zelkova Keaki*, Maximowicz, *Mél. Biol.* ix. 21 (1872); Sargent, *Garden and Forest*, vi. 323, fig. 49 (1893); Shirasawa, *Icon. Ess. Forest. Japon*, text 65, t. 36, figs. 1-17 (1900); Mayr, *Fremdländ. Wald- u. Parkbäume*, 525, figs. 247-249 (1906).

*Zelkova serrata*, Makino, in Tokyo *Bot. Mag.* xvii. 13 (1903).

*Zelkova hirta*, Schneider, *Laubholzkunde*, i. 806 (1906).

*Corchorus hirtus*, Thunberg, *Fl. Jap.* 228 (1784).

*Ulmus Keaki*, Siebold, *Verh. Bat. Gen.* xii. 28 (1830).

*Planera acuminata*, Lindley, *Gard. Chron.* 1862, p. 428

*Planera japonica*, Miquel, *Ann. Mus. Bot. Lugd.* iii. 66 (1867).

*Planera Keaki*, Koch, *Dendrol.* ii. 1. 427 (1872).

*Abelicea Keaki*, Schneider, *Dendrol. Winterstudien*, 238 (1903).

*Abelicea hirta*, Schneider, *Laubholzkunde*, i. 226 (1904).

A tree attaining in Japan 120 feet in height and 15 feet or more in girth. Bark smooth, greyish, resembling that of a beech, but on old trees dividing on the surface into irregular rounded scaly plates. Branchlets slender, at first with a scattered slight pubescence, but becoming glabrous in summer. Leaves (Plate 267, Fig. 7), membranous, thinner than those of the Caucasian species, about 3 inches long and $1\frac{1}{4}$ inch broad, ovate, with a long acuminate apex, slightly cordate and unequal at the base; nerves, nine to twelve pairs, each ending in the long, sharp, often recurved tip of a serration; margin ciliate at first, the cilia falling off in summer; upper surface dark green, with scattered, short, stiff hairs; lower surface light green, at first slightly pubescent, becoming glabrous in summer; petiole $\frac{3}{16}$ to $\frac{1}{4}$ inch, glabrescent.

Fruit about $\frac{1}{8}$ inch broad, sub-globose, oblique, with a concave depression on either side of a transverse ridge on the upper surface, which bears the remains of the styles; prominently veined, glabrescent.

In winter the twigs are glabrous, and the buds are minute, ovoid, uniformly brown in colour, with glabrous slightly ciliate scales.

### DISTRIBUTION

*Zelkova acuminata* occurs in China and Korea, as well as in Japan, but appears to be only common in China in the Tsin-Ling mountain in Shensi, where it was collected by Père Giraldi.[1] It has also been found[2] in the province of Chekiang, on the hills near the Taihoo Lake, and on the mountains inland from Ningpo. Carles[2] collected it near Seoul in Korea.

It is much more widely distributed in Japan, occurring throughout Kiushiu, Shikoku, and Honshu, forming forests in mixture with maple, beech, *Quercus grosseserrata*, and other broad-leaved trees. According to Shirasawa, it is usually of branching habit in the forest, the tallest and stoutest trees being those cultivated

---

[1] Diels, in Engler, *Jahrb.* xxxvi. No. 5, p. 33 (1905).    [2] Hemsley, *Journ. Linn. Soc. (Bot.)*, xxvi. 449 (1894).

near houses and temples, which often have clean stems due to pruning. Mayr, however, records a forest tree, 123 feet in height, with a stem 33 inches in diameter, and clean of branches to 57 feet. Dupont,[1] who gives many interesting particulars concerning this species, states that it ascends in Kiushiu to 3000 feet on northern slopes, and in Honshu to 4000 feet on southern slopes ; and that it requires for its best development a deep, permeable, and rich soil, such as is found on alluvial tracts. It thrives well also on volcanic soils and sandy loams, but does not succeed on poor sands or on stiff clays. Dupont states that on suitable soils and situations it grows remarkably straight, whether isolated or crowded in the forest, attaining on soil of middling quality at 1600 feet elevation in the latitude of Fuji-yama, 5 feet in girth at 60 years old, 9 feet at 120 years, and 12½ feet at 180 years. The growth of isolated trees on alluvial soil is still more rapid, the annual rings averaging ⅓ inch in width. He advocates the planting of this tree on account of its rapid growth and the value of its timber, which he considers to be superior to that of oak. (A. H.)

It seems strange that this tree, whose wood is more highly valued by the Japanese than any other hard wood, should be planted on so small a scale in Japan. Probably it requires too many years to come to maturity to induce private persons to plant it when bamboo, Cryptomeria, and pine offer so quick and certain a return. But though I saw no plantations of Keaki,[2] I believe the Government are making efforts to preserve and increase the area under this species. It is said to be found wild in the south up to about 5000 feet, and in the north up to about 2000 feet. It also grows wild in Hokkaido, but not to so great a size as in the north-eastern districts of the main island which are famous for their large trees. I heard of, but was not able to see, one said to be the largest in Japan at Sendai. Rein[3] speaks of one which was felled at Meguro, near Tokio, in 1874, and measured 11.7 metres in girth at one metre high.

The largest I measured myself was growing by the side of the Nakasendo road, at a place called Hideshiwa, near the village of Sooga in Shinano, in a grove of trees just below the road, and may have been wild or planted. It was about 115 feet high and 20 feet in girth, dividing at about 20 feet, into two tall upright limbs, each 10 feet or more in girth, and seemed to contain about 800 to 1000 cubic feet of timber. Close to it was a very large Æsculus, and on the other side of the road another Keaki, 113 feet high, by 13 feet 6 inches in girth, clean and straight to about 70 feet high.

Such trees as these are found only where they have been crowded when young, the tendency of the species being to assume a branching and spreading habit, so that most of these which are seen planted singly are thick and spreading rather than tall.

In the forest the Keaki grows scattered among other trees, and is said by the author of the handbook on Japanese forestry to love calcareous soils. I never saw any such soil in Japan, but it seemed to grow equally well on all kinds of soil provided it is deep and moist. As to the age to which the tree attains I cannot speak positively, but it looks like a very long-lived tree. Its bark is smooth and greyish in colour, somewhat like that of the beech. It seeds freely and reproduces itself

---

[1] *Essences Forest. du Japon*, 45 (1879).
[2] The Japanese name of this species is *Keyaki*, occasionally spelled *Keaki*.    [3] *Industries of Japan*, 225 (1889).

easily, the seedlings being hardy and fast growing and bearing shade well. In appearance the Zelkova is not an ungraceful tree, resembling a beech perhaps more than an elm, but its small leaves make it a poor shade tree, and its habit of growth varies very much according to the situation in which it grows.

## CULTIVATION

This species was introduced into cultivation in England by J. Gould Veitch,[1] who sent seeds from Japan in 1862. It was apparently introduced on the Continent a few years earlier by Siebold; and Koch, writing in 1872, mentions that it had been cultivated previously for several years in the Botanic Garden at Berlin, where it had sustained severe frost without injury.

I raised a quantity of seedlings in 1901 from Japanese seed, which grew rapidly at first, and seemed quite hardy; but those which I have planted out grow slowly, and, where not protected from spring frost, have been killed back every winter, so that they produce bushy shrubs. I should, therefore, suppose that it requires more summer heat and moisture than most parts of England afford, and that it should be planted only in rich, deep soil, where it can be shaded and drawn up by other trees. · Careful pruning is also evidently necessary to check its tendency to produce lateral branches when young, and I do not anticipate that it will ever attain large dimensions in Great Britain or be worth planting for its timber.

So far as we have ascertained there are no large trees existing in England, the best specimens we have seen at Kew Gardens, at Tortworth, and in Lord Kesteven's woods at Casewick, Lincolnshire, not exceeding 20 to 30 feet in height.

Mr. C. Palmer tells me that in November 1864, he planted a specimen received from Veitch, then 3 feet high, in an exposed situation, at about 500 feet above the sea, near Stukeley Grange, Leighton Buzzard. This tree in 1874 was 18½ feet high by 6 inches only in girth. In 1892 it had increased to 2 feet in girth.

A good-sized and healthy-looking tree, of whose age no record can be found, grows near the pond below the entrance from Woburn village to Woburn Park on the right-hand side of the drive.

At Kilmacurragh, Co. Wicklow, a healthy tree measured, in 1904, 41 feet in height and 3 feet 1 inch in girth.

The oldest and largest tree of this species that we know of in Europe is growing in the Botanic Gardens at Carlsruhe. I am indebted to Herr Max Leichtlin, of Baden-Baden, for a photograph of this tree (Plate 250), and Herr Gräbener has also sent us one. He informs us that it is one of three seedlings which was raised at St. Petersburg (or brought from Leyden). It was planted sometime between 1859 and 1861, and has never suffered from frost, having sustained the severe winter of 1879-1880 without injury.[2] In 1904 it measured at 1½ metre from the ground 3.10 metres in girth.

---

[1] *Hortus Veitchii*, 386 (1906). Mr. H. J. Veitch informs me that the plants raised at Coombe Wood only attained 3 to 4 feet in height, and gradually died out, the whole being finally lost during a severe winter.

[2] Mayr says that at Grafrath it has endured −25° Cent. without injury.

This species [1] has been planted pure in small experimental plots at five different forestry stations in Germany ; but the results have not been encouraging. Both Schwappach and Mayr consider that it probably would succeed if planted in mixture with other hardwood trees.

In the Botanic Garden at Copenhagen, Henry saw, in 1908, a fine tree, about 40 feet high, but dividing into three stems at 2 feet from the ground. This tree was planted about 1870, and exceeds considerably in size a *Z. crenata* planted beside it the same year. At Christiania, *Z. acuminata* remains a bush about a foot high, being repeatedly cut by frost.

The largest specimens in the United States are two trees, growing in Dr. Hall's gardens at Warren, Rhode Island, which were raised from seed sent home in 1862. According to Sargent,[2] who says that this species is probably the only Japanese tree worth introducing into North America, on a large scale for timber, these have received no special care, the soil is not particularly good, and their growth has been checked by overcrowding. They were about 50 feet high in 1893, with trunks about a foot in diameter, and had produced flowers and fruit. There appear [3] to be several other trees of this species in the same garden, and hundreds of self-sown seedlings were observed near them in 1893. The only drawback [4] to the cultivation of this species in America is that it is subject to the attack of the elm-leaf beetle.

## TIMBER

The wood [5] resembles in structure that of the elm, the vessels being disposed in similar broken lines. The sapwood, according to Mayr, about $1\frac{1}{2}$ inch in thickness, and white in colour, is separated from the brown heartwood by a narrow pink zone. According to Dupont,[6] the wood, while like that of the elm in appearance, is more like the ash in working, as it bends readily, and is of great strength, surpassing even the oak in this quality. It resists exposure to moisture, and is very durable for building purposes. It is much used in Japan for making furniture, an especial kind, cut from burrs, and called *jo-rin-moku*, being especially esteemed. The most beautiful trays and cabinets [7] which come from Japan are made of dark, irregularly grained, and wavy-lined wood of the Zelkova. Many of these trays are ornamented with the bark of Pterocarya. Dupont says that the wood contains an oil, with a disagreeable odour, which prevents it from being made into articles used for containing liquids.

The wood is the most valuable in Japan for building, for furniture, and for all purposes where a strong, tough, durable wood, not liable to warp, crack, or decay, is required ; and it is also valuable for carving and lacquering. It is the highest priced wood in Japan, worth in ordinary sizes up to 4s. per cubic foot, and for finely grained or very wide planks much more. I was told that a single large plank of

[1] Schwappach, *Anbauversuche mit Fremdländischen Holzarten*, 79 (1901).

[2] *Garden and Forest*, vi. 324 (1893).     [3] *Ibid.* 468.     [4] *Ibid.* 369.

[5] Figured in Mayr, *op. cit.* t. xx. fig. 44.     [6] *Ess. Forest. du Japon*, 45 (1879).

[7] Sargent, *Garden and Forest*, x. 40 (1879).

finely grained wood of this tree shown at the Osaka Exhibition in 1903 was priced at 800 yen (over £80); and I saw beautifully figured pieces myself of a peculiar reddish tint which were held for fancy prices, such pieces being much valued for the construction of the dais which is a marked feature in Japanese rooms, and on which is the seat of honour. Most of the pillars, beams, gateways, gates, and carved roofs, which are so striking a feature in Japanese temples, old and new, are made of Keaki wood, which seems indestructible by time or damp when covered in, and I was told that some of these which looked sound, though much weather worn, were 1000 years old. It seemed to me, however, that 1000 years is simply a convenient expression in Japan for anything very old, though no doubt historical evidence could be found if wanted as to the durability of this fine wood. The fancy grained varieties are known as Jorin, Uzura (partridge), Tama (gem), or Botan (peony), and these are used for cabinetmaking and fancy work. The colour, according to Rein, is deepened by long submersion in water. Rein gives the specific gravity of the wood at 0.682. I bought some of the wood in Japan, and have used it in making a large wardrobe; it takes polish well, makes good joints, and seems equal to mahogany for furniture making, but so far as I can learn has not as yet been imported, and is unknown in the trade.                    (H. J. E.)

# CELTIS

*Celtis*, Linnæus, *Gen. Pl.* 337 (1837); Bentham et Hooker, *Gen. Pl.* iii. 354 (1880).
*Mertensia*, Humboldt, Bonpland et Kunth, *Nov. Gen. et Spec.* ii. 30 (1817).
*Momisia*, F. G. Dietrich, *Lexic. Garten.* v. 128 (1819).
*Solenostigma*, Endlicher, *Prod. Fl. Norf.* 41 (1833).

THE genus Celtis, belonging to the order Ulmaceæ, comprises about sixty species, spread over the temperate and tropical regions of the northern hemisphere; and was divided into four sub-genera by Planchon.[1] In the following account, the characters of one of the sub-genera, *Euceltis*, are given, as all the species in cultivation belong to it.

Deciduous trees, without spines. Leaves stipulate, alternate, distichous, simple, stalked, serrate or entire, usually oblique and three-nerved (rarely four- or five-nerved) at the base, the midrib and basal nerves giving off pinnately secondary nerves.

Flowers minute, pedicellate on the branchlets of the year, polygamo-monœcious. Staminate flowers in few-flowered fascicles from the axils of caducous bud-scales. Perfect flowers solitary or in two- to three-flowered fascicles in the axils of the lower leaves. Calyx four- or five-lobed, imbricate in æstivation, deciduous. Corolla absent. Stamens, four or five, inserted under the margin of a pubescent disc; filaments, subulate, erect and exserted in the staminate flowers, shorter and included, occasionally absent, in the perfect flowers; anthers two-celled, extrorse, opening longitudinally. Ovary sessile, one-celled, crowned with a short style, divided into two divergent, elongated, reflexed lobes, papillo-stigmatic on the inner surface; ovule solitary, suspended. Fruit, a fleshy drupe, with a firm epicarp, a succulent thin mesocarp, and a thick-walled bony stone, containing one seed. Cotyledons emarginate at the apex, raised above ground in germination.

Seven species of Celtis are in cultivation in this country, which may be distinguished as follows :—

I. *Leaves ovate.*

* *Leaves quite glabrous.*

1. *Celtis Davidiana*, Carrière. China. See p. 929.
   Leaves shining on both surfaces, toothed in the upper third, shortly acuminate, minutely punctate when viewed with a lens.

[1] De Candolle, *Prod.* xvii. 169 (1873).

2. *Celtis glabrata*, Steven.   Asia Minor, Caucasus.   See p. 929.
    Leaves bluish green, serrate except near the base, acute or very shortly acuminate, conspicuously punctate when viewed with a lens.

#### ** *Leaves pubescent.*

3. *Celtis occidentalis*, Linnæus.   North America.   See p. 930.
    Leaves caudate-acuminate, serrate in the upper half or two-thirds, smooth to the touch above, pubescent on the nerves beneath.

4. *Celtis crassifolia*, Lamarck.   North America.   See p. 932.
    Leaves shortly acuminate, serrate in the upper half or two-thirds, scabrous to the touch above, pubescent on the nerves beneath.

II. *Leaves lanceolate.*

#### * *Leaves usually entire.*

5. *Celtis mississippiensis*, Bosc.   North America.   See p. 933.
    Leaves, entire in margin, rarely dentate at the apex, glabrous except for axil tufts at the base beneath.

#### ** *Leaves serrate.*

6. *Celtis australis*, Linnæus.  Southern Europe, North Africa, Caucasus.  See below.
    Leaves caudate-acuminate, covered beneath throughout with a soft pubescence.

7. *Celtis caucasica*, Willdenow.   Caucasus, Persia, Afghanistan, Baluchistan, N. India.   See p. 928.
    Leaves shortly acuminate, pubescent beneath only on the midrib and nerves.

(A. H.)


## CELTIS AUSTRALIS, Nettle Tree

*Celtis australis*, Linnæus, *Sp. Pl.* 1043 (1753); Loudon, *Arb. et Frut. Brit.* iii. 1414 (1838); Planchon in DC. *Prodr.* xvii. 169 (1873); Boissier, *Fl. Orientalis*, iv. 1156 (1879); Willkomm, *Forstliche Flora*, 545 (1887); Mathieu, *Flore Forestière*, 293 (1897).

A tree, usually attaining 50 to 70 feet in height, and 10 feet in girth, but in rare cases becoming as much as 20 feet in girth.   Bark thin, greyish, smooth, somewhat resembling that of the beech, but on old trunks sometimes covered with warty excrescences.   Young branchlets pubescent.   Leaves (Plate 267, Fig. 5), about 4 inches long by $1\frac{1}{4}$ inch broad, oval-lanceolate, unequal and cuneate at the base, contracted above into a very long caudate-acuminate apex, serrate except near the base ; upper surface dark green, scabrous, shortly-pubescent ; lower surface greyish, covered with a soft tomentum ; petiole greyish-tomentose, about $\frac{1}{2}$ inch long.   Fruit globose, up to $\frac{1}{2}$ inch in diameter, at first whitish, then red, and finally dark-brown ; with a scanty sweetish flesh.   Fruit pedicel, very slender, an inch or more in length.   The seedling[1] is similar to that of *C. occidentalis* ; but the cotyledons are wider, rhomboidal in shape, and with a shallower emargination ; and the primary leaves are longer, narrower, and more acuminate.   The seedling attains about 8 inches in height in the first year.

[1] Lubbock, *Seedlings*, ii. 495 (1892).

In winter, the twigs are slender, tomentose. Leaf-scars crescentic, 3-dotted, on prominent pulvini. Stipule-scars linear, one on each side of each leaf-scar. Terminal bud not formed and scar present at the apex of the twig, as in *C. occidentalis*. Lateral buds, appressed to the twig, compressed, ovoid, acute, with 2 to 3 pairs of loose, tomentose, ciliate, imbricated scales.

No varieties of *C. australis* have been described; but in the eastern part of its area there are forms connecting it with *C. caucasica*, which is a very closely allied species.

*Celtis australis* is widely distributed throughout the Mediterranean region, and extends into Asia Minor and probably farther eastward, in the Caucasus and north Persia. In France, it is common in Provence and Languedoc, where it is often cultivated as coppice, and is met with as a rare tree as far north as Poitiers and Lyons. It is usually an inhabitant of the plains and low hills, but occasionally ascends to 3000 feet in the mountains. In the north of France, it is scarcely hardy, at least when young. It occurs in Switzerland in the canton of Tessin. Farther east, its northern limit is the southern parts of Tyrol, Styria, and Hungary, whence it extends southward through the Balkan States to Greece and Crete. In Banat, Istria, and Dalmatia it often forms small woods, and ascends to 1600 feet. It is in all these regions extensively planted, and has become naturalised in many districts. It also occurs in Spain and Portugal, Sardinia, Italy, Sicily, Morocco, Algeria, and Tunis; and is said to grow in the Madeira Islands.

*Celtis australis* attains a great age, and trees of extraordinary size are recorded, there being one[1] in the public square at Aix in Provence, estimated to be 500 years old, which is 19 feet in girth, and higher than any of the adjoining houses. There are fine specimens also in the Botanic Gardens at Montpellier, which are 10 to 12 feet in girth. Willkomm saw very large trees in the Balearic Islands, and says that enormous trees are to be seen in Istria and Dalmatia, one at Pisino being supposed to be 1000 years old.

*Celtis australis* produces suckers from the roots, and when cut gives good coppice shoots. In the south of France, coppice woods of this species are very valuable, as the shoots[2] attain about 4 inches in diameter in ten or twelve years, and are worth one to two francs each. The wood resembles much that of the ash, of which it has all the good qualities, and is used in carriage-building, and for making numerous kinds of small articles, as tool-handles, hay-forks, trenails, tent-pegs, etc. Whip handles in France are almost universally made of this wood, and are called "perpignans," because the chief place of manufacture is at Perpignan. The foliage is given to cattle as fodder, the seed contains a sweet oil, and the bark and root yield a yellow colouring matter.

*Celtis australis* was introduced[3] in 1796, according to Loudon, who mentions a

---

[1] Mouillefert, *Traité des Arbres*, ii. 1207 (1898).      [2] Jolyet, *Les Forêts*, 226 (1901).

[3] In the *London Catalogue of Trees*, 18 (1730), three species of Celtis are mentioned as being in cultivation in England :—

    "I. Virginian Nettle Tree. Red fruit. In several gardens near London there are large trees.

    II. Nettle Tree, with black fruit. European sort, is most rare in England.

    III. *Celtis* with large yellow fruit. Has been grown in Devonshire many years, where there are some large trees, which produce ripe fruit, from which many plants have been raised ; but we know not where it came from originally."

tree at Mitcham, which was 6 feet 8 inches in girth, and had a spread of 60 feet; and a tree at Kew, 40 feet in height, which no longer exists, there not being at present a single specimen there of this species. It appears to be very rare in cultivation at the present time, the only trees which we have seen being one at Liphook, and another at Hursley Park. The latter is a small unhealthy-looking tree about 20 feet high, though of considerable age. There is also a small tree at Tortworth.

According to Bureau,[1] this species supports at Paris the severest winters without injury; but according to Pardé,[2] it bears with difficulty severe frosts in the north of France. The seedlings which Elwes raised at Colesborne were killed by 20° of frost, and though the tree may succeed in the warmest and driest parts of the south-east of England, it seems hardly worth planting elsewhere.      (A. H.)


## CELTIS CAUCASICA

*Celtis caucasica*, Willdenow, *Sp. Pl.* iv. 994 (1805); Loudon, *Arb. et Frut. Brit.* iii. 1415 (1838); Boissier, *Flora Orientalis*, iv. 1156 (1879).
*Celtis australis*, Brandis, *Forest Flora of N.W. India*, 428 (1874), and *Indian Trees*, 595 (1906) (not Linnæus); Hooker, *Flora Brit. India*, v. 482 (1888); Gamble, *Indian Timbers*, 629 (1902).

A tree of moderate size, very similar to *C. australis*, of which it is possibly only a geographical form. It differs in the following characters :—Leaves ovate-lanceolate, broader in proportion to their length, and more rhomboidal, with a shorter and non-caudate acuminate apex; upper surface glabrescent, scarcely scabrous; lower surface with slight pubescence, confined to the nerves and midrib. Drupes yellow.[3]

This species, which is connected with the European species by var. *cuspidata*,[4] with long-acuminate leaves, is widely spread through the Caucasus, Persia, Afghanistan, Baluchistan, and northern India. In the Caucasus, it is associated with *C. australis*; but farther east the latter species is scarcely met with. In Afghanistan, according to Aitchison,[5] it is usually a planted tree near shrines and in graveyards; but it is quite wild along the Darban and Shendtoi rivers; and in Baluchistan, its leaves, according to Lace,[6] are often used, as they are in India, for feeding sheep and goats, the trees being pollarded for this purpose. It occurs in India in the north-west Himalaya, as far east as Nepaul ascending to 8000 feet, where it is a common tree, wild in the forests, and around villages. According to Webber,[7] in Gorakhpur, it reverses the season of casting its leaves, which wither and fall off in the hot weather, and it flowers in the early months of the cold season. The wood is tough and strong, and is used for oars, tool-handles, sticks, and other purposes requiring toughness and elasticity.

This species, though mentioned by Loudon, was not in cultivation in England in his day; and Schneider[8] doubts if it has yet been introduced on the Continent.

---

[1] *Nouv. Arch. Mus. Hist. Nat.* vi. 181 (1894).
[3] Brandis mentions a variety with purplish-black fruit.
[5] *Journ. Linn. Soc.* (*Bot.*) xviii. 93 (1880).
[7] *Forests of Upper India*, 232 (1902).

[2] *Arb. Nat. Des Barres*, 242 (1906).
[4] Planchon, in DC. *Prod.* xvii., 170 (1873).
[6] *Ibid.* xxviii. 305 (1891).
[8] *Laubholzkunde*, 231 (1904).

There is, however, a tree of this species in the Kew Collection, which is marked "Aitchison" on the label, and was probably raised from seeds sent by Aitchison from Afghanistan, about the year 1881. (A. H.)

## CELTIS GLABRATA

*Celtis glabrata*, Steven, *ex* Planchon, in *Ann. Sc. Nat.* sér. 3, x. 285 (1848).
*Celtis Tournefortii*, Lamarck, var. *glabrata*, Boissier, *Fl. Orient.* iv. 1157 (1879).

A shrub or small tree. Young branchlets with a minute scattered pubescence. Leaves (Plate 267, Fig. 10), about 2 inches long, 1¼ inch broad, quite glabrous, ovate, unequal and rounded or broadly cuneate at the base, acute or very shortly acuminate at the apex, coarsely serrate except near the base ; upper surface bluish green, roughened with minute papillæ; lower surface lighter green in colour ; punctate with numerous translucent minute dots, when viewed with a lens ; petiole glabrous, ⅛ inch. Fruit pedicels about an inch. Drupes globose, reddish brown.

This species occurs in Asia Minor, in Lycia and Cilicia, and in the Caucasus. Schneider doubts if it has been introduced into cultivation ; but there is at Kew a small tree, undoubtedly of this species. *C. Tournefortii*, Lamarck, a closely allied species, occurring in Sicily, Greece, and Asia Minor, is mentioned by Loudon, as having been introduced into England in 1739, and cultivated in 1838 in the London Horticultural Society's Garden ; but I have seen no specimens in this country.

(A. H.)

## CELTIS DAVIDIANA

*Celtis Davidiana*, Carrière, *Rev. Hort.* 1868, p. 300.
*Celtis Bungeana*, Blume (in part) ; Hemsley, *Journ. Linn. Soc. (Bot.)* xxvi. 449 (1894).
*Celtis sinensis*, Persoon (in part) ; Maximowicz, *Mél. Biol.* ix. 27 (1872) ; Bretschneider, *Botanicon Sinicum*, i. 117 (1882).

A small tree. Young branchlets slightly pubescent. Leaves (Plate 267, Fig. 11), about 2½ inches long, 1¼ inch broad, ovate or ovate-lanceolate, base rounded, contracted above into a short acuminate apex, rarely entire, usually slightly toothed in the upper third ; glabrous and shining on both surfaces, dark green above, light green beneath, punctate when viewed with a lens ; petiole ⅓ inch, pubescent. Fruit-pedicels, slender, ¾ inch long. Drupes small, ovoid, black in colour.

This species occurs in north China, in the hills around Peking, and in the mountains of the province of Shingking ; and was found by me growing as a small tree, about 20 feet in height, in the mountains of Hupeh. It has been confused with two other Chinese species, *C. Bungeana*[1] and *C. sinensis*,[2] which do not appear to be in cultivation in Europe. It is readily distinguishable from all the other cultivated species by the very shining glabrous leaves; and is a very distinct and handsome tree.

[1] Blume, *Mus. Bot. Ludg. Bat.* ii. 71 (1852).     [2] Persoon, *Syn. Pl.* i. 292 (1805).

It was introduced into Kew Gardens, where there is a small tree about 15 feet high, by seeds sent from Peking by Bretschneider in 1882. It had, however, been previously introduced into France by Père David, who sent seeds to Carrière in 1868, from which a tree was raised in the Jardin des Plantes. This tree, according to Franchet[1] had become with age identical in character with *C. Bungeana*; but this is incorrect. It fruited for the first time[2] at Paris in 1894. Schneider[3] mentions trees of this species in the Botanic Gardens at Strassburg and Darmstadt.

(A. H.)

## CELTIS OCCIDENTALIS, Hackberry

*Celtis occidentalis*, Linnæus, *Sp. Pl.* 1044 (1753); Michaux, *Hist. Arb. Am.* iii. 225, t. 8 (1813); Loudon, *Arb. et Frut. Brit.* iii. 1417 (1838); Sargent, *Silva N. Amer.* vii. 67 (in part), t. 317 (1895), and *Trees N. Amer.* 299 (1905) (in part).

A tree, attaining in America, 100 feet in height and 9 feet in girth. Bark grey, broken on the surface into appressed scales, and often roughened on old trees with thick discontinuous corky ridges. Young branchlets glabrous or pubescent. Leaves (Plate 267, Fig. 4), uniform in size, about $2\frac{1}{2}$ inches long and $1\frac{1}{4}$ inch wide, ovate, unequal and rounded or shortly cuneate at the base, with a long caudate-acuminate usually non-serrated apex; serrate in the upper half or two-thirds; upper surface smooth to the touch; lower surface pubescent on the nerves; petiole $\frac{1}{4}$ inch or more, glabrous or pubescent. Fruiting pedicels short, about $\frac{3}{8}$ inch. Drupe, purplish-black or orange when ripe, globose or ovoid, about $\frac{2}{5}$ inch in diameter.

In winter the twigs show the following characters :—Branchlets slender, zigzag, reddish-brown, shining, glabrous. Leaf-scars oblique on prominent pulvini, three-dotted. Stipule scars minute, linear, one on each side of each leaf-scar. Terminal bud not formed, the end of the branchlet falling off in summer, and leaving a minute orbicular scar at the apex of the twig. Buds[4] all axillary, uniform in size, about $\frac{3}{16}$ inch long, alternate, distichous, appressed to the twig, ovoid, acute, compressed, covered by three pairs of pubescent, ciliate, imbricated scales.

*Seedling.*[5]—Primary root long, tapering, flexuose, with numerous lateral fibres. Caulicle erect, pubescent, about $1\frac{1}{2}$ inch long. Cotyledons oblong, cuneate and three-nerved at the base, emarginate at the apex, green above, pale beneath, about $\frac{3}{4}$ inch long. Stem hispid. First pair of leaves opposite, ovate, acuminate, serrate, three-nerved, covered in the young stage with clear dot-like glands. Succeeding leaves similar, but alternate.

Scarcely any varieties are known, unless *C. crassifolia* be considered a geographical form of this species. *C. pumila*, Pursh,[6] a low shrub, of xerophytic

---

[1] *Plantæ Davidianæ*, i. 269 (1884).     [2] *Rev. Hort.* 1894, p. 97.     [3] *Laubholzkunde*, 228 (1904).

[4] Sometimes in this species, the axil of the leaf produces three buds side by side. The middle bud sends out a shoot in the following year, whilst the lateral ones are left as a reserve. If the shoot happens to die in the year after, one of the two accessory buds develops. Cf. Kerner, *Nat. Hist. Plants*, Eng. transl. ii. 32 (1898).

[5] Cf. Lubbock, *Seedlings*, ii. 493, fig. 646 (1892).

[6] *Fl. Amer. Sept.* i. 200 (1814); Hill, *Bull. Torr. Bot. Club*, xxvii. 496 (1900).

habit, usually growing on sand dunes, rocky places, and dry hills and mountains, and widely distributed in the United States from Delaware and Pennsylvania to Utah and Colorado, is considered by some to be a variety of this species, but is probably distinct.

This species is widely distributed, occurring in the north from Quebec to Manitoba, and extending southward to Louisiana, Missouri, Kansas, and North Carolina. Its exact distribution is not clearly known, as it has been confused with other species; but it appears to be commoner in the north and east, while *C. crassifolia* is most prevalent in the Mississippi valley and west of the Alleghanies. It is replaced by *C. mississippiensis* in the extreme south, and by *C. reticulata* in the dry regions of the far west.

In New England it is a low round-headed tree, and is well depicted by Sargent in *Garden and Forest*, iii. 39, fig. 43 (1890), which represents a tree growing close to the seashore in Massachusetts. It is not common east of the Hudson,[1] but on the banks of this river grows with a slender trunk and long graceful pendulous branches. Amongst North American species none, perhaps, retains its foliage green and fresh so late in the season.                                                      (A. H.)

The American Nettle Tree or Hackberry was introduced into England in 1656 by Tradescant, and the first description of the species, made from a tree cultivated in England, was published in Ray's *Historia Plantarum*, ii. 1917. It seems to be the only species of Celtis which bears our climate well enough to be worth planting, but is so rare in cultivation that very few people know it, and it is rarely found in nurseries.

I have found it easy to raise from seed, and though the seedlings grow slowly and are somewhat susceptible to frost when young, it will, as it grows older, endure a greater degree of cold in winter than *C. australis*. The largest tree mentioned by Loudon was one at Syon, which, in 1838, was 54 feet high and 7 feet in girth. This cannot be the same as one which now grows there, which, according to the garden catalogue, was 50 feet by 3 feet 3 inches in 1849, and when I measured it in 1903, was 60 feet by 4 feet 4 inches. There is a fine tree in Kew Gardens, which, owing to its being crowded on one side by an evergreen oak, leans considerably, and is about 45 to 50 feet by 6 feet 4 inches. This is probably one of the original trees of the Kew Arboretum, though not mentioned by Loudon, unless, as is possible, he mistook it for *C. australis*.

The finest and best-shaped tree, however, that we know of, is one at West Dean Park (Plate 251), which is probably the one mentioned by Loudon as then fourteen years planted and 19 feet high. Now it is 50 feet by 5½ feet, and when I saw it in July 1906 was bearing fruit abundantly. This tree evidently lives to a considerable age, as I saw one in the Botanic Gardens at Padua which was planted in 1760, and measures no less than 32 metres high by 2 metres in girth, with a fine clean bole.

Henry saw a good specimen in the Botanic Garden at Copenhagen.

                                                                                (H. J. E.)

---

[1] *Garden and Forest*, i. 465 (1888).

## CELTIS CRASSIFOLIA, Hackberry

*Celtis crassifolia*, Lamarck, *Encycl.* iv. 138 (1797); Michaux, *Hist. Arb. Amer.* iii. 228, t. 9 (1813); Loudon, *Arb. et Frut. Brit.* iii. 1418 (1838); Britton, *Man. Fl. North. States and Canada*, 339 (1901).
*Celtis cordata*, Persoon, *Sp. Pl.* i. 292 (1805).
*Celtis Audibertiana*, Spach, *Ann. Sc. Nat.* sér. 2, xvi. 41 (1841).
*Celtis occidentalis*, Sargent, *Silva N. Amer.* vii. 67 (1895), and *Trees N. Amer.* 299 (1905). (In part.)

A tree, attaining rarely as much as 130 feet in height, and 10 feet in girth. Bark as in *C. occidentalis*. Young branchlets pubescent. Leaves usually about $2\frac{1}{2}$ inches long by $1\frac{1}{2}$ inch broad, but on isolated branches, commonly found in the inner part of the crown, often 6 inches long and 3 inches broad; ovate, unequal, and shortly cuneate at the base, shortly acuminate at the apex, serrate, as a rule, only in the upper half; upper surface scabrous to the touch; lower surface pubescent on the nerves; petiole, $\frac{1}{4}$ inch or more, pubescent. Fruiting pedicels long, $\frac{3}{4}$ inch or more. Drupe purple, red or black when ripe, globose or ovoid, about $\frac{2}{5}$ inch in diameter.

This species, which is not distinguished by Sargent, even as a variety, from *C. occidentalis*, is remarkably distinct in foliage, and appears to be a more upright and faster-growing tree in cultivation than that species. Michaux states that it is one of the finest species of the genus, remarkable for its great height and straight trunk, and that it is common in the states west of the Alleghanies, especially in Ohio and Kentucky, where, however, its timber was little esteemed on account of its weakness and liability to speedy decay on exposure to the weather. Its distribution has been confused with that of *C. occidentalis*; but, according to Britton, it occurs from New York to South Carolina, Ohio, Illinois, Missouri, and Tennessee.

(A. H.)

Ridgway[1] speaks of this species of Hackberry under the name of *C. occidentalis*, as "a very tall and beautiful tree in rich bottoms, growing frequently 120 to 130 feet high and 3 feet in diameter, with a tall straight trunk of 60 to 70, or even 80 feet to the first limb. When growing to its full perfection in a dense forest, there is an individuality about the aspect of this tree which it is difficult to describe, owing to the extreme slenderness and great length of the trunk, which not unfrequently comprises three-fourths of the total height of the tree; and the smooth grey bark conspicuously clouded on the north side with blackish moss or lichen for its entire length. This striking appearance is sometimes increased by vines of the Virginia creeper ascending to the topmost branches, which are wreathed and matted with its foliage. One tree was seen whose silvery shaft gleamed among the surrounding tree tops, in a wood where the summit level was considerably more than 100 feet aloft, and though only 10 feet in circumference must have been upwards of 90 feet

[1] *Proc. U.S. Nat. Museum*, 1882, p. 72.

to the first limb, which grew not more than 25 feet from the extreme summit of the tree."

Though I was not fortunate enough to find any such trees standing, when I visited the remains of this forest in 1904, yet I saw enough to make me wish that an area of this unique forest could be preserved to show what the virgin forests of the Wabash valley were once like; for there is no other part of the temperate world where so many species of hardwood trees grow to such a size as they formerly did here.

This species appears to have been introduced about the beginning of the nineteenth century; several trees, 10 to 15 feet in height, being mentioned by Loudon.

At Kew it appears to be straighter and more vigorous in growth than *C. occidentalis*; and all the specimens have a few branches mainly in the upper and inner parts of the tree, which bear very large leaves. One of the trees, growing on the walk behind the Aroid House, is 38 feet high by 3½ feet in girth.

(H. J. E.)

## CELTIS MISSISSIPPIENSIS

*Celtis mississippiensis*, Bosc, *Dict. Agric.* x. 41 (1810); Sargent, *Silva N. Amer.* vii. 71, t. 318 (1895), and *Trees N. Amer.* 300 (1905).

*Celtis lævigata*, Willdenow, *Berlin Baumz.* 81 (1811); Loudon, *Arb. et Frut. Brit.* iii. 1420 (1838).

*Celtis occidentalis*, Linnæus, var. *integrifolia*, Nuttall, *Gen.* i. 202 (1818).

*Celtis occidentalis*, Sargent, *Forest Trees N. Amer.*, *10th Census U.S.* ix. 125 (1884) (in part); and *Garden and Forest*, iii. 39 (in part), ff. 9, 10, 11 (1890).

A tree, attaining in America, 80 feet high and 9 feet in girth. Bark bluish-green, and covered with prominent excrescences. Young branchlets glabrous. Leaves (Plate 267, Fig. 9), up to 3 inches long and 1¼ inch wide, ovate-lanceolate or lanceolate, unequal and rounded or broadly cuneate at the base, long-acuminate at the apex; margin usually entire, occasionally irregularly serrate towards the apex; light green and glabrous, except for slight axil tufts at the base beneath; petiole, about ½ inch, glabrous. Fruiting-pedicels, about ¾ inch. Drupes, ovoid, ⅛ to ¼ inch, bright orange red, with thin dry flesh and a smooth light brown stone.

This species is distributed from southern Indiana and Illinois, through Kentucky, Tennessee, and Alabama to Florida, and through Missouri, Arkansas, and Texas to Nuevo Leon. It is also a native of the Bermudas. It is very abundant and of its largest size in the basin of the lower Ohio River, a tree measured by Schneck in Richland County, Illinois, being 95 feet high and 5½ feet in girth. Here it is often associated with *C. crassifolia*, from which it may be distinguished[1] by its usually smaller size, shorter trunk, entire leaves, and bright orange-red fruit. It is the most common species in Kentucky and Tennessee; but is rare in the Gulf States. Though apparently found in Texas and Nuevo Leon, it is replaced to the

---

[1] Elwes noticed that the wrinkled bark of this species easily distinguished it in the forest from *C. crassifolia*.

westward by *Celtis reticulata*, Torrey,[1] which occurs in Texas, New Mexico, Arizona, Utah, Nevada, and Lower California, and has not yet apparently been introduced.

*C. mississippiensis* was probably introduced, by the elder Michaux, into France, where the species was first described in 1810 from a tree cultivated at Paris. It was not introduced into England in Loudon's time. The only cultivated specimen known to us is a small tree in Kew Gardens.                    (A. H.)

[1] *Ann. Lyc. N. York*, ii. 247 (1828). *C. occidentalis*, var. *reticulata*, Sargent, *Forest Trees N. Amer.*, 10*th Census, U.S.* ix. 126 (1884); *Garden and Forest*, iii. 40 f. 12 (1890). *C. mississippiensis*, var. *reticulata*, Sargent, *Silva N. Amer.* vii. 72, t. 319 (1895), and *Trees N. Amer.* 301 (1905).

# ALNUS

*Alnus*, Linnæus, *Gen. Pl.* 285 (1737); Bentham et Hooker, *Gen. Pl.* iii. 404 (1880); Winkler, in
   Engler, *Pflanzenreich*, iv. 61, *Betulaceæ*, 101 (1904).
*Betula*, Linnæus, *Gen. Pl.* 485 (1764) (in part).
*Alnaster* and *Clethropsis*, Spach, *Ann. Sc. Nat.* sér. 2, xv. 200, 201 (1841).
*Semidopsis*, Zumaglini, *Fl. Pedem.* i. 249 (1849).
*Alnobetula*, Schur, *Verhand. Siebenb. Ver. Naturw.* iv. 68 (1858).

DECIDUOUS trees or shrubs, belonging to the order Betulaceæ. Leaves alternate,
simple, stalked, usually serrate or dentate, rarely entire, penninerved. Stipules[1]
enclosing the leaf in the bud, caducous or deciduous. Flowers opening either
in early spring before or with the unfolding of the leaves, or in two species in
autumn, monœcious, unisexual, without petals, in few-flowered cymes in the
axils of short-stalked peltate scales of pedunculate catkins. Staminate catkins,
at first naked and erect, afterwards pendulous, in the axils of the last leaves or of
leafy bracts; scales three-flowered; bracteoles, three to five, adnate to the base
of the scale; calyx, four-partite; stamens four, filaments short, undivided; anthers
dorsi-fixed, not pilose at the apex. Pistillate catkins erect, solitary or racemose, in
the axils of the leaves and produced in autumn, or terminal on a short leafy branch
and produced in spring; scales two-flowered; bracteoles, two to four, adnate to the
base of the scales; calyx absent; ovary two-celled; styles two, stigmatose at the
apex; ovule, solitary in each cell, suspended. Fruit, a strobile or cone, formed by
the scales of the pistillate flower becoming, when ripe, thick, woody, obovate, three-
to five-lobed or truncate at the thickened apex. Cones persistent on the branch
after the opening of the closely imbricated scales and the escape of the nutlets.
Nutlet, compressed, minute, bearing at the apex the remains of the styles, marked at
the base by a scar, with or without lateral wings. Seed, solitary by abortion, filling
up the cavity of the nutlet.

About twenty-five species are distinguished, inhabiting Europe, Algeria, extra-
tropical Asia, North America, Central America, and the Andes of South America
from Colombia to Peru.

The following key includes all the species in cultivation, with the exception of
*Alnus serrulata*, Willdenow, a North American shrub.

I. *Buds sessile, with several (two to six) outer scales, which are unequal in length.*
   1. *Alnus firma*, Siebold et Zuccarini. Japan. See p. 952.

---

[1] Cf. Lubbock, "On Stipules of the Alder," in *Journ. Linn. Soc. (Bot.)*, xxx. 527 (1895).

Leaves ovate-lanceolate, green beneath, with fifteen or more pairs of lateral nerves.

2. *Alnus viridis*, De Candolle.[1]    Mountains of Central Europe, plains of Northern Russia, Siberia, Labrador, and Greenland.

Leaves broadly oval, green beneath, with eight to ten pairs of lateral nerves.

This species is a shrub, rarely attaining more than 10 feet in height, and will not be further dealt with by us.

II. *Buds stalked, with two outer scales, almost equal in length.*

### A. *Leaves white or grey beneath.*

3. *Alnus incana*, Mœnch.[2]    Europe, Caucasus, North America.    See p. 945.
Branchlets pubescent, buds rounded at the apex.    Leaves grey beneath, with the lateral nerves running to the tips of serrated lobules ; margin not revolute.

4. *Alnus rubra*, Bongard.    Western North America.    See p. 956.
Branchlets glabrous, buds beaked at the apex.    Leaves whitish beneath, with nerves as in *A. incana* ; margin revolute.

### B. *Leaves green beneath.*

\* *Leaves simply serrate, not lobulate, rounded or acute at the apex.*

5. *Alnus elliptica*, Requien.    A hybrid between *A. cordata* and *A. glutinosa*, occurring wild in Corsica.    See full description under *A. cordata*, p. 949.
Branchlets glabrous and covered with wax glands.    Leaves elliptic, rounded at both base and apex, with prominent brown axil-tufts beneath.

6. *Alnus rhombifolia*, Nuttall.    Western North America.    See p. 958.
Branchlets pubescent.    Leaves ovate or oval, rounded at the base, acute or rounded at the apex, pubescent throughout beneath.

\*\* *Leaves simply serrate, not lobulate, conspicuously acuminate at the apex.*

7. *Alnus cordata*, Desfontaines.    Corsica, Southern Italy.    See p. 949.
Branchlets glabrous, covered with wax glands.    Leaves broadly ovate, cordate at the base, cuspidate-acuminate at the apex, with conspicuous axil-tufts beneath.

8. *Alnus japonica*, Siebold et Zuccarini.    Japan, Manchuria, Korea, Formosa. See p. 953.
Branchlets glabrous.    Leaves lanceolate, cuneate at the base, long-acuminate at the apex, with minute axil-tufts beneath.

9. *Alnus maritima*, Nuttall.    Delaware, Maryland, Indian Territory.    See p. 955.
Branchlets glabrous.    Leaves ovate or obovate, cuneate at the base, shortly acuminate at the apex, glabrous beneath.

10. *Alnus nitida*, Endlicher.    North-west Himalaya.    See p. 954.
Branchlets densely and minutely pubescent.    Leaves ovate-elliptical, cuneate

---

[1] *Fl. Franç.* iii. 304 (1805).    *Alnus Alnobetula*, Hartig, *Naturges. Forst. Kulturpfl.* 372 (1851).    *Betula Alnobetula*, Ehrhart, *Beit.* ii. 72 (1788).

[2] *Alnus tinctoria*, Sargent, a closely allied species, occurring in Japan and Manchuria, is distinguished by its large, broadly ovate leaves, acuminate at the apex.    *A. incana* has small, narrowly ovate leaves, acute at the apex.    Cf. p. 946.

or rounded at the base, short- or long-acuminate, with slight axil-tufts beneath.

*** *Leaves with serrate lobules.*

11. **Alnus glutinosa**, Gærtner.   Europe, Siberia, Western Asia, North Africa. See below.

Branchlets usually glabrous.   Leaves obovate, cuneate at the base ; obtuse, truncate or rounded at the apex ; with prominent axil-tufts beneath.

12. *Alnus glutinosa*, Gærtner, var. *barbata*, Ledebour (*Alnus barbata*, C. A. Meyer). Caucasus.   See p. 938.

Branchlets pubescent.   Leaves elliptical, rounded at the base and apex, covered on both surfaces with pubescence, densest on the midrib and nerves beneath.

13. *Alnus tenuifolia*, Nuttall.   Western North America.   See p. 957.

Branchlets glabrous.   Leaves ovate, broad and rounded at the base, acute or shortly acuminate at the apex, pubescent on the midrib beneath with inconspicuous axil-tufts.

14. *Alnus subcordata*, C. A. Meyer.   Caucasus, North Persia.   See p. 951.

Branchlets pubescent.   Leaves ovate-oblong, unequal and rounded or sub-cordate at the base, cuspidate-acuminate at the apex, pubescent on the midrib and nerves beneath.

## ALNUS GLUTINOSA, Common Alder

*Alnus glutinosa*, Gaertner, *De Fruct.* ii. 54 (1791) ; Loudon, *Arb. et Frut. Brit.* iii. 1678 (1838) ;
Willkomm, *Forstliche Flora*, 339 (1887) ; Mathieu, *Flore Forestière*, 421 (1897) ; Winkler,
*Betulaceæ*, 115 (1904).
*Alnus nigra*, Gilibert, *Exerc.* ii. 401 (1792).
*Alnus communis*, Desfontaines, *Tabl. Hort. Paris*, 213 (1804).
*Alnus vulgaris*, Persoon, *Syn.* ii. 550 (1807).
*Betula Alnus glutinosa*, Linnæus, *Sp. Pl.* 983 (1753).
*Betula Alnus*, Scopoli, *Fl. Carn.* ii. 233 (1772).
*Betula glutinosa*, Lamarck, *Dict.* i. 454 (1783).
*Betula palustris*, Salisbury, *Prod.* 395 (1796).

A tree, occasionally attaining 100 feet in height and 12 feet or more in girth. Bark of young trees smooth and greenish ; after twenty years old becoming brownish-black and divided on the surface into broad flattened plates.   Young branchlets, three-angled at the tip, usually glabrous, occasionally pubescent, covered with glands, which secrete a waxy resin, often seen on the dried twigs as a bluish bloom.   Leaves (Plate 268, Fig. 6) averaging $3\frac{1}{2}$ inches long and 3 inches broad, variable in shape, but nearly always broadest above the middle, obovate, sub-orbicular or elliptical ; cuneate at the base ; obtuse, truncate, or retuse at the apex ; margin entire in the basal third, elsewhere lobulate, each lobule serrate or dentate ; upper surface dark green, shining, glabrous ; lower surface light-green, pubescent along the midrib and nerves, with conspicuous tufts of rusty-brown hairs in the axils; petioles, glabrous or pubescent, $1\frac{1}{2}$ to 1 inch long; stipules conspicuous, deciduous, ovate to lanceolate, obtuse, fringed with glandular hairs. The leaves turn blackish in autumn.

Flowers appearing very early, before the leaves, in February or March. Catkins, three to six in a raceme, at the tip of a branchlet. Staminate catkins, 2 to 4 inches long, at first erect and rigid, afterwards lax and pendent; stamens,[1] four, yellow, opposite the segments of the four-lobed calyx. Pistillate catkins, always erect, at first about $\frac{1}{4}$ inch long, smooth, with reddish-brown stigmas; afterwards $\frac{1}{2}$ inch long, ovoid, cone-like, the scales ending in purple shield-like expansions, each with a central brown point. Cones at first green, ultimately black, persistent on the tree after the escape of the nutlets. Nutlets obovate, blunt-angled, wingless or with a very narrow coriaceous wing.

The nutlets[2] are gradually shaken out of the cones by the wind during autumn and winter. Their walls are provided with small air-tight cavities, which enable them to float in water, and secrete an oil, which protects them from being wetted. Usually falling into streams and ditches, they float undamaged and unchanged during winter, and germinate in the water in early spring. The young seedlings, drifting to the bank, establish themselves where they happen to be stranded in a suitable place.

In winter, the twigs are glabrous and usually covered with a thin waxy secretion. Leaf-scars pentagonal or rhomboid, parallel to the twig on a projecting cushion, five-dotted, the lowermost three dots coalesced together. Stipule-scars linear, one on each side of a leaf-scar. Terminal bud present, similar to the lateral buds; all conspicuously stalked, ovoid, obtuse, with two external scales, viscid-glandular, and often covered with a purplish bloom. Pith triangular in section.

The common alder coppices freely from the stool; but rarely if ever produces root-suckers.

### VARIETIES

The common alder, distributed over a wide area, shows considerable variation in the wild state, and several varieties have been described.

1. Var. *barbata*, Ledebour, *Fl. Rossica*, iii. 657 (1851); Winkler, *Betulaceæ*, 118 (1904). *Alnus barbata*, C. A. Meyer, *Verz. Pfl. Kauk.* 43 (1831).

This variety is remarkably distinct in the foliage, but has the flowers and fruit of typical *A. glutinosa*. Young branchlets pubescent. Leaves, about 3 inches long and 2 inches wide, rounded at the base, rounded or occasionally acute at the apex, margin with serrate lobules, ciliate; nerves eight or nine pairs, running parallel and curved to the margin; upper surface dark green, shining, minutely pubescent; lower surface pubescent, the pubescence densest along the midrib and nerves; petiole $\frac{3}{4}$ inch, pubescent. Buds pubescent.

---

[1] Kerner, in *Nat. Hist. Plants*, Eng. trans. ii. pp. 119, 133, 135, gives an elaborate account of the way in which pollination is effected by the wind, and of the devices for the protection of the pollen in rainy weather.

[2] Cf. Miall's account of the fruit of the alder in *Round The Year*, p. 279.

This **variety** occurs in the Caucasus, and is very similar to var. *denticulata*,[1] occurring in the same region and in north Persia, which is less pubescent. Var. *barbata* is in cultivation at Kew.

2. Var. *quercifolia*, Willdenow, *Berlin Baumz.* 44 (1796). Oak-leaved alder. Leaves obovate, lobed like the common oak. This variety has been found wild in Sweden.

3. Var. *sorbifolia*, Dippel, *Laubholzkunde*, ii. 161 (1892). Service-leaved alder. Leaves oval, lobed like those of *Pyrus intermedia*. This variety has been found wild in Finland.

4. Var. *laciniata*, Willdenow, *loc. cit.* Cut-leaved alder. Leaves (Plate 268, Fig. 7), divided half-way to the midrib into three to six pairs of non-serrated triangular segments ; petiole slender, about an inch long.

The cut-leaved alder, according to Duhamel, occurs wild in the north of France, particularly in Normandy, and in the woods of Montmorency near Paris. Thouin states, according to Loudon, that it was first found by Trochereau de la Berlière, and planted by him in his garden near St. Germain, where the stool remained in 1838, from which all the nurseries of Paris were supplied with plants. The largest trees we have seen of this variety are described on p. 942.

5. Var. *imperialis*, Petzold and Kirchner, *Arb. Musc.* 599 (1864). *Alnus imperialis*, Desfossé-Thuillier, *Illust. Hort.* vi. 97, fig. (1859).

Leaves (Plate 268, Fig. 8) oval, divided more than half-way to the midrib, into six or seven pairs of long narrow lanceolate non-serrated curved segments. This variety, so far as we know, does not attain to as large a size as the ordinary form of the cut-leaved alder. A specimen at Ponfield, Hertford, is 25 feet high by 1 foot 8 inches in girth.

6. Var. *incisa*, Willdenow, *Sp. Pl.* iv. 335 (1805) (var. *oxyacanthæfolia*, Loddiges, *Catalogue*, 1836). Thorn-leaved alder. Leaves (Plate 268, Fig. 10) small, deeply incised, resembling those of the common hawthorn. A fine specimen, 44 feet high by 2 feet 8 inches in girth, is growing in the arboretum at Barton, near Bury St. Edmunds.

7. Var. *rubrinervia*, Dippel, *loc. cit.* A tree, pyramidal in habit, with large and shining leaves, furnished with red petioles and nerves, vigorous in growth and handsome in appearance. In cultivation at Aldenham.

8. Var. *pyramidalis*, Dippel, *loc. cit.* Branches erect, leaves as in the type.

9. Var. *aurea*, Verschaffelt, *ex* Dippel, *loc. cit.* Lemaire, *Illust. Hortic.* 1866, t. 490. Leaves yellow. Found as a seedling in Vervaene's nursery at Ledeberg-les-Gand. In cultivation at Aldenham.

10. Var. *maculata*, Winkler, *loc. cit.* Leaves variegated with yellow. There is a small specimen at Aldenham, which is slow in growth.

Hybrids[2] between *Alnus glutinosa* and *Alnus incana* are common in the wild state, where the two species are growing together, and have been observed in

---

[1] Ledebour, *loc. cit. Alnus denticulata*, C. A. Meyer, *loc. cit.*

[2] *A. glutinosa* x *incana* ; *A. spuria*, Callier. Schneider, *Laubholzkunde*, 130 (1904), distinguishes three forms of this hybrid.

Sweden, Norway, Russia, Denmark, Bosnia, and Montenegro. These hybrids are intermediate in the characters of the leaves and fruit; and so far as we know, are not in cultivation in England.

A hybrid[1] between *A. glutinosa* and *A. serrulata* has also arisen in cultivation in Silesia and Brandenburg.

## DISTRIBUTION

The common alder is distributed through nearly the whole of Europe, Siberia, Western Asia, and North Africa. In Europe, its northerly limit extends from lat. 63° 52′ in southern Norway at Anderoen, to lat. 63° 20′ on the west side of the Gulf of Bothnia in Sweden, reaching Uleaborg in Finland in lat. 65°, where, however, it only exists as a shrub, and is continued through the interior of Finland and Russia along the parallel of 62°. In Siberia, its distribution is not accurately known; but it occurs in the Ural and Altai mountains, and in the district around Lake Baikal. Its southern limit, commencing in the province of Talysch in the Caucasus, between 39° and 40° lat., extends through Asia Minor and Greece to Sicily, where it reaches at Catania, lat. 37° 25′, its most southerly point in Europe. It occurs in Spain and Portugal, as far south as the Sierra Morena, about lat. 38°. It is also distributed through the mountains of Algeria and Morocco. A variety,[2] recorded for Japan, is probably a distinct species.

According to Sir Herbert Maxwell[3] the Anglo-Saxon name for alder was *alr*, in Norse *olr* (now, according to Schübeler, *aar*, *older*, and *or*); and the Gaelic name *fearn*, the names surviving in place names such as Allerton, Allerbeck, Ellerslie, Balfern, Farnie, Glenfarne.

It is generally distributed throughout the British Isles, growing usually on river banks, along the sides of lakes, and in wet or marshy places; and ascends in the Grampians to 1600 feet.

It is common throughout France, in similar situations, and ascends to 5700 feet in the Pyrenees. Although most usual on siliceous soils, it grows on chalk in Champagne; and in Germany, has been shown to be indifferent to the mineral constituents of the soil, provided a sufficient quantity of moisture is present. It is met with as pure woods, on peat-bogs and marshy places, in north Germany, in the Baltic provinces and Lithuania in Russia, and also in Hungary; but more usually is mixed with birch and aspen, and more rarely grows in company with other hardwoods. It ascends in the mountains of Norway to 1100 feet, in the Central Alps in Switzerland to 3200 feet, and in the Carpathians to 3800 feet. The alder is susceptible to late frosts and is injured by cold dry winds, and on this account thrives best in the colder parts of Europe on slopes with a westerly aspect.

It is naturalised in various localities in the eastern United States, particularly in southern New York and in New Jersey. It succeeds well in North America, when

---

[1] *A. glutinosa* × *serrulata*; *A. silesiaca*, Fiek. Cf. Schneider, *loc. cit.*
[2] Var. *japonica*, Matsumura, *Journ. Coll. Sci. Tokyo*, xvi. 2, p. 9 (1902).
[3] Green's *Encyclopedia of Agriculture*, i. 62.

planted in wet situations; but is not, as a rule, a long-lived tree, and never grows to be so large as it does in England.[1]                              (A. H.)


## PROPAGATION AND CULTURE

Though old writers tell us that the alder was often planted by inserting long cuttings, or by burying pieces of the root in the soil; and though layering is the mode usually adopted by nurserymen for propagating the varieties; yet, as a general rule, it is best raised from seed. The cones ripen in autumn and are freely produced almost every year so far as I have observed. As soon as they begin to open, they should be gathered when quite dry; and though it is said that the seeds will keep for two years or more in the cone, yet, as a rule, they should be sown in autumn or in early spring and lightly covered with earth. Though I have not tried it myself, I believe that the germination is more rapid and regular if the seed is soaked in water before sowing, but seedlings can be procured so cheaply from nurserymen that I have always bought them at one or two years old. They are easy to transplant and grow fast if the soil is moist, being fit to plant out at three or at most four years old; and I have had as good or better success by planting them in spring as in autumn. The alder bears coppicing well, if not cut too near the ground, but the stools have a tendency to decay in the centre and to spread outwardly. True suckers are not produced, though the roots when exposed by running water will throw up shoots. The usual age for coppicing is fifteen to twenty years, and I am informed by Sir Hugh Beevor that he obtained a yield of 1700 poles per acre, which at seventeen years' growth from the stool averaged 20 feet long with a girth of 7 to 11 inches, giving a yield of about 1100 cubic feet per acre. If the trees are allowed to stand for timber they should be cut at fifty to seventy years, when they may average 50 to 70 feet high by 4 to 5 feet in girth. The only lot of alders I ever sold standing, 300 in number, realised £100, being at the rate of 4d. or 5d. per foot. Sir Herbert Maxwell states, that as long as clogs remain in common use, there will be little difficulty in realising £40 per acre for mature alder coppice, and this on land so wet as to be worthless for any other purpose. Except in localities where a good and regular market is assured, I should not recommend the planting of alder except in places too cold, wet, and marshy for willow or poplar to thrive; but Selby,[2] whose opinion of the tree as an ornamental one was better than my own, states as the result of his own experience, that the nature of the roots of the alder causes the tree to attract and retain the moisture in the soil, to such an extent that it will convert into a morass, land which, if drained and planted with other trees, might be rendered dry and productive. He adds that from experiments he has made he is "fully convinced that a plantation of alders would soon render the ground (even if previously of tolerably sound and dry quality) soft and spongy, and in time convert it into a decided bog." I cannot learn that this observation has been confirmed by others, and am inclined to doubt its being of general application.

[1] Hough, *Trees N. States and Canada*, 131 (1907).          [2] *British Forest Trees*, p. 218 (1842).

The common alder is rarely planted as an ornamental tree, and only on wet situations or on the banks of ponds and streams is it able to attain its full dimensions. But if desired for ornament in such a situation, I should recommend the cut-leaved alder in preference to the common one; and for drier ground either the grey or the Italian alder.

### REMARKABLE TREES

Among the finest trees that we have seen or heard of, those at Pain's Hill take a high place. One of these on an island is probably over 90 feet, but we could not measure it. Another growing by the lake has seventeen stems about 75 feet in height, growing from a stool which is 19 feet in girth. I estimated the contents of these poles at about 200 cubic feet. At Whitton there is a tall slender tree about 90 feet by 6 feet without a branch for 46 feet, and with a clean stem to 70 feet up. At Betchford Park, Surrey, Henry measured a tree 90 feet high by 11 feet 4 inches at 3 feet from the ground, dividing into two stems at 4 feet. At Enville Hall, Stourbridge, he saw one which was 87 feet by 8 feet 2 inches. Sir Hugh Beevor tells us of a large one at Shottisham, Norfolk, 70 feet high and 18 feet in girth near the ground, out of the base of which a mountain ash of large size is growing.

At Holme Lacy there is a large tree near the home farm, which has a short bole 6 feet high by 18 feet in girth, dividing into four main stems about 60 feet high; I estimated the contents of this tree to be not far short of 300 cubic feet.

On the banks of the Nene near Lilford Hall, Northamptonshire, there is a fine row of large and picturesque alders, of which Plate 252 gives a good representation, but I was unable to measure them on account of the water.

In Boughton Park, near Kettering, the property of the Duke of Buccleuch, a remarkable alder is growing near the Broad Avenue, in comparatively dry ground. It measures 70 feet by 9½ feet, and has the lower part of the trunk covered with bark so like that of an elm that it was difficult to recognise it as an alder, by the trunk alone.

At Aldermaston Park there is a very large old tree which looks as if it had been pollarded, and which in 1906 was 17 feet 4 inches in girth.

At Elvaston Castle, Derby, Mr. A. B. Jackson has seen a remarkably fine tree, which he estimates at 90 feet by 7½ feet, with a clean bole 60 to 70 feet long.

In Wales the finest alder I have seen is in a wood at Penrhyn Castle which is about 75 feet by 6 feet 9 inches, with a clean bole 40 feet long. The contents of this tree were estimated by me at over 100 feet of timber.

Of the cut-leaved alders the finest I have seen is about 68 feet by 10½ feet on the banks of the lake at Syon. At Melbury there is a tree very similar in size and appearance to that at Syon; and Henry measured, in 1905, one at Cassiobury Park, 85 feet by 11½ feet; and another at Belton, 85 feet by 10 feet 2 inches. Colonel Birch Reynardson sends me a photograph of a tree at

Holywell Hall, Stamford, which is 65 feet high and no less than 17 feet in girth at two and a half feet, dividing into two trunks of equal height. At Audley End, Henry measured, in October 1908, a fine specimen, 65 feet high, and 15 feet in girth at a foot from the ground, dividing above into three great stems, with a spread of branches 80 feet in diameter.

In Scotland the alder attains as large a girth and perhaps a greater age than in England. One of the oldest and largest on record is in the flower gardens at Gordon Castle, and was figured in 1881.[1] It was only 35 feet high but 14 feet 9 inches in girth and with a spread 63 yards in circumference. The late Mr. J. Webster supposed it to be nearly three centuries old, but when I saw it in 1907, nothing remained but a hollow stump 16½ feet in girth. At Fasnakyle, in Strathglass, there is a very large old alder, of which Mr. Stevenson Clarke sends me a photograph, and which in 1904 measured 18 feet in girth at two feet from the ground. It has a rowan tree 2 feet in diameter growing on it.

At Shanbally, near Dumfries, Henry measured an alder in 1904 which was 65 feet by 10 feet 5 inches, dividing at eight feet into two stems; and at Scone he saw a cut-leaved alder, which was 66 feet by 6 feet 3 inches in the same year.

Some of the most remarkable alders in Ireland grow in the old Deer Park at Kilmacurragh on strong wet land covered with tall bracken and rushes. Henry thinks that in former times this park may have been part of the virgin forest of Wicklow. The trees are scattered in groups as if self-sown; and though fully mature, are in most cases sound. They average 55 to 60 feet in height, and in some cases have clean boles 30 to 40 feet high and 8 to 10 feet in girth. The one figured (Plate 253), was about 60 feet by 11 feet 4 inches, and stands near the top of the park.

At Powerscourt, an immense alder was felled in 1902, the butt being 20 feet in girth near the ground. At Churchill, Armagh, a tree growing in peat soil, measured, in 1904, 94 feet in height by 6 feet 4 inches in girth, with a clean stem to 60 or 70 feet.

## Timber

Alder is now a wood of third-rate importance in the English timber trade, and though still used for making the clogs[2] which are worn in Lancashire, is so low in value that it will not bear much cost for carriage. On this account it is usually worked up on the ground where it grows into pieces suitable for clog soles by men who travel about from place to place, and this work is carried on by preference in summer.

In Scotland, however, alder is still used for making herring barrels, in districts where it is plentiful, though imported staves as usual are taking its place. The timber is said to be most valuable for piles, and to be certainly durable under water,

---

[1] *Trans. Scot. Arbor. Soc.* ix. pl. 1 (1881).

[2] A complete set of specimens, illustrating the manufacture of clog-soles from alder wood, was sent to the Kew Museum, in 1904, from Enniscorthy, Ireland. Cf. *Kew. Bull.* 1904, p. 6, where the clog-sole industry is described.

so that where milldams, weirs, or similar work is being done on an estate, it may be profitable to use it in preference to more costly wood.

It is useless for posts or fencing, as it decays quickly when exposed to wet and dry conditions. It makes very good panelling, and is strong enough for inside work such as window-sills, and may be used for cheap furniture, but is said to be subject to the attacks of wood-boring beetles, unless previously steeped for some time in lime water. The colour is a pale reddish brown; and the large burrs which are commonly found on old trees show a very pretty figure when cut in slices, but are usually too small and full of flaws to have any marketable value, though Sir Thomas Dick Lauder says that handsome tables can be made of them.[1] Loudon quotes Mitchell to the effect that in Dorsetshire the local saying used for willow and poplar in the midland counties, is applied to alder poles when peeled, viz.—

> Thatch me well and keep me dry
> Heart of oak I will defy;

but, according to Cobbett, the bark must be taken off with a draw knife as soon as possible after the poles are cut, and even then they will only last a year or two as hop poles.

Alder wood which is dug up from peat bogs is said to become as black as ebony, but I have seen none large or sound enough to be used like bog oak. As fuel the wood is little valued in England, though Mouillefert says that in France it is considered specially suitable for heating ovens and glass works, though considerably inferior in heating power to beechwood. The charcoal made from it was at one time in great demand for making gunpowder, but, so far as I can learn, is now little used for that purpose. The wood is said to be used on the continent for making cigar boxes; this may be the case for very cheap cigars, but all the cigar boxes I have seen appear to be made of the wood of the West Indian "cedar," *Cedrela odorata*. The bark was used for tanning in the north, but only contains about 16 per cent of tannin, and Mouillefert says that in France a black dye used for felt is made from the bark and sulphate of iron.

Another use for alder wood which seems to be little known is the making of hat-blocks, an industry carried on in Dunstable, Luton, and other towns near the principal hat factories. For this purpose the larger sized trees are preferred, cut into plank of not less than 10 inches wide and 3 inches in thickness. I am informed that in consequence of the increasing difficulty in procuring this wood in England, it is now imported from the continent, and as much as 1s. 3d. a foot is paid at the ports on the north-east coast.

The following details of the cost of making clog soles from alder were taken on my own estate in 1908 :—

About 100 trees, estimated at 56 years old, growing on the bank of a stream at Colesborne, on an area of about one-third of an acre, were sold at 7d. and 8d. a

---

[1] Alder burrs seem to have been a favourite wood with cabinet-makers in Sweden in former times, as I saw several handsome cabinets veneered with this wood in the Northern Museum at Stockholm.

foot, and as measured down to 3 inches diameter, produced 817 cubic feet and
realised . . . . . . . . . £24 10 0
Deduct expenses of cutting and hauling out . . £2 16 0
„ „ loading and delivering to station at
  6s. per ton . . . . . 2 8 0
                                                              ————— 5 4 0

Leaving a net return of . . . . . . . £19 6 0
     Four men were occupied for 42 days in working the timber up on the ground
and produced :

| | | | | | | |
|---|---|---|---|---|---|---|
| 196½ dozen pair 1st size, men's, at a cost of 1s. 4d. per dozen | . | £13 | 2 | 0 |
| 209 „ „ 2nd „ women's, „ 1s. 2d. „ | . | 12 | 3 | 10 |
| 98½ „ „ 3rd „ boys', „ 10d. „ | . | 4 | 2 | 1 |
| 119½ „ „ 4th „ children's, „ 8d. „ | . | 3 | 19 | 8 |

                     Total for labour . . . . £33 7 7

     The maker informed me that the cost of carriage to Oldham was £8, 2s. at the
rate of £1 per ton, and that the sum realised was £72, in addition to which he had
the whole of the waste and chips to sell for firewood.


## ALNUS INCANA, GREY ALDER

*Alnus incana*, Mœnch, *Meth.* 424 (1794); Willdenow, *Sp. Pl.* iv. 335 (1805); Loudon, *Arb. et Frut.*
     *Brit.* iii. 1687 (1838); Willkomm, *Forstliche Flora*, 349 (1887); Mathieu, *Flore Forestière*, 426
     (1897); Winkler, *Betulaceæ*, 120 (1904).
*Alnus lanuginosa*, Gilibert, *Exercit. Phyt.* ii. 402 (1792).
*Alnus glauca*, Michaux, f., *Hist. Arb. Amer.* iii. 322 (1813).
*Betula Alnus incana*, Linnæus, *Sp. Pl.* 983 (1753).
*Betula incana*, Linnæus, f., *Suppl.* 417 (1781).

     A tree, attaining about 70 feet in height and 6 feet in girth. Bark smooth and
silvery grey, only fissuring slightly at the base of old trunks. Young branchlets
greyish pubescent. Leaves (Plate 268, Fig. 1) about 3 inches long and 2 inches wide,
ovate or oval, rounded or cuneate at the base, acute or slightly acuminate at the
apex; lateral nerves nine to twelve pairs, running straight to the margin, each
ending in a short acute lobe, which is finely serrate and ciliate; upper surface dull,
dark green, pubescent; lower surface greyish, covered with soft hairs, densest on
the midrib and nerves, without axil-tufts; petiole, ¾ inch long, pubescent.
     Catkins in number and position like those of *A. glutinosa*; but male catkins
looser, with distant shining red-brown scales and yellow anthers. Cones smaller
than in *A. glutinosa*, with more numerous scales, thinner and less distinctly five-
lobed. Nutlets depressed, pentagonal, reddish-brown, with wing almost as broad as
the body.
     In winter the twigs are three-angled at the tip, and densely covered with a fine

pubescence.   Leaf-scars and stipule-scars as in the common alder.   Buds reddish-brown, ovoid, conspicuously stalked, with two external scales, finely pubescent on the surface and only slightly glandular.   Pith triangular or three-lobed.

The grey alder exhibits in the wild state considerable variation in the shape and pubescence of the leaves, and the cones may be sessile or shortly stalked.   Many varieties are mentioned by Winkler and Schneider, most of which are scarcely worth discriminating ; but the following are noteworthy :—

1. Var. *argentata*, Norrlin.   Leaves silvery on both surfaces, and covered with a dense silky pubescence.   Observed in Finland, Silesia, Saxony, and Switzerland.

2. Var. *glauca*, Regel.   Leaves bluish-green and nearly glabrous beneath.

3. Var. *orbicularis*, Callier.   Leaves small, almost orbicular, with five pairs of lateral nerves, wild in Silesia.   This is occasionally cultivated under the name of var. *parvifolia* ; but var. *parvifolia*, Regel, which occurs in Sweden and Finland has still smaller leaves, only ½ inch in length, and ovate in shape.

4. Var. *acuminata*, Regel.[1]   Leaves (Plate 268, Fig. 9) divided more than half-way to the midrib, into three to six pairs of long, narrow, triangular, serrate segments. This form has been observed wild in Sweden, and has been much confused with another wild variety in the same country, var. *pinnatifida*, Wahlenberg,[2] which resembles in the shape of the leaves *A. glutinosa*, var. *incisa* ; and has not been seen by us in cultivation.   Var. *acuminata* is common in gardens, and is usually known as var. *incisa* or var. *pinnatifida*.

5. Var. *aurea*, Schelle.   Leaves and fruit yellow.   This variety is growing well at Aldenham, and is striking in appearance.

6. Var. *montrosa*, Dippel.   A dwarf shrub, with the tips of the branches ribbon-like and fasciated, which originated in Spath's nursery.   In cultivation at Aldenham.


### DISTRIBUTION

The grey alder is widely distributed throughout the greater part of Europe and the Caucasus.   It is also met with in North America, where, however, it is only a shrub, commonly growing in swamps and on river banks, and forming dense thickets rarely more than 10 or 12 feet high, and is spread throughout British territory from Newfoundland to the eastern base of the Rocky Mountains, descending in the United States to New York, Pennsylvania, Wisconsin, and Nebraska.

*Alnus incana* is replaced in northern and eastern Asia by two closely allied species : *A. hirsuta*,[3] Turczaninow, not in cultivation, a native of Siberia, Kamtschatka, Manchuria, Saghalien, and Japan ; and *A. tinctoria*, Sargent,[4] which is confined to Manchuria and Japan.   The latter species differs mainly from *A. incana*, in the larger size and different shape of the leaves, which are broadly

---

[1] *Mem. Soc. Nat. Mosc.* xiii. 158, t. 17, f. 8 (1861).                    [2] *Fl. Suec.* 622 (1824).

[3] *Bull. Soc. Nat. Mosc.* 1838, p. 101.   *Alnus incana*, var. *hirsuta*, Spach.

[4] *Garden and Forest*, x. 472, f. 59 (1897).   *Alnus incana*, var. *glauca*, Shirasawa, *Icon. Ess. Forest. Japon*, t. 19, ff. 1-17 (1900).   This species is known in Japan as *yama-harinoki*, or mountain alder, and is much used for making small articles in the Hakone mountains.   Cf. Rein, *Industries of Japan*, 239, 336 (1889).

ovate, 4 to 6 inches long, 3½ to 5 inches wide ; base broad and rounded or truncate, occasionally cuneate ; apex acuminate or cuspidate ; pubescent on both surfaces, glaucous or brownish beneath ; with 9 to 12 pairs of nerves, each ending in a triangular serrated lobule ; petiole an inch or more in length. The amount of pubescence on the branchlets, petioles, and leaves is variable ; but the buds appear to be always densely pubescent. The cones are much larger than those of *A. incana*, attaining about ¾ inch in length and ½ inch in diameter. *A. tinctoria* grows in Yezo, according to Sargent, on low slopes in rich moist ground, usually at some distance from the banks of streams, which are generally occupied by *A. japonica*. *A. tinctoria* attains in Japan 60 feet in height, and 6 feet in girth ; and was collected by Elwes at Asahigawa in Yezo. It was formerly in cultivation at Coombe Wood, where it was probably raised from seed sent by Maries ; but no specimens can now be found there ; and the only one which we have seen in England is a tree at Aldenham,[1] about 15 feet in height, which is reported to be growing vigorously. There are trees of *A. tinctoria* in the Arnold Arboretum, Massachusetts, which were raised from seed collected by Sargent in Japan in 1892.

The grey alder extends in Europe much farther to the northward than the common alder, its northern limit in Scandinavia being about lat. 70° 30'. In Finmark it reaches the mouth of the river Tana, and following the shore of the Arctic Sea, the northern limit extends throughout Russia along the Arctic Circle. Its distribution is divided into two areas, a northern one extending southward in the plains of Russia to the 55th N. parallel ; and a southern area, which comprises the mountain ranges of the Carpathians, Alps, Jura, and Apennines, where the tree grows at high elevations in the mountains, and descends along the river valleys to lower altitudes, as along the Rhone, Isère, Drôme, Durance, and Var in France, and along the Rhine and its tributaries in Germany, and along the Danube in Austria. Its southern limit passes westwards from Russia through Transylvania to Banat and Servia ; but the tree is not found in Croatia, Dalmatia, or Istria. In Italy it descends along the Apennines as far south as lat. 43° 40', and grows as a rule between 4000 and 6000 feet, occasionally as low as 3000 feet. It ascends in the Erz mountains to 2100 feet, in the Swiss Alps and the Tyrol to 5000 feet, and in France thrives at 6000 feet altitude near Barcelonette and Briançon.

In Scandinavia the grey alder is common in the pine and spruce forests, usually occurring as underwood ; but in favourable situations near streams attaining a considerable size. There are many fine specimens in the beautiful natural park, close to Gefle on the Baltic. These trees, many of which are suckers from the roots of old trees that had been felled, are narrowly pyramidal in habit. The largest measured 75 feet in height and 5 feet in girth. In Denmark, Mr. Prytz of the forest service, who has measured trees 65 feet high and 7½ feet in girth, informed me that the wood had been tested, and clogs made of *A. incana* had worn as well as those manufactured from the common alder. I saw several trees in a beech forest near Nykjöbing averaging 60 feet in height and 3 or 4 feet in girth. In Denmark it grows better on dry soil than the common alder.

---

[1] At Aldenham, and in gardens on the Continent, this species is cultivated under the erroneous name, *A. incana*, var. *hirsuta*. Cf. Schneider, *Laubholzkunde*, i. 134 (1904).

In the Baltic Provinces of Russia it is common as coppice treated with a short revolution ; and often takes possession of forests, when the larger trees have been cut away, and succeeds in doing so, as it is able to grow very well on dry soil.   In Germany and Austria it grows chiefly on the banks of streams and rivers, but is also met with on hilly ground and on mountain precipices.   It is very rarely met with on peat-bogs. In the Alps it is especially common on gravelly soil, and it is the most common species in many places, where the mountain torrents form vast areas of gravel and sand, through which their branches spread in all directions.   One of the most remarkable and beautiful of these woods is situated at 2500 feet elevation on the river Romanche near Bourg d'Oisans in Isère.   The whole area is about 200 acres, one-half of which is composed of a dense wood of grey alder, mixed with a small number of aspens and ashes, the other half being more open and consisting of a mixture of grey alder and white willow.   The dense wood is treated as coppice, with a revolution of thirty years, forty standards per acre being reserved each time of felling.   When cut, the grey alder produces vigorous shoots, which grow rapidly till they are thirty-five or forty years old ; after which time growth ceases and the shoots begin to die.   At Bourg d'Oisans natural seedlings are very numerous.

The grey alder, unlike the common alder, suckers freely from the root, often at a great distance from the parent stem.   It layers easily, and can also be propagated by cuttings.   This facility of reproduction renders it of great service for the re-afforestation of the mountains in France, especially in the difficult work of planting trees on the sides of the torrents, where the soil is easily washed away.

*Alnus incana* is not a native of the British Isles, and has not yet been discovered in the fossil state there.                                                    (A. H.)

## CULTIVATION

Though the tree is hardly known to English foresters, I believe that it may become an exceedingly useful one on account of its extreme hardiness, rapidity of growth and ability to thrive in very cold heavy soil, and in places subject to late and early frosts.   I have used it with great success as a nurse to trees like *Thuya plicata*, in situations which were too wet and cold for that tree when young, and believe that it might be economically used for quickly suppressing rank herbage which would smother more tender and slower-growing trees in low and damp situations.   It can be procured quite cheaply from French nurseries as one- or two-year seedlings, and grows with extraordinary rapidity on any soil, providing a dense cover, and rendering the land fit for planting.   It soon overtops other trees, and if left standing requires the branches to be lopped so as to allow their heads to get up.   It seems to thrive equally well on wet ground, and to grow much better than the common alder on soil too dry for that tree.   I believe that the wood is at least as good, and according to Mouillefert is less brittle, than that of the common alder.

Though Loudon says that it was introduced as long ago as 1780, I have never seen a tree of any size in England ; but Sir Hugh Beevor has sent me a photograph of one at Hargham in Norfolk, which measures about 72 feet high by 3 feet in girth, being drawn up in a wood by other trees. It is growing on good loam over clay, and is the only big pole of the species in the wood; others which have been cut throw up many shoots from the stool, of which the majority die. On Lord Castletown's property at Doneraile, Co. Cork, there is a wood,[1] partly composed of grey alder, which has in places covered the ground with its suckers. (H. J. E.)

## ALNUS CORDATA, ITALIAN ALDER

*Alnus cordata*, Desfontaines, *Tabl. Hort. Paris*, 244 (1815); Winkler, *Betulaceæ*, 110 (1904).
*Alnus cordifolia*, Tenore, *Flor. Neap.* i. *Prod.* p. lxiv. (1811), and ii. 340 (1820); Loudon, *Arb. et Frut. Brit.* iii. 1689 (1838); Baillon, *Nat. Hist. Plants*, vi. 223, figs. 158-164 (1880); Masters, *Gard. Chron.* xix. 284 f. 42 (1883); Mathieu, *Flore Forestière*, 428 (1897).
*Betula cordata*, Loiseleur, *Notice*, 139 (1810), *ex* Loiseleur, *Fl. Gall.* ii. 317 (1828).

A tree, attaining 80 feet in height. Bark greyish-brown, smooth or slightly warty. Young branchlets, three-angled at the tip, stout, glabrous. Leaves (Plate 268, Fig. 4) about 4 inches long and 3 inches broad, oval or ovate, cordate at the base, shortly and abruptly acuminate at the apex; margin not lobulate, regularly serrate; nerves six to ten pairs, looping before reaching the margin; upper surface dark green, shining, glabrous; lower surface light green, glabrous, except for axil-tufts of rusty brown pubescence; petiole 1 to 2 inches, glabrous. Male catkins, three to four in a terminal raceme. Cones, solitary or two to three in an erect terminal raceme, 1 to 1¼ inches long, ovoid. Nutlet, sub-orbicular, with a thin narrow wing.

In winter the twigs are glabrous, with leaf-scars and stipule-scars like those of the common alder. Buds long-stalked, arising from the twigs at a wide angle, ovoid, beaked at the apex, glabrous and covered with wax glands; scales ciliate in margin.

*Seedling* [2]:—Cotyledons oblong-oval, slightly fleshy, pale green, about ⅓ inch long, with a very short grooved petiole. Caulicle pubescent, about ½ inch long, ending in a tapering flexuose tap-root. Young stem brown, pubescent. Leaves alternate; first pair broadly ovate, acute or cuspidate, irregularly and acutely serrate, with pubescent petioles; ultimate leaves cordate, cuspidate.

This species shows no variation in the wild state, except that the leaves are occasionally rounded and not acuminate at the apex. It differs considerably from *A. subcordata*, which has been supposed to be a variety of it; and is readily distinguished from all other species by the conspicuous cordate base of the leaves.

*Alnus elliptica*, Requien, *Ann. Sc. Nat.* v. 381 (1825) is a remarkable natural hybrid between *A. cordata* and *A. glutinosa*, which was originally found growing on the banks of the river Salenzara in Corsica. It has leaves, similar in size to those of *A. cordata*, but thinner in texture, oval or elliptical, rounded at the base and apex; margin not lobulate, finely and equally serrate; glabrous on both surfaces, except for axil-tufts beneath. The fruits are not so large as in *A. cordata*, and are inter-

---

[1] Described by Prof. Fisher in *Quarterly Journal of Forestry*, ii. 95 (1908).
[2] Cf. Lubbock, *Seedlings*, ii. 531, f. 666 (1892).

mediate between that species and *A. glutinosa*. The branchlets are glabrous and covered with wax glands. This hybrid, which in general aspect strongly resembles *A. cordata*, but is readily distinguished by the thinner leaves, not cordate at the base, appears to be very vigorous in growth at Kew, where there is a tree growing beside the lake, which is 72 feet in height and 5 feet in girth. The bark is like that of *A. cordata*, being greyish in colour and slightly warty on the surface.

*Alnus cordata* has a very restricted distribution, being confined to Corsica and southern Italy. In Corsica it ascends to 3000 feet, as at Vizzavona, where I saw it in a beech forest, growing not only beside a stream, but also on the side of the hill at some little distance off. Here the trees were about 70 feet high and 5 feet in girth, with clean timber to 50 feet, and were narrowly pyramidal in habit, with ascending branches. It grows in southern Italy from the Bay of Naples southwards; and according to Tenore occurs both on marshy ground and in the mountains. It forms woods on Mt. Serino.

This elegant species, with foliage somewhat resembling at a distance that of the Caucasian lime, which is retained late in the autumn, was introduced, according to Loudon,[1] in 1820. It flowers in March, before the leaves appear; and seems to grow as fast and to be as hardy as the common alder.

It supports well the climate of the north of France; and at Nancy, where the winters are severe, flowers and fruits regularly, and has attained 7 inches in diameter after twelve years' growth. According to Mouillefert,[2] it succeeds better on dry soils than either the common or the grey alder; and has been planted on the chalky soil of Champagne, where it is treated as coppice with a short revolution. At Grignon it has borne −4° Fahr. without injury, but suffered in 1880, when the temperature fell to −13° Fahr. Here on poor chalky soil it has attained, at thirty-five years old, 48 feet in height and 2 feet 8 inches in girth; and on better soil, 64 feet by 3 feet 1 inch.

The finest tree that we have seen of this species grows on the lawn at Tottenham House, Savernake, Wilts, and is a well-shaped tree, measuring no less than 69 feet high by 9 feet 3 inches in girth at four feet from the ground (Plate 254). When Elwes found it on April 3, 1908, it was in full flower, and covered with the cones of the previous year. It does not appear to be a very old tree, and is growing in a deep and rather heavy soil overlying chalk, at an elevation of about 400 feet.

In the new park at Merton Hall, Thetford, a tree, growing in a wind-swept situation, on very dry, light, sandy soil, measured in 1908, 50 feet high and 10 feet in girth, with a spread of branches 56 feet in diameter. Lord Walsingham believes that this tree was planted about 1843, as the new park was enclosed in the preceding year. The bark at the base is deeply fissured and scaly.

---

[1] *Arb. et Frut. Brit.* 1689 (1838); but in Loudon, *Gard. Mag.* 1837, p. 143, and 1839, p. 39, a tree at Britwell House, Bucks, growing on gravelly soil, was reported to be 60 feet high; and this would show that the date of introduction was earlier than 1820. So far as we can learn, this tree no longer exists.

[2] *Essences Forestières*, 252 (1903). However, two trees at Verrières near Paris, about 80 years old, have only attained 60 feet in height and 5 feet 8 inches in girth; and M. Philippe L. de Vilmorin states (*Hortus Vilmorinianus*, 54 (1906)), that their growth seems to have come long ago to a standstill.

A specimen is growing at Milford House, near Godalming, which was planted by the famous botanist and traveller, Phillip Barker Webb. The present owner, R. W. Webb, Esq., informed us in 1905 that it was very healthy, measuring 8 feet in girth, and estimated to be about 50 feet in height.

There is a fine tree growing near the pond in front of the palm-house in Kew Gardens, which is 71 feet high by 5 feet 8 inches in girth. At Tortworth, a tree measures 60 feet high by 6 feet in girth; and at Waterer's Nursery, Knaphill, Woking, another is 50 feet by 5 feet 10 inches.

At Nuneham Park, Oxford, a tree, growing on hilly dry ground, on the green-sand formation, measures 51 feet by 5 feet 5 inches, and is very thriving. Elwes has seen a tree at Bicton, measuring 65 feet by 5 feet 10 inches, and another at Melbury, where it grows vigorously and fruits.

In the playing fields at Eton, on the banks of the Thames, there are two trees, the larger of which is 40 feet high by 6 feet 4 inches in girth. These were in full foliage on 17th November 1907, having scarcely lost a leaf, and were bearing fruit. They have not developed tall straight stems, as in the other places where the tree is thriving; and this is probably owing to their position being exposed to easterly and north-easterly winds. At Ponfield, Herts, a young tree 35 feet by 2 feet 4 inches in 1906 is doing well on dry soil; and there is a good specimen in the Cambridge Botanic Garden. Another at Yattendon Court, Berks, is 50 feet by 3 feet 8 inches.

In Scotland it also grows well at the Botanic Gardens, Edinburgh, where there is a tree 60 feet by 5 feet 10 inches; and in the west at Castle Kennedy, and at Monreith, where Elwes saw a tree 30 feet high, bearing cones in September 1906.

There is a fine specimen in the Glasnevin Botanic Garden, which is 64 feet high by 5 feet 4 inches in girth. (A. H.)

## ALNUS SUBCORDATA, CAUCASIAN ALDER

*Alnus subcordata*, C. A. Meyer, *Verz. Pfl. Kauk.* 43 (1831); Winkler, *Betulaceæ*, 112 (1904).
*Alnus cordifolia*, Tenore, var. *subcordata*, Regel, in *Mém. Soc. Nat. Mosc.* xiii. 170 (1861), and in DC. *Prod.* xvi. 2, p. 185, (1868).

A tree, attaining about 60 feet in height. Bark grey, warty on the surface, ultimately scaling at the base of old trunks. Young branchlets pubescent. Leaves (Plate 268, Fig. 5) about 4 inches long and 2½ inches broad, ovate-oblong, rounded and unequal or subcordate at the base, cuspidate-acuminate at the apex; coarsely serrate or bi-serrate in the upper half, finely serrate in the lower half; nerves, about eight pairs, running to the margin; upper surface dark green, slightly pubescent; lower surface light green, pubescent throughout, the pubescence densest along the nerves and in the axils; petiole, ¾ inch, pubescent. Staminate catkins, three to five in a raceme. Cones solitary or several, ovoid-elliptic, about an inch long; nutlets broadly ovoid, with a very narrow wing.

This is a moderate-sized tree, occurring in the province of Talysch in the Caucasus, and in the provinces of Asterabad and Ghilan in Persia, where Dr. Stapf informs me that he has seen large trees south of the Caspian Sea.

It is closely allied to, if not identical with *Alnus orientalis*, Decaisne, a native of Asia Minor and Cyprus. The latter has not been introduced, so far as we know, into English or continental gardens.

It was known in cultivation a good many years ago in France, as Gay records[1] a tree 30 feet high growing at Verrières in 1861 ; but we are unaware of the exact date of its introduction into England. It appears to grow as well and to be as hardy in England as *A. cordata* ; and a fine tree, growing near the lake in Kew Gardens, is 52 feet high and 4 feet 10 inches in girth. There is a small specimen at Aldenham.

(A. H.)

## ALNUS FIRMA

*Alnus firma*, Siebold et Zuccarini, *Abh. Akad. München*, iv. 3. p. 230 (1845) ; Sargent, *Forest Flora Japan*, 63 (1894) ; Winkler, *Betulaceæ*, 102 (1904).

*Alnus Sieboldiana*, Matsumura, *Journ. Coll. Sci. Tokyo*, xvi. 5, p. 3 (1902).

*Alnus yasha*, Matsumura, *op. cit.* p. 4 (1902).

*Alnus pendula*, Matsumura, *op. cit.* p. 6 (1902).

*Alnus multinervis*, Schneider, *Laubholzkunde*, 123 (1904).

A tree, attaining in Japan a height of 30 feet, but usually smaller. Young branchlets three-angled at the tip, pubescent. Leaves (Plate 268, Fig. 2) about 4 inches long and $1\frac{3}{4}$ inch broad, plicate ; nerves deeply immersed above and very prominent beneath, about fifteen to eighteen pairs, running parallel and straight to the margin ; ovate-lanceolate, rounded and unequal at the base, acuminate at the apex ; upper surface dark green, shining, glabrous ; lower surface light green, pubescent, the pubescence strongest on the midrib and nerves ; margin finely and regularly serrate, ciliate ; petiole $\frac{1}{2}$ inch, pilose ; stipules often persistent, ovate-lanceolate, $\frac{1}{2}$ inch long, membranous, glabrous. Buds sessile, conical, long-pointed, curved, green, glabrous, with two external scales.

Flowers appearing in spring. Staminate catkins terminal or lateral, 1 to $2\frac{1}{2}$ inches long. Pistillate catkins, one, two, or three to five, arising from one bud. Cones, solitary or racemose, variable in size, $\frac{1}{2}$ to 1 inch long, in the different varieties. Nutlets, obovate-oblong or sub-rhomboid with a membranous wing.

This alder displays great variation in Japan, no less than three distinct species being recognised by Matsumura and Schneider. These appear to be geographical varieties :—

1. Var. *multinervis*, Regel in *Bull. Soc. Nat. Mosc.* xxxviii. 2. 423 (1865), and in DC. *Prod.* xvi. 2, p. 183 (1868). *Alnus pendula*, Matsumura. *Alnus multinervis* Schneider.

Branchlets pubescent. Leaves with numerous lateral nerves, eighteen pairs or more. Cones in pendulous racemes, $\frac{2}{5}$ inch long. This form is the

---

[1] Note with a specimen in Kew Herbarium. This tree is not mentioned in *Hortus Vilmorinianus* (1906).

only one in cultivation, and grows on the banks of streams in sub-alpine regions in Yezo and Hondo.

2. Var. *Sieboldiana*, Winkler, *loc. cit.*  *Alnus Sieboldiana*, Matsumura.

Branchlets glabrous.  Lateral nerves twelve to fifteen pairs.  Cones solitary, 1 inch long.  A native of the sea-coast in Hondo.

3. Var. *yasha*, Winkler, *loc. cit.*  *Alnus yasha*, Matsumura.

Branchlets pubescent.  Lateral nerves twelve to fifteen pairs.  Cones solitary or racemose, ¾ inch long.  Occurs in mountain woods in Kiusiu, Shikoku, and Hondo.

According to Sargent, *Alnus firma* is largely planted along the borders of rice-fields near Tokyo, to afford support for the poles on which the freshly cut rice is hung to dry.  He observed var. *multinervis* on the mountains of Hondo, where it grows on dry rocky soil and reaches 5000 feet elevation, and describes it as a graceful tree 20 to 30 feet in height.  The species, as mentioned above under the varieties, is widely distributed throughout the whole of Japan.

It was introduced by Sargent into New England in 1892 ; and, according to Winkler, was brought by Zabel into the forest garden of Münden in Germany. There are trees 6 to 10 feet in height in the collection at Kew.  It is a remarkably distinct species, with plicate many-nerved leaves, recalling those of two other Japanese trees, viz. : *Carpinus japonica* and *Acer carpinifolium* ; and is worthy of a place in collections of shrubs, as it scarcely can be considered to be a tree.

(A. H.)

## ALNUS JAPONICA, Japanese Alder

*Alnus japonica*, Siebold et Zuccarini, *Abh. Akad. München*, iv. 3, p. 320 (1845); Sargent, *Garden and Forest*, vi. 343, f. 53 (1893), and *Forest Flora, Japan*, 63, t. 20 (1894); Shirasawa, *Icon. Ess. Forest. Japon*, text 38, t. 19, ff. 18-34 (1900); Winkler, *Betulaceæ*, 114 (1904).

*Alnus maritima*, Nuttall, var. *japonica*, Regel, in DC. *Prod.* xvi. 2, p. 186 (1868).

*Alnus maritima*, Nuttall, var. *formosana*, Burkill, *Journ. Linn. Soc. (Bot.)* xxvi. 500 (1899).

A tree attaining about 80 feet in height.  Young branchlets usually glabrous. Leaves (Plate 268, Fig. 12) about 4 inches long and 1½ to 1¾ inch wide, lanceolate or narrowly elliptical, cuneate at the base, long-acuminate at the apex ; margin not lobulate, finely serrate ; nerves, about twelve pairs, mostly running to the margin ; upper surface dark green, shining, pubescent on the midrib and nerves ; lower surface light green, glabrous except for minute axil-tufts ; petiole about ½ inch, slightly pubescent.  Buds minute, stalked, glabrous, glandular.

Flowers,[1] appearing in spring, the fruit ripening in autumn ; otherwise similar to *Alnus maritima*.

This species occurs in Japan, Manchuria, Korea, and Formosa.  In Manchuria[2] it grows along the sea-coast from St. Olga Bay southwards, and also inland, either solitary or in groups, in sandy soil along the rivers.  It has been collected in Korea

---

[1] In Formosa, according to Burkill, the flowers are produced later, in summer ; and he adduces this as a reason for uniting this species with the American *A. maritima*.      [2] Komarov, *Flora Manshuriæ*, ii. 60 (1904).

at Port Chusan; and in Formosa, it grows near Tamsui at the north end of the island. According to Sargent, it is the most beautiful and largest of the alders in Japan, forming a pyramidal tree, often 70 or 80 feet in height, and well furnished to the ground with branches clothed with large dark green lustrous leaves. It differs from *A. maritima* in the larger, differently shaped and coloured leaves, and in the time of flowering.

Sargent states that it is perfectly hardy in New England, where it grows rapidly and promises to become a large and handsome tree. It was introduced, according to Nicholson,[1] in 1886. There are small healthy trees in the collection at Kew; and at Aldenham a specimen is about 14 feet high. (A. H.)

## ALNUS NITIDA, Himalayan Alder

*Alnus nitida*, Endlicher, *Gen. Pl. Suppl.* IV. ii. 20 (1847); Brandis, *Forest Flora N.-W. India*, 460, t. 57 (1874), and *Indian Trees*, 623 (1906); Hooker, *Flora Brit. India*, v. 600 (1888), and *Bot. Mag.* t. 7654 (1899); Gamble, *Indian Timbers*, 670 (1902); Winkler, *Betulaceæ*, 108 (1904).

*Clethropsis nitida*, Spach, *Ann. Sc. Nat.* sér. 2, xv. 202 (1841); Cambessedes in Jacquemont, *Voy. dans l'Inde, Bot.* 159, t. 159 (1844).

A large tree attaining 100 feet in height and 15 feet in girth. Bark blackish, with thin quadrangular scales. Young branchlets densely and minutely pubescent. Leaves (Plate 268, Fig. 3) about 5 inches long, and $2\frac{1}{2}$ inches broad, ovate-elliptical or elliptical, base rounded or cuneate, apex acuminate; margin entire, obscurely crenate, or remotely and slightly serrate; thin in texture; nerves, nine to twelve pairs, looping before reaching the margin; upper surface dark green, glabrous, shining; lower surface light green, glabrous except for slight axil-tufts; petiole, $\frac{3}{4}$ to 1 inch, minutely pubescent. Male catkins, 2 inches long, in terminal erect, often leafy racemes. Cones, $\frac{3}{4}$ to $1\frac{1}{2}$ inch long, three to five in erect lateral racemes; nutlet with a narrow thickened margin. The flowers in this species open in September.

*Alnus nitida* occurs in the north-western Himalaya, from Kashmir to Kumaon, usually at low elevations, 2000 to 4000 feet, fringing the banks of rivers, and not uncommonly descending with them into the plains. It occasionally, however, ascends as high as 9000 feet; and is common on the Sutlej in the dry region of Kunawar, as far as Spui on the right bank and Namgia on the left bank, according to Brandis, who states that it attains 100 feet high and 15 feet in girth; but Gamble has never seen it so big, and says it is usually crooked and branching. An attempt was made recently to float out the wood from the Tons river forest, but failed, as the timber quickly became waterlogged. The bark is used for tanning and dyeing.

*Alnus nitida* is one of the few Himalayan broad-leaved trees which have

---

[1] *Dict. Gardening, Suppl.* 34 (1900).

succeeded in this country, where, however, it is little known, the only trees in cultivation[1] that we know of being three thriving specimens which are growing near the lake in Kew Gardens. The largest of these is now 40 feet by 2 feet 3 inches in girth. They were raised from seed sent by Mr. R. E. Ellis of the Indian Forest Department in 1882. (A. H.)


## ALNUS MARITIMA

*Alnus maritima*, Nuttall, *Sylva*, i. 34 (1842); Sargent, *Garden and Forest*, iv. 268, t. 47 (1891), *Silva N. Amer.* ix. 81, t. 458 (1896), and *Trees N. Amer.* 215 (1905); Winkler, *Betulaceæ*, 114 (1904).
*Alnus oblongata*, Regel, *Mém. Soc. Nat. Mosc.* xiii. 171 (1861) (in part).
*Betula-Alnus maritima*, Marshall, *Arb. Am.* 20 (1785).

A tree attaining in America 30 feet in height and 1 foot in girth. Bark smooth, greyish-brown. Young branchlets slightly pubescent, three-angled at the tip. Leaves (Plate 268, Fig. 11) in cultivated specimens 2½ inches long, 1¾ inch wide, somewhat larger in wild specimens, ovate or obovate, cuneate at the base; acute, slightly acuminate, or rounded at the apex; nerves, eight to twelve pairs, running to the margin; margin not lobulate, remotely serrate in the upper two-thirds with minute incurved glandular teeth; upper surface dark green, shining, glabrous; lower surface light green, glabrous; petioles ½ inch, slightly pubescent. Buds minute, stalked, ovoid, glabrous, slightly glandular.

Flowers appearing in July on the branches of the year, and opening in September. Staminate catkins in scurfy pubescent racemes in the axils of the upper leaves. Pistillate catkins usually solitary from the axils of the lower leaves. Cones ripening in the following September, so that both flowers and ripe fruit occur simultaneously on the tree, ovoid, ⅝ inch long; nutlet obovate, narrowed and apiculate at the apex, with a thin membranous border.

This alder grows on the banks of streams and ponds in Delaware and Maryland, usually near, but not immediately upon the sea-coast, as its name would seem to imply. However, it abounds on the banks of the Nanticoke and Wicomico rivers in Maryland, near the high-water mark. What appears to be the same species was collected by Hall on the Red River in Indian Territory.

It was introduced into cultivation by Thomas Meehan, who sent it in 1878 to the Arnold Arboretum, where it is tolerably hardy, flowering and fruiting freely, though it was killed to the ground in 1885. There are now two trees, about 6 feet high, growing in the nursery at Kew, which were sent by Prof. Sargent in 1899. These flower in September, and produce fruit in quantity. (A. H.)

[1] Mr. A. B. Jackson has lately seen a tree at Grayswood, Haslemere, which is 18 feet high and 9 inches in girth.

## ALNUS RUBRA, Oregon Alder

*Alnus rubra*, Bongard, *Mém. Acad. St. Pétersb.* ii. 162 (1833); Winkler, *Betulaceæ*, 124 (1904); Sargent, *Bot. Gazette*, xliv. 226 (1907.)

*Alnus oregona*, Nuttall, *Sylva*, i. 28, t. 9 (1842); Sargent, *Silva N. Amer.* ix. 73, t. 454 (1896), and *Trees N. Amer.* 210 (1905).

*Alnus incana*, Moench, var. *rubra*, Regel, *Mém. Soc. Nat. Mosc.* xiii. 157 (1861).

A tree attaining 80 feet in height and 10 feet in girth. Bark greyish or whitish, thin, roughened by minute wart-like excrescences. Young branchlets glabrous, three-angled at the tip, scarcely viscid except at the beginning of the season. Leaves (Plate 268, Fig. 16) about 4 or 5 inches long and $2\frac{1}{2}$ inches wide, ovate or elliptical, rounded or cuneate at the base, acute at the apex; nerves, about 15 pairs, each running straight and parallel to the apex of a lobule, which is furnished with minute gland-tipped serrations; margin slightly revolute and ciliate; upper surface dark green, slightly pubescent; lower surface whitish or greyish, covered with a minute brown pubescence; petiole, $\frac{3}{4}$ inch, with a few scattered hairs. Buds beaked at the apex, glabrous, stalked. Stipules ovate, acute, tomentose, $\frac{1}{8}$ to $\frac{1}{4}$ inch long.

Flowers opening in spring before the leaves. Staminate catkins, three to six in a raceme, 4 to 6 inches long when fully opened. Cones, three to six in a raceme, $\frac{1}{2}$ to 1 inch long, with truncate scales, much thickened at the apex; nutlet orbicular or obovate, surrounded by a membranous wing.

This species can only be confused with *A. incana*, from which it differs in the glabrous branchlets and the usually larger leaves with revolute margins. The buds also differ, those of *A. rubra* being elongated, pointed, and glabrous, whilst those of *A. incana* are shorter, rounded at the apex, and pubescent.

*Alnus rubra*, according to Sargent, ranges from Sitka, where it often clothes mountain-sides to elevations of 3000 feet above the sea, southwards through the islands and coast ranges of British Columbia, and through western Washington and Oregon, and the cañons of the Californian coast ranges, to the Santa Inez mountains near Santa Barbara. It grows to its largest size in the neighbourhood of Puget Sound, where it commonly fringes the banks of streams and grows in wet places.[1]

This species was introduced into cultivation a few years ago, and there are two trees in Kew Gardens about 15 feet in height. Elwes has raised seedlings from these trees, which grow very rapidly in heavy soil at Colesborne, but being planted in a situation very subject to late frosts, have suffered on several occasions, when the grey and common alders standing near were quite untouched.

The wood is light, soft, brittle, and not strong, but close-grained and takes a fine polish; and is now largely used in Washington and Oregon for making furniture.　　　　　　　　　　　　　　　　　　　　　　　　　　　　　　　(A. H.)

---

[1] In Vancouver Island, the stem and branches are often covered with *Polypodium falcatum*, the creeping rhizomes of which find anchorage in its moss-covered bark. Cf. *Postelsia*, 1906, p. 76. A figure of the tree, growing in a moist part of the forest and surrounded by devil's club (*Echinopanax horridum*), is given in Piper, *Flora of the State of Washington*, plate vii. (1906.)

## ALNUS TENUIFOLIA

*Alnus tenuifolia*, Nuttall, *Sylva*, i. 32 (1842); Sargent, *Silva N. Amer.* ix. 75, t. 455 (1896), and
*Trees N. Amer.* 211 (1905); Winkler, *Betulaceæ*, 124 (1904); Schneider, *Laubholzkunde*, 133 (1904).
*Alnus incana*, Moench, var. *glauca*, Regel, *Mém. Soc. Nat. Mosc.* xiii. 154 (1861) (in part).
*Alnus incana*, Moench, var. *virescens*, Watson, in Brewer and Watson, *Bot. Calif.* ii. 81 (1880).
*Alnus occidentalis*, Dippel, *Laubholzkunde*, ii. 158 (1892); De Wildeman, *Icon. Select. Hort. Thenensis*,
ii. 147, t. 75 (1901).

A tree attaining 30 feet in height and 2 feet in girth. Bark bright red-brown, broken on the surface into small scales. Young branchlets glabrous. Leaves (Plate 268, Fig. 15) about 3 inches long by 2 inches wide, ovate, broad and rounded at the base, acute or shortly acuminate at the apex; nerves, nine or ten pairs, running parallel and straight to the margin, and ending in acute triangular lobes, which are finely serrate; upper surface dark green, pubescent on the midrib and nerves; lower surface yellowish green, glandular, pubescent on the midrib with slight axil-tufts; petioles pubescent, $\frac{3}{4}$ to 1 inch long. Buds stalked, pubescent at the base. Stipules deciduous, lanceolate, acute, pubescent. Staminate catkins, three to four, in slender-stemmed racemes; stamens four. Cones, ovoid-oblong, $\frac{1}{3}$ to $\frac{1}{2}$ inch long, three to four in a raceme; scales thickened, three-lobed and truncate at the apex; nutlets nearly circular, surrounded by a thin membranous border.

This species, distributed over a wide area, shows two well-marked geographical varieties :—

1. Var. *virescens*, Callier. This is the commonest form, and has been described above (Plate 268, Fig. 15).

2. Var. *occidentalis*, Callier. *Alnus occidentalis*, Dippel. Leaves (Plate 268, Fig. 14) larger, 4 to 5 inches long and 3 to 4 inches wide; nerves, twelve pairs; slightly bluish-green and pubescent throughout beneath; stipules ovate, broad, obtuse. This variety is rare, and has only been observed in British Columbia and Oregon.

This species is widely distributed in western North America. It occurs in British Columbia, from Francis Lake in lat. 61° to the valley of the Lower Fraser River, and extends eastward along the Saskatchewan River to Prince Albert. It extends southwards along the Rocky Mountains to northern New Mexico, and is the common species in the northern interior region, east of the divide of the Cascade Mountains, in eastern Washington, Oregon, Idaho, and Montana. It is very abundant on the southern California Sierra, forming great thickets at 6000 to 7000 feet above the sea, along the head-waters of the rivers of southern California flowing to the Pacific Ocean. It is equally abundant and attains its largest size in Colorado and northern New Mexico, and is met with in Nevada and Utah.

This species is rare in cultivation. There are two or three trees of each variety in the alder collection at Kew, which are about 15 feet in height, and show no special beauty or vigour. Var. *virescens* is thriving at Aldenham; and a fine specimen at Grayswood, Haslemere, planted in 1888, is now about 30 feet in height and 1 foot 8 inches in girth.

According to De Wildeman, var. *occidentalis* was introduced from British Columbia in 1891, by Dieck and Purpus, into the arboretum at Zoschen.

## ALNUS RHOMBIFOLIA

*Alnus rhombifolia*, Nuttall, *Sylva*, i. 33 (1842); Sargent, *Silva N. Amer.* ix. 77, t. 456 (1896), and *Trees N. Amer.* 212 (1905); Winkler, *Betulaceæ*, 115 (1904).
*Alnus oblongifolia*, Watson, in Brewer and Watson, *Bot. Calif.* ii. 80 (1880) (in part).

A tree attaining 80 feet in height and 9 feet in girth. Young branchlets pubescent. Leaves (Plate 268, Fig. 13) on young trees up to 5 inches long and 3 inches broad, on old trees 3 inches long and $1\frac{1}{2}$ inch broad, ovate or oval, rounded and unequal at the base, acute or rounded at the apex; margin slightly thickened and reflexed, finely and irregularly serrate, and ciliate; nerves, ten or eleven pairs, running parallel and slightly curved to the margin; upper surface dark green, shining, glabrous; lower surface light green, pubescent, the pubescence strongest on the midrib and nerves; petiole, $\frac{1}{2}$ inch, pubescent. Buds stalked, pubescent. Staminate catkins, in pubescent racemes, deciduous before the opening of the leaves; stamens, two or three, rarely one. Cones oblong, $\frac{1}{3}$ to $\frac{1}{2}$ inch long, fully grown at midsummer, but remaining closed till the trees flower in the following year; nutlet broadly ovate with a thin, acute margin.

This species grows on the banks of streams from northern Idaho to the eastern slope of the Cascade Mountains in Washington and Oregon, extending southward over the coast ranges and along the western slopes of the Sierra Nevada to the mountains of southern California. It is the common alder of central California, and the only species at low altitudes in the southern part of this state.

It is extremely rare in cultivation, the only specimen which we have seen being a small plant in Lord Aldenham's remarkable collection of shrubs and trees at Aldenham, Herts. According to Nicholson,[1] it was introduced into cultivation in 1888.

[1] *Dict. Gardening, Suppl.* 34 (1900).

# BETULA

*Betula*, Linnæus, *Sp. Pl.* 982 (1753); Bentham et Hooker, *Gen. Pl.* iii. 404 (1880); Winkler, in
　　Engler, *Pflanzenreich*, iv. 61, *Betulaceæ*, 56 (1904).
*Betulaster*, Spach, in *Ann. Sc. Nat.* sér. 2, xv. 198 (1841).
*Apterocaryon* and *Chamæbetula*, Opiz, in *Lotos*, v. 258 (1855).

DECIDUOUS trees or shrubs belonging to the order Betulaceæ. Bark smooth with longitudinal lenticels, often peeling off in papery strips, and becoming on old trunks thick and furrowed near the base. Branchlets of two kinds: long shoots with several leaves and axillary buds, no true terminal bud being formed; and short shoots or dwarf spurs, each with two (rarely one or three) leaves and a terminal bud. Buds viscid, elongated, ovoid, fully grown and green at midsummer, composed of imbricated scales, but with the two basal ones short and lateral, usually only four scales being visible externally; inner scales accrescent, and marking in falling the base of the shoots with ring-like scars. Leaves alternate, simple, stalked, penninerved; serrate, dentate or incised. Stipules lateral, enclosing the leaf in the bud, fugacious. Flowers monœcious, fertilised by the wind, in cylindrical catkins, composed of closely imbricated three-lobed scales, with three flowers on each scale. Male catkins,[1] formed in the preceding autumn, clustered in the axils of the upper leaves of a long shoot, erect and naked during winter, pendulous in spring. Staminate flowers, with a one- to four- lobed calyx; stamens two, with short bifurcated filaments, each of the four branches bearing an erect half-anther, there being thus apparently twelve stamens on each scale. Pistillate catkins, solitary, or two to four in a raceme, terminal on the short shoots, and appearing with the leaves in spring. Pistillate flowers, without a calyx, two-celled, with one ovule in each cell; styles two, stigmatic at the apex. Cones, ripening[2] usually in autumn, composed of woody three-lobed scales and small fruits, deciduous together; nutlets oval or obovate, compressed, bearing the persistent styles at the apex, and with the outer shell produced into a marginal transparent wing, interrupted at the apex; seed solitary, pendulous, without albumen.[3]

In winter, species of Betula are readily distinguished by the short shoots on the older wood, which end in a terminal bud, and are densely clothed with scars, as each season's growth is very short and marked by two crescentic leaf-scars in addition to the ring-like scars left by the fall of the scales of the bud of the previous spring. The long shoots show similar ring-like scars at the base, and bear axillary buds

---

[1] In some of the shrubby species the male catkins are solitary on the ends of the short shoots, and remain enclosed in the buds during winter, appearing in spring.　　　　[2] In *B. nigra* the fruit ripens in May or June.
[3] The cones, scales, and fruits shown in Plates 269 and 270, were all drawn by Miss F. H. Woolward, except in the case of Figs. 8 and 16.

arranged alternately. The leaf-scars, semicircular or crescentic, and three-dotted, have on each side a linear stipule-scar. The pith of the twigs is oblong in section.

About thirty species of Betula are known, all natives of the northern hemisphere, extending from the Arctic circle to Texas in the New World, and to southern Europe, the Himalayas, China, and Japan in the Old World. A considerable number are shrubs, the treatment of which does not come within the scope of our work. Of the arborescent species, a few, either not introduced or imperfectly known, are not included in the following account.

*B. corylifolia*, Regel et Maximowicz, though not yet introduced, is included in the key and fully described below, as it is very distinct and has been much confused with other Japanese species.

There are young plants in the nursery at Kew, received from the Arnold Arboretum, as *B. globispica*, Shirai,[1] which appear to be a very distinct species ; but, as there is no authentic material in the Kew Herbarium with which to compare them and ascertain if they are correctly named, it is unadvisable to deal with this species at present.

Similarly, young plants of *B. alnoides*, Buchanan-Hamilton, var. *pyrifolia*, Franchet, growing at Coombe Wood, which were raised from seed sent from central China by Wilson in 1901, are left undescribed, as they show considerable variation, and we cannot be certain, until they have borne fruit, of their identification.

### KEY TO ARBORESCENT SPECIES OF BETULA IN CULTIVATION

I. *Branchlets and leaves quite glabrous.*
   1. *Betula verrucosa*, Ehrhart.   Europe, Northern and Eastern Asia.   See p. 966.
      Leaves bi-serrate, shortly acuminate.
   2. *Betula populifolia*, Marshall.   North America.   See p. 987.
      Leaves lobulate and irregularly serrate, ending in a long caudate acumen.
II. *Branchlets or leaves or both pubescent.*
      * *Leaves cordate at the base.*
   3. *Betula Maximowiczii*, Regel.   Japan.   See p. 976.
      Leaves 5 or 6 inches long, broadly ovate.
   4. *Betula ulmifolia*, Siebold et Zuccarini.   Japan.   See p. 979.
      Leaves, 3 inches long, narrowly ovate.
      ** *Leaves cuneate at the base.*
   5. *Betula pubescens*, Ehrhart.   Europe, Northern Asia, Greenland.   See p. 962.
      Leaves light green beneath, rhombic-ovate, bi-serrate, with six pairs of nerves. Branchlets not glandular, clothed with minute dense erect pubescence.
   6. *Betula davurica*, Pallas.   Manchuria, Korea, North China.   See p. 974.
      Leaves light green beneath, narrowly ovate, bi-serrate, with six to eight pairs of nerves.   Branchlets glandular, with minute erect pubescence interspersed with a few long hairs.

---

[1] The Japanese name of this species, according to Matsumura and Goto, is *Jizo-kamba*.

7. *Betula nigra*, Linnæus.   North America.   See p. 988.

Leaves greyish beneath, ovate, acute, with large serrated teeth, and six to eight pairs of nerves.   Branchlets glandular, tomentose.

*** *Leaves rounded or truncate at the base.*

A. *Leaves orbicular or oval.*

8. *Betula corylifolia*, Regel et Maximowicz.   Japan.   See p. 975.

Leaves acute, pale beneath with conspicuous silky hairs on the midrib and nerves; nerves twelve to fourteen pairs.

B. *Leaves ovate-oblong, considerably longer than broad, and widest near the middle.*

9. *Betula lutea*, Michaux.   North America.   See p. 990.

Leaves, with nine to twelve pairs of nerves, pilose on the midrib and nerves of both surfaces, and on the petiole.   Branchlets pilose.

10. *Betula lenta*, Linnæus.   North America.   See p. 991.

Leaves with nine to twelve pairs of nerves, pilose on the midrib and nerves of both surfaces; petiole glabrescent.   Branchlets glabrous, except for a few hairs above the leaf-insertions.

11. *Betula utilis*, Don.   Himalayas, China.   See p. 980.

Leaves, with nine to twelve pairs of nerves; lower surface with dense axil-tufts of pubescence; petiole tomentose.   Branchlets tomentose.

C. *Leaves ovate, not much longer than broad, widest near the base.*

† *Branchlets very glandular.*

12. *Betula Ermani*, Chamisso.   Eastern Siberia, Manchuria, Japan.   See p. 977.

Leaves truncate at the base, with ten to twelve pairs of nerves, glabrescent and conspicuously glandular beneath.   Branchlets glabrous.

13. *Betula fontinalis*, Sargent.   Western North America.   See p. 992.

Leaves thin in texture; rounded, truncate, or subcordate at the base, with six pairs of nerves; both surfaces minutely glandular, with scattered long hairs.   Branchlets with long pale hairs.

‡ *Branchlets not conspicuously glandular.*

14. *Betula papyrifera*, Marshall.   North America.   See p. 983.

Leaves about 3 inches long, with six to eight pairs of nerves; lower surface glandular and with conspicuous axil-tufts.   Branchlets pubescent or glabrous.

15. *Betula utilis*, Don, var. *Jacquemontii*, Regel.   Himalayas.   See p. 981.

Leaves about 3 inches long, with seven or eight pairs of nerves; lower surface glandular, with long hairs on the midrib and nerves, and without conspicuous axil-tufts.   Branchlets with short, erect, dense pubescence.

16. *Betula luminifera*, Winkler.   Central China.   See p. 980.

Leaves, 5 or 6 inches long, with ten to fourteen pairs of nerves, pubescent on both surfaces.   Branchlets with dense erect pubescence.

(A. H.)

## BETULA PUBESCENS, COMMON BIRCH

*Betula pubescens,* Ehrhart, *Beit. Naturk.* vi. 98 (1793); Mathieu, *Flore Forestière,* 415 (1897); Winkler, *Betulaceæ,* 81 (1904).

*Betula tomentosa,* Reitter and Abel, *Abbild. Holzart,* i. 17 (1790).

*Betula alba,* Linnæus, *Sp. Pl.* 982 (1753) (in part); Roth, *Tent. Fl. Germ.* i. 404 (1788); Willkomm, *Forstliche Flora,* 302 (1887); Schneider, *Laubholzkunde,* i. 116 (1904).

*Betula alba,* Linnæus, var. *pubescens,* Loudon, *Arb. et Frut. Brit.* iii. 1691 (1838).

*Betula odorata,* Bechstein, *Diana,* i. 74 (1797).

A tree, usually attaining 70 or 80 feet in height, and 5 or 6 feet in girth, occasionally larger. Branches ascending or spreading, branchlets usually not pendulous. Bark smooth, white, and papery, often peeling off in transverse shreds, with black triangular markings below the insertion of the branches; thick and deeply furrowed at the base of old stems. Young branchlets covered with short, erect pubescence, often minute and only discernible with a lens, usually retained in the second year. Leaves (Plate 269, Fig. 1), 1½ inch to 2 inches long, ¾ inch to 1½ inch wide, rhomboid-ovate or ovate, usually cuneate at the base, acute or acuminate at the apex; margin ciliate and coarsely serrate; nerves, five or six pairs; upper surface with scattered pubescence; lower surface pubescent on the midrib and nerves; petiole, ½ inch, pubescent and glandular.

Fruiting-catkins (Plate 269, Fig. 1), cylindrical, about 1 inch long, ⅓ inch wide, at first erect, afterwards pendulous, on long pubescent stalks; scales, puberulous, ciliate, with the central lobe more prolonged than is the case in *B. verrucosa,* and with the lateral lobes angular and usually erect, but occasionally recurved.[1]

In winter the twigs are slender, dark brown, densely covered with short, erect pubescence. Buds, ⅕ inch long, ovoid, rather blunt at the apex, viscid, with glabrous, ciliate scales.

Seedling.[2]—Cotyledons, about ⅕ inch long, oblong-ovate, obtuse, glabrous, with pubescent petioles, about ⅛ inch long. Caulicle short, pubescent, raising the cotyledons above the ground. Stem pubescent and non-glandular, bearing primary leaves, which are alternate, ovate, cordate at the base, simply and coarsely serrate, and pubescent.

The birch,[3] under ordinary conditions of growth, does not produce root-suckers; however, when cut down, although coppice shoots are not given off from the stool, the roots give rise to numerous tomentose shoots, which bear leaves larger than those of ordinary branches, cordate at the base, dentate in margin, and pubescent on both surfaces.

[1] When the lateral lobes are recurved, the scales are similar in shape to those of *B. verrucosa*; and in such cases we may suspect a hybrid between the two species or an intermediate form. The pubescent stalks of the catkins and the puberulous scales are, however, apparently characteristic of *B. pubescens.*

[2] Cf. Lubbock, *Seedlings,* ii. 541, fig. 672 (1892).

[3] Cf. Dubard, in *Ann. Sc. Nat.* xvii. 169, plate 2, fig. 4 (1903).

The witches' brooms, which are so common on birch trees, are generally supposed to be due to a fungus, *Exoascus betulinus*, the threads of which penetrate the young growing twigs, causing them to branch repeatedly and thus form large, irregular, nest-like clusters. Miss Ormerod,[1] however, states that these abnormal growths are caused by the development of unhealthy buds, which have been attacked by a gall-mite.[2] It is possible that in some cases it is the fungus, and in other cases the gall-mite, which is the cause of these witches' brooms.

A large number of birches were killed in 1900 in Epping Forest by a fungus, identified, by Paulson,[3] with *Melanconis stilbostoma*, Tulasne, which attacks the young growing branches.

## Varieties

This species is very variable in the wild state, both as regards the stature of the tree and the shape, size, and pubescence of the foliage. A large number of varieties have been distinguished by Continental botanists, of which the nomenclature is very confused; and as most of these are separated by inconstant characters and are of no value from the cultivator's point of view, it will be sufficient here to refer the reader to the works of Willkomm, Winkler, and Schneider, where the different forms are fully dealt with. The following varieties are, however, worthy of note :—

1. In Alpine and northern localities this species is often met with as a small shrub with twisted branchlets, but with leaves very variable in character. This group of forms may be distinguished as var. *tortuosa*, Koehne, *Deut. Dend.* 109 (1893).

2. Var. *Murithii*, Gremli, *Excursionsfl. f. d. Schweiz*, 365 (1893).

B. *Murithii*, Gaudichaud, *Fl. Helv.* vi. 178 (1830); Christ, *Ber. Schweiz, Bot. Ges.* v. 16 (1895).

An Alpine shrub, occurring in the Bagnes valley, near Mauvoisin (Valais), in Fribourg, and in the Joux valley (Vaud) in Switzerland. This has broadly ovate or ovate-triangular leaves, with large simple serrations, and prominent reticulate venation beneath.

3. Var. *denudata*, Grenier et Godron, *Fl. France*, iii. 147 (1855). This name may be given to a series of forms, characterised by rhombic leaves, cuneate at the base, and glabrous beneath or with only slight axil-tufts. This is often cultivated as var. *pontica*,[4] var. *carpatica*,[5] var. *odorata*, etc., and is usually a tree of considerable size.

4. Var. *urticifolia*, Spach, in *Ann. Sc. Nat.*, sér. 2, xv. 187 (1841); Schneider, *Laubholzkunde*, i. 117 (1904).

Betula *urticifolia*, Regel, in *Mém. Soc. Nat. Mosc.* xiii. 115 (1860); Willkomm, *Forstliche Flora*, 313 (1887); Winkler, *Betulaceæ*, 80 (1904).

[1] *Injurious Insects*, 212 (1890).

[2] Gillanders, *Forest Entomology*, 25 (1908), identifies the gall-mite with *Eriophes rudis*, Canestrini; and gives a figure of swollen buds on the branch of a birch tree. These had been found in close proximity to a witches' broom.

[3] *Essex Naturalist*, xi. 1, p. 273 (1901). Cf. also *Nature*, lxii. 599 (1900). Mr. Massee thinks that root-rot or unsuitable soil conditions, rather than the fungus, were the cause of death of these trees.

[4] There is a good-sized healthy tree in the Botanic Garden at Glasnevin, under the name of B. *alba*, var. *pontica*, which was mentioned by Loudon as being, in 1838, thirty-five years old and 35 feet high.

[5] The true var. *carpatica* (B. *carpatica*, Waldstein and Kitaibel, in Willdenow, *Sp. Pl.* iv. 464 (1805)) is a low tree allied to var. *tortuosa*, the distinctive characters of which are given in Schneider, *op. cit.* 119.

This variety, which was long considered to be a distinct species, has been shown by Beissner and Schneider to be an abnormal form of *B. pubescens*. It differs from the type in the longer, more acuminate, slightly lobed leaves; and in the fruiting-catkins, which are very long (1½ inch or more) and slender, with pubescent ciliate scales, the middle lobe of which is elongated. The leaves and fruit are shown in Plate 269, Fig. 2.

This peculiar form has been found wild in the province of Wermland, in Sweden, and is only a tree of small dimensions. It is often planted in botanical gardens.

5. Some peculiar forms have arisen in cultivation, as var. *aurea*, young foliage tinted with yellow, sent out a few years ago by G. Paul, Cheshunt Nurseries; and var. *nana*, a dwarf form.

## HYBRIDS

1. Hybrids have often been observed between this species and *B. verrucosa*, and have received various names, as *B. hybrida*, Bechstein, in *Diana*, i. 80 (1797); *B. aurata*, Borkhausen, *Forstbot.* i. 498 (1800); *B. glutinosa*, Wallroth, *Sched. Crit.* 497 (1822); and *B. ambigua*, Hampe, in Reichenbach, *Fl. Sax.* 120 (1842). These hybrids are intermediate in the characters of the branchlets, foliage, and fruit; and may be suspected in cases where the branchlets are more or less glandular and show slight pubescence.

2. *B. intermedia*, Thomas, in Gaudichaud, *Fl. Helvet.* vi. 176 (1830), a hybrid between *B. pubescens* and *B. nana*, is a shrub about 10 feet high, with small leaves, which is found in the Swiss Jura, Greenland, Iceland, northern Europe, and Siberia. This has been collected[1] in a few localities in the Highlands of Scotland.

## DISTRIBUTION

This species is widely distributed through Europe and northern Asia, extending farther northward than *B. verrucosa*, but not descending so far south. It is the most northerly tree in Europe, growing on the shores of the Arctic Sea from North Cape (lat. 71°) to the mouth of the White Sea; its northern limit eastward through north-eastern Russia and Siberia to Kamtschatka being near the Arctic circle, though in some localities it ascends a degree or two higher. Its southern limit in Russia and Siberia appears to be the edge of the steppes, on which the tree does not grow; but it occurs in the Caucasus and Armenia. Farther westward the southern limit is the Carpathians and the Alps; and the tree is not found in the Apennines or in the Pyrenees, its most southerly point in France being near Grenoble. It is also a native of Iceland and of south-western Greenland. The distribution may then be roughly described as the northern hemisphere, from Greenland in the west to Kamtschatka in the east, between the parallels of 45° and 71°.

As compared with *B. verrucosa*, this species is found on the continent of Europe on wetter soils and in moister climates, and is the birch which grows on marshy ground and on undrained peat-mosses.

[1] Cf. E. S. Marshall, in *Journ. Bot.* xxxix. 271 (1901), and *Bot. Exchange Club Report* for 1904, p. 33.

In Norway, Schübeler distinguishes this species as the highland birch, and speaks of *B. verrucosa* as the lowland birch, the latter not being found north of Snaasen in lat. 64° 12', or on the fells higher than 1600 to 1800 feet.

In Russia it forms large pure forests in the provinces of Olonetz and Vologda; while in Esthonia, Livonia, and Finland, it is mixed with pine, spruce, and aspen; farther south it is gradually replaced by *B. verrucosa*, with which, however, it is sometimes associated. Von Sivers[1] states that in the Baltic provinces this species is the characteristic tree of the low-lying moors, and on account of its resistance to May frosts, holds its own with the spruce and common alder. On better soils it forms immense forests, where *Betula humilis* and *Rhamnus frangula* are the underwood, and which are the favourite summer resort of the elk. It never attains such large dimensions as *B. verrucosa*, scarcely ever surpassing 100 feet in height.

In northern Germany, large forests, composed mainly of this species and common alder, are common on marshy ground. In France it is usually met with in the moister parts of the forests or on peat-mosses.

It appears to be much more common, as a wild tree, in the British Isles, than *B. verrucosa*, the moist climate being favourable to its development; and the extensive birch forests of the Highlands of Scotland are usually *B. pubescens*. As the two species have not as a rule been distinguished by collectors, and no discrimination has been made between planted and wild trees, it is impossible at present to give an accurate account of the distribution of the two species in this country.                                                                                      (A. H.)

The distinction between the two forms or species of common birch which is almost universally admitted by Continental botanists and foresters appears to have been generally overlooked in Great Britain; and though most local floras admit both, yet, after much inquiry and investigation, I have found it impossible to define their distribution as indigenous trees. In many districts where the birch now reproduces itself by seed abundantly, the original parents were of both forms, which are not distinguished by nurserymen, though the name "silver birch" is supposed to be, and should correctly be used for the rough-twigged form. The bark of this being more silvery—though this character is variable and disappears with age—and the habit more pendulous and graceful, it should be chosen as an ornamental tree. But, where birch is planted on peat bogs or wet moors in order to produce a timber crop, and to prepare the ground for planting other trees, as has been recommended by Mr. G. U. Macdonald,[2] or to act as a nurse for other trees, it would be preferable to use the downy-twigged form, which is considered to be more naturally at home and to thrive better on wet than on dry rocky soil.[3]

But both forms as well as their hybrids grow together in many parts of England and Scotland, and my experience in planting them does not justify me in saying that there is a marked difference in their relative growth or size. However,

---

[1] *Forst. Verhält. Balt. Prov.* 18 (1903).

[2] "Protection of Young Spruce from Frost," in *Trans. Roy. Scot. Arb. Soc.* xix. 287 (1906).

[3] M. Bommer, Director of the Botanic Garden at Brussels, pointed out to me in the Museum there, characters in the bark by which he could distinguish the two common birches. It seems to me, however, that bark is, especially in the birch, so much influenced by climate, soil, and the age and vigour of the trees, that these characters were not reliable.

when visiting the nursery at Balmoral in 1904, I noticed self-sown seedlings of both forms, in which the distinction was very striking, and took some away with me, which I planted on my trial beds at Colesborne. Of these the growth of *B. verrucosa* has been immensely superior; and Mr. J. Michie, Commissioner to His Majesty at Balmoral, writing to me on the subject, says, "I have no doubt about the rough and smooth twigged birches remaining constant through life; they are distinct varieties, and in nature grow side by side frequently. I do not say that the rough-twigged variety always develops a pendulous habit with age, but it grows larger and has a lighter colour of bark. It generally weeps, on the same ground where the other remains rigid, of less size, and with darker bark."

Birch is the most Alpine tree in Great Britain, and ascends in the Highlands to about 3000 feet.

The Gaelic name of the birch is *beith* (pronounced bey), and according to Sir Herbert Maxwell,[1] is found in various forms in Scottish place-names, as Drumbae, the birch-ridge, Auchenvèy and Largvey in Galloway (*achadh-na-beith*, birch field, and *learg-bheith*, birch hill-side). Beòch in Ayrshire, Galloway, and Dumfriesshire is *beitheach* (beyagh), birch-land.

In Ireland, this word [2] occurs in many names of places, as Ballybay in Monaghan (mouth of the ford of the birch), Kilbeheny, a village in Tipperary (birch-wood), Aghavea in Fermanagh (birch-field). Beagh is also a common place-name in different parts of Ireland.

(H. J. E.)

## BETULA VERRUCOSA, Silver Birch

*Betula verrucosa*, Ehrhart, *Beit. Naturk.* vi. 98 (1791); Willkomm, *Forstliche Flora*, 314 (1887); Mathieu, *Flore Forestière*, 407 (1897); Winkler, *Betulaceæ*, 75 (1904).
*Betula pendula*, Roth, *Tent. Fl. Germ.* i. 405 (1788); Schneider, *Laubholzkunde*, i. 113 (1904).
*Betula alba*, Linnæus, *Sp. Pl.* 982 (1753) (in part).
*Betula alba*, Linnæus, var. *pendula*, Aiton, *Hort. Kew*, iii. 336 (1789); Loudon, *Arb. et Frut. Brit.* iii. 1691 (1838).
*Betula rhombifolia*, Tausch, in *Flora*, xxi. 2, p. 752 (1838).
*Betula lobulata*, Kanitz, in *Linnæa*, xxxii. 351 (1863).
*Betula odorata*, Reichenbach, *Icon. Fl. Germ.* xii. 2, t. 626, f. 1288 (1850) (not Bechstein).

A tree, attaining in Russia 120 feet in height and about 12 feet in girth. Bark like that of *B. pubescens*, but more silvery white in colour. Main branches ascending, smaller branches and branchlets pendulous. Young branchlets glabrous, with scattered minute glands, which are persistent in the second year.

Leaves (Plate 270, Fig. 9) about 1½ to 2 inches long, and 1 to 1½ inch broad, deltoid, with a broadly cuneate base and an acuminate apex; margin biserrate; nerves five or six pairs; both surfaces glabrous and glandular; petiole ¾ inch, glabrous, glandular.

Fruiting-catkins (Plate 270, Fig. 9) cylindrical, about 1 inch long and ⅓ inch wide, directed towards the apex of the branchlet, on a slender, glabrous, glandular

---

[1] *Scottish Land Names*, 109 (1894).          [2] Joyce, *Irish Names of Places*, i. 506 (1883).

stalk, about ½ inch long ; scales glabrous, ciliate, with rounded, recurved lateral lobes larger than the middle lobe.

The twigs in winter are slender, shining, glabrous, covered with scattered glands and waxy patches. Buds ovoid, acute, ¼ inch long, appressed to the branchlet, with brown glabrous scales.

The shoots, which spring, as in *B. pubescens*, from the roots, after a tree is felled, are covered with numerous glandular warts and layers of wax, and bear large, incised, pubescent leaves. The seedlings of this species have simply serrate, pubescent leaves resembling those of seedlings of *B. pubescens*,[1] but conspicuously glandular on both surfaces ; the stem is pubescent, but bears numerous glands.

## Varieties

1. Var. *dalecarlica*, Linnæus, f., *Suppl.* 416 (1781), Fern-leaved Birch. Leaves (Plate 270, Fig. 10) produced into a long acuminate apex, and with the margin cut into pinnatifid serrated lobes. This variety has been found growing wild in the provinces of Dalecarlia and Wermland in Sweden, and is occasionally seen in cultivation, there being a good specimen in Kew Gardens.[2] It is sometimes known in nurseries as *var. laciniata*.

2. Shrubby forms, with leaves smaller than in the type, have been distinguished as var. *oycowiensis*, Regel, in DC. *Prod.* xvi. 2, p. 164 (1868), found growing wild in Silesia and Galicia ; and var. *arbuscula*, Winkler, observed by Fries in the wild state in Dalecarlia.

3. Var. *japonica*, Rehder, in Bailey, *Cycl. Am. Hort.* i. 159 (1900) ; Schneider, *Laubholzkunde*, i. 112 (1904).

> *Betula japonica*, Siebold, *Verh. Batav. Gen.* xii. 25 (1830) ; Winkler, *Betulaceæ*, 78 (1904).
> *Betula latifolia*, Tausch, *Fl. Ratisb.* 751 (1838) ; Komarov, *Act. Hort. Petrop.* xxii. i. p. 38 (1904).
> *Betula alba*, Linnæus, sub-species *latifolia*, Regel, in *Bull. Soc. Nat. Mosc.* xxxviii. 399 (1865), and in DC. *Prod.* xvi. 2, p. 165 (1868).
> *Betula alba*, Linnæus, var. *Tauschii*, Shirai, in *Tokyo Bot. Mag.* viii. 319 (1894).

In eastern Asia, in Manchuria, Saghalien, and Japan, the common birch is represented by a series of forms which have been grouped together by Winkler under the name *B. japonica*, Siebold. In some respects they approach more closely *B. verrucosa* than *B. pubescens*, and are perhaps best treated as a geographical variety of the former species. Trees of Japanese origin cultivated in Kew Gardens show the following characters :—Young branchlets sparingly glandular, glabrous or with a few scattered hairs. Leaves, 2 inches long, 1¼ inch wide, ovate, cuneate at the base, acuminate at the apex ; margin ciliate, sharply and simply serrate ; nerves seven or eight pairs ; upper surface with scattered pubescence ; lower surface light green, glandular, glabrous except for slight pubescence on the midrib and nerves ; petiole, ¾ inch, glabrous, glandular. Fruiting-catkins about 1 inch long, ⅓ inch wide,

---

[1] Cf. Watson, *Compendium*, 560 (1870) ; and Kerner, *Nat. Hist. Plants*, Eng. Trans. ii. 514 (1898).

[2] According to Schübeler, p. 461, this beautiful variety was first found in 1767 at Lilla Ornäs, about seven English miles south of Falun in Sweden, when it was quite a small tree, 6 feet high, but grew to be in 1878, 64½ feet high, with a trunk 6 feet 8 inches in girth. An excellent illustration of it is given by Schübeler (Fig. 86) with outlines of the leaves (Fig. 87).— (H. J. E.)

cylindrical, on long, slender, minutely pubescent, glandular stalks; scales long-stalked, veined, puberulous, ciliate, with the triangular central lobe slightly longer than the broad, rounded, recurved, lateral lobes; nutlets with rather broad wings.

This variety[1] is distinct in the larger number of nerves in the narrower, longer, slightly pubescent leaves, which are simply serrate in margin; and in the characters of the fruit-scales. The bark of cultivated trees is more like that of *B. Ermani* than the common birch, as it is uniformly white in colour, with raised whitish lenticels, and scales off in transverse shreds. According to Sargent,[2] it is a slender tree, attaining about 80 feet in height in Yezo.

There are three trees of this variety in Kew Gardens, about 20 to 25 feet in height, which were raised from Japanese seed sent by Sargent in 1891 under the erroneous name *B. ulmifolia*. A similar tree,[3] 25 feet high, cultivated at Kew as *B. alba*, var. *latifolia*, was obtained from Madrid in 1887. These trees are narrowly pyramidal in habit, and very ornamental on account of their beautiful white bark; and appear to be fast in growth and very thriving.

4. Several varieties have arisen in cultivation, of which the most noteworthy are:—

Var. *Youngi*, Schneider, Young's weeping birch; and var. *elegans*, Schelle, Bonamy's[4] weeping birch. Both these forms have long, slender, pendulous branchlets; and are usually grafted on stems 6 to 8 feet high, when they assume the habit of the weeping sophora. A fine specimen is growing in Smith's nursery at Worcester.

Var. *fastigiata*, Schelle, is characterised by its upright branches, the tree resembling in its appearance a Lombardy poplar. According to a writer in *Woods and Forests*, this variety retains its foliage later in autumn than any other form of the silver or common birch.

Var. *purpurea*.[5] Leaves purple, resembling in colour those of the purple beech, valuable for ornamental planting.

### DISTRIBUTION

This species is widely distributed in Europe, and in northern and eastern Asia. The northern limit, beginning in Scotland, crosses Norway in lat. 64°, Sweden in lat. 65°, and ascends in Russian Lapland to Lake Ruanjärvi; and thence, crossing Lake Onega, passes through the province of Vologda to Siberia, where its exact distribution has not been made out. In eastern Asia, var. *japonica* is met with in Manchuria, Saghalien, and Japan. The type occurs in the mountains of north China, and was found near Lake Kokonor (lat. 37° 50') by Przewalski; and it appears to be the common birch in the Altai and Ural mountains. It is not found in Persia or Afghanistan, but occurs on the higher mountains of the Caucasus and in Armenia and Asia Minor. In Europe the southern limit extends from the

---

[1] The Japanese name for this variety is *Shira-Kamba*.    [2] *Forest Flora of Japan*, 61 (1894).

[3] This tree has broadly ovate leaves, subcordate or rounded at the base, and larger than those of the trees raised from Japanese seed, sent by Sargent; but in other respects is identical, and is probably also of Japanese origin.

[4] This originated in Bonamy's nursery at Toulouse, and is usually known in gardens as *B. alba pendula elegans*. Cf. *Rev. Hort.*, 1869, p. 135, fig. 33, and *Gard. Chron.*, 1869, p. 1278.

[5] Probably identical with var. *atrosanguinea*, stated by Schübeler to have originated in France, and to be growing in the Botanic Garden at Christiania.—(H. J. E.)

Rhodope mountains in Bulgaria, through Servia, Bosnia, and Istria to the Venetian Alps; and following the southern limit of the Alps in northern Italy, it extends from the maritime Alps along the Apennines to Aspromonte in Calabria, crossing over to Sicily, where it reaches its extreme southerly point on Mount Etna in lat. 37° 40'. It does not occur in Corsica or Sardinia. It is common in the forests of the plains and lower mountains of France in the north, east, and west; but towards the south only grows at high elevations, as in the mountains of Auvergne and in the Pyrenees. It grows in Portugal in the Sierra de Gerez; and in Spain throughout the northern mountains, in Catalonia, Aragon, Navarre; and also in the Sierra Guadarrama, the mountains of Toledo, and the Sierra de Gata.

The largest forests of the species occur on the plains of the Baltic and central provinces of Russia, where it grows either pure or in mixture with aspen and grey alder, or with the common pine and spruce. Von Sivers[1] points out that the two species of birch occur in the Baltic provinces in different soils and situations. *B. verrucosa* grows on the glacial drifts, where it reaches large dimensions, and often forms pure forests of clean, straight stems, which on the better class of soil, amongst spruce, often reach 100 to 130 feet in height.

There are also extensive pure woods in the plains of northern and central Germany; but farther south the tree is more at home in the mountains, as in the Alps and Carpathians, and only forms small woods, or grows scattered or in groups amidst other trees.

This species is most common in continental Europe on dry soils, thriving best in localities where the common pine does well, as in loamy sands with a moderately moist subsoil; but dwindles and ceases to grow on marshy ground or on undrained peat-mosses. It requires more light than the other species, and in woods of *B. verrucosa* the soil is usually covered with grass; the leaf mould and moss, so common on the ground in woods of *B. pubescens*, being usually absent.

(A. H.)

### CULTIVATION OF THE COMMON AND SILVER BIRCH

After the oak, there is perhaps no tree which has been so generally attractive to artists and lovers of the picturesque as the birch, which will grow almost anywhere, and is often looked on by English foresters and woodmen as a weed. This it may be on land fit to grow fine timber; where, however, it is not often so prevalent as on poor dry soils, or on wet, boggy land; but when the question of covering waste land with timber of some sort at a low cost has to be considered, there is no tree that will do it so cheaply and so surely as the birch.

It seeds very profusely, and the seed is so light that it spreads with great rapidity, and germinates in places where hardly any other tree will live. It is absolutely the hardiest tree we have, and though its economic value is low at present, yet probably it will, when our coal gets scarce and dear, be looked on as the cheapest and best of firewood.

[1] *Forst. Verhält. Balt. Prov.* 18 (1903).

In north Russia it is the usual fuel for railway engines, and all over Scandinavia is the principal firewood, but though our climate does not produce the tree as well as that of more northern regions, where the bark is almost the only material used for covering the roofs of common buildings; yet, having regard to the great beauty of the tree in landscape, it should be much more generally grown than it is. All this was well brought out by Loudon many years ago in his great work, and yet the birch remains a neglected tree. But it has another virtue which must appeal to many in these days. It is of all trees the one most distasteful to rabbits, and on my property is the only tree which grows up from self-sown seed, on land which on account of its poverty has been treated as a rabbit warren.

As a nurse for other trees on poor land, whether of a dry and rocky nature, or wet and peaty, the birch seems to me to have a greater value than most writers on forestry have admitted. For wherever the soil is naturally covered by self-sown birch, or a fairly thick crop can be obtained by sowing, the land will be made fit for the planting of more valuable trees, such as larch or Corsican pine, at a lower cost; and after providing shelter for smaller plants than could otherwise be used, it can, when the permanent crop has been established, be cut and sold at an age when spruce or Scots pine of the same age would be worthless.

A very successful instance of its use in this way has been described[1] by G. U. Macdonald, forester at Raith, Fifeshire, the object here having been to plant moorland with spruce, in a locality where the late and early frosts were so severe that the spruce would hardly grow at first without some protection.[2]

On very dry oolite soil, birch is the only tree which reproduces itself naturally among long coarse grass, which it will, if thick enough, eventually suppress; and, though a large quantity of small birch wood may not be always saleable at as good a price as at Raith (20s. per ton for crate-wood), yet it is such excellent firewood, even when quite small, that having regard to the low cost of its seed, I can suggest no means whereby the desired result could be obtained so cheaply.

If desired to establish a birch covert by sowing, I would advise the careful selection of seed from trees naturally growing on land of similar character, because though foresters in this country have not yet realised the preference shown by the rough twigged birch for dry rocky land, it is universally accepted as a fact in Germany, whilst for wet or boggy land the downy twigged birch is preferable.

To raise birch from seed is not always easy; and whether it is better to sow in autumn directly the seed is ripe, or in spring, is a question which, after trying both plans, I have not yet decided to my own satisfaction. But, as a rule, I would follow nature and sow in autumn, not attempting to cover the seed with earth, but covering with some fir boughs, fern, or leaves, until it began to germinate. So far

[1] *Trans. Roy. Scot. Arb. Soc.* xix. 287 (1906).

[2] It has been used with great success as a nurse for beech in some of the plantations, which were made by the Danish forester, Ulrich, near Copenhagen, and which were shown to the Royal English Arboricultural Society, in August 1908. The object here was to protect young beech trees from spring frosts, and afford shade during their youth. Birches were planted in lines about 8 feet apart, and the beech planted between the rows of birch eight or ten years later. When the beech are sufficiently tall, the birch are thinned and finally cut out entirely and used as firewood. This system seemed to me to be one well worthy of adoption in England for other trees which require shade in youth.

as I have seen the germination is slow and irregular, and seems to depend a good deal on the maturity and age of the seed. For though birch seed is best sown the first year, I have had fair results from seed which had been kept a year in a bag, whilst I have sometimes obtained poor results from fresh seed sown in spring. I always sow the seed of exotic birches in pots or under glass, and prick out the seedlings at a year old.

As a rule the seedlings grow fast, and must not be left long in the nursery, as they do not transplant well when old, but there is much variation among the different species; and it seems that some of the American birches do not grow well or live long in this country, unless grafted on the roots of the common birch.

As a rule the birches seem to grow best in nature when unmixed with other trees; and in some of the best birch woods I have seen in Norway, Japan, and America are almost pure, but if mixed with conifers or other hardwoods, and not suppressed by their shade, they often attain large dimensions.

### REMARKABLE TREES, COMMON AND SILVER BIRCH

Among the numerous birches which I have measured in England, I cannot mention any tree which is equal in height to some that I have seen in Norway and Russia, as the tallest do not exceed about 90 feet. Along the road which passes through Savernake Forest from Marlborough to Andover there are a great number of large and beautiful trees, planted as an avenue on both sides; but of their age I can obtain no record. Near the school is one of the finest (Plate 225), which measured in 1908 about 90 feet high by 8 feet in girth. Another near it was covered with large witches' brooms; and a third, from which a large branch had been recently torn off by the wind, was pouring out sap in such quantity that a pool of it had formed on the ground. In this park a birch, which was considered by the woodman to be the largest, and which like the rest appeared to be *B. pubescens*, measured 77 feet by 10 feet 7 inches.

Plate 256 shows the graceful habit of a fine birch in front of Lord Walsingham's house at Merton, Norfolk. Plate 257 shows a group of self-sown birches in Sherwood forest, close to the Queen oak, described on page 322 of this work.

At Dropmore there is a birch (*B. pubescens*) about 55 feet high, with an immense bole 21 feet in girth at a foot from the ground, and dividing a little way up into three main stems, 9 feet 10 inches, 8 feet 2 inches, and 6 feet 3 inches in girth respectively. One large limb has rested on the ground for many years, but does not seem to have rooted.

In Windsor Park there are numerous fine birches, one of the handsomest of which, growing by Prince Albert's chapel, was in 1904 71 feet by 8 feet 8 inches. At Longleat there is a fine tree of which Colonel Thynne has sent me a photograph, and which measured 76 feet by 9 feet 4 inches in 1906. At Barton there is a tree from 80 to 85 feet high, with a clean stem about 50 feet long and 5 feet 7 inches in girth, and drooping branches. This tree is still young and vigorous, and is one of the finest we have seen anywhere.

At Beauport, Henry saw one dividing at 5 feet into five stems, and girthing below the fork 11 feet.   At Arley Castle he measured one 76 feet by 8 feet 1 inch in 1905.   At Croome Court there is a tree 77 feet high, of which one large stem has been broken off short.   Mr. J. Smith mentions [1] a birch growing at Embley Park, Hants, in 1887, which was 85 feet by 6 feet 7 inches, with a bole of 25 feet.

In Wales the birch, so far as I have seen, does not attain so large a size as in England and Scotland, the finest I know of being at Ogwenbank, near the entrance to the great slate quarry at Penrhyn.   This, though only about 50 feet high, spreads over an area 25 paces in diameter, and has two main stems which are 13 feet and 12 feet 2 inches in girth respectively.

In Scotland, Mr. Renwick considers a birch (*B. verrucosa*) at Auchendrane, in Ayrshire, to be the finest in the west of Scotland.   He gives an account of this tree, with a photograph, in *Trans. Nat. Hist. Soc. Glasgow*, vii. 262 (1905).   It is 67 feet in height, with a bole of 13 feet, girthing 10 feet 8 inches ; and was planted, according to Miss Cathcart, by her mother in 1818, having been purchased from Booth's nursery near Hamburg as a cut-leaved weeping birch.

A still larger tree at Newton Don, near Kelso, which was cut down in 1901, measured, in 1893, 80 feet high, with a short bole, 13 feet in girth at 1 foot 7 inches above the ground, and dividing at 3 feet up into two main stems.   Captain C. B. Balfour informs us that Jeffrey, in his *History of Roxburghshire*, describes it in 1859 as being then 74 feet high and 14 feet in girth at the base.

At Monzie there are some tall birches, drawn up by other trees, one measuring 90 feet high by 8 feet in girth.   At Blair Drummond there are several old birches, all with large boles, some with remarkable burrs, and one with low spreading branches layering.   One of these measured 60 feet by 13 feet 10 inches ; and another 70 feet high is 10 feet 8 inches in girth.

In Darnaway Forest there are many fine birch trees on the banks of the Findhorn, one of which was stated by Sir Thomas Dick Lauder to girth 13 feet ; but I am informed by Mr. D. Scott, forester to the Earl of Moray, that many of these have died, and those remaining are fast decaying.   The largest measure 9 to 10 feet in girth, and many of them contain 70 to 75 cubic feet of timber.   The rough twigged birch predominates here, but does not as a rule assume a pendulous habit until it is of some age.

At Gordon Castle there is a fine tree in the park, which in 1904 was 68 feet by 9 feet ; and at Murthly, in the drive from Dunkeld, I measured in 1906 a very tall and slender birch (*B. verrucosa*) which was 89 feet high and only 3 feet 9 inches in girth.

In the Pass of Killiecrankie and many other Highland glens the birch grows freely mixed with oak on the rocky slopes, and in the wide valley of the Spey there are beautiful open woods of pure birch, covering a large extent of the gravelly flats and terraces which every traveller on the Highland Railway between Kingussie and Grantown must have admired.   In the swampy flat at the head of Loch Morlich, in Glenmore, there is an open wood of curiously distorted, twisted, and stunted

[1] *Trans. Scot. Arb. Soc.* xi. 532 (1887).

birches, mixed with alders, in the hollow trunks of which the goosander occasionally breeds, and the goldeneye duck may also sometimes do so (Plate 258).

In no part of Great Britain do the birches assume the same tall, clean growth, or have the same smooth, silvery bark that they do in the forests of Scandinavia, where pure woods of birch are in some districts very prevalent on the lower land, whilst on the fjelds and mountains it ascends as a scrubby and stunted tree to a greater elevation than any other. Schübeler[1] describes and figures some instances of abnormal growths in birch. His figure 88 shows a fallen trunk from which six healthy-looking young trees are growing in a line, and I have a similar though less striking instance in my own woods. The tallest birch he knew in Norway was at Drobak in Sœterdalen, 30 kilometres south of Christiania, and measured 100 feet by 5 feet 9 inches. Another at Sondre Tveten, in Eidanger, was 79 feet by 11 feet. He figures a remarkable tree at Dunserud, in Eker, which divides into six large trunks, and measures 75 feet in height. Perhaps the most shapely and beautiful of those which he figures, are a tree at the farm of Hohls, in south Trondhjem Amt, measuring 80 feet by 11 feet, and another at Gravrok, 18 kilometres south of Trondhjem, which measured 80 feet by 16 feet. It seems from these particulars that the birch is the largest deciduous tree in Norway, and attains greater dimensions than in this country; but though I have spent several months in the forests of north and south Trondhjem in the pursuit of my favourite quarry, the elk, I never saw such trees as those above mentioned, and believe that they all grow in the neighbourhood of farms on unusually fertile soil.

## TIMBER AND BARK

The timber of the birch can hardly be said to have any general recognised value in England, though in some districts it can be sold to coopers, chairmakers, and clogmakers;[2] in others, especially since charcoal burning has ceased in most places to be a profitable industry, it can only be looked on as firewood. It is so perishable in contact with the soil, that it is of no use for fencing unless creosoted; and though in former times,[3] according to Sang, the Highlanders made everything that they used of it—rafters, ploughs, harrows, carts, and fences—yet now it would only be used as a makeshift, when other wood could not be had. I have seen large old burry birches, which when cut into boards, were fit for small cabinet panels; but the wood twists a good deal in drying, and is usually inferior, in grain, texture, colour, and figure to the wood of several species of American birch which can be imported in larger size and at a low price. As underwood it has an uncertain value for making brooms and tool handles and is also used for bobbins.

In Sweden and Russia the burrs found on the trunks of this tree are converted into many ornamental articles of great beauty. Beer-pots carved out of these burrs, and hooped with wood or silver, are often heirlooms in Scandinavian houses;

---

[1] *Viridarium Norvegicum*, 469.
[2] Birch clog soles are used in the Yorkshire manufacturing towns, whilst in Lancashire alder is preferred.
[3] Pennant, *Tour in Scotland*, 112 (1772), says that wine was extracted from the live tree.

and as the wood takes a high polish, it is highly valued in Siberia for work-boxes, cigarette cases, and other small fancy articles.

In northern Sweden and Russia the wood is sometimes found full of undulations, which make it very ornamental for furniture, and some bedroom furniture made by the Nordiska Kompaniet (Lundberg and Laja) of Stockholm, was almost equal in beauty to satinwood. I have also seen it used as veneer with the best effect for decorating cabins in steamers built in Denmark and Sweden. This is known in Sweden as " Flammig björk." Another curious form of birch wood is that known in Finland and Sweden as " Masur." I was informed by Mr. Jacobssen, Swedish Vice-Consul at Åbo, that this variety in Finland is only found in certain places, Karku, Tyrois, and Kalvola. A number of logs which I saw in the works of the Finska Colorit Aktiebolag at Åbo were covered on the outside with small pitted depressions, somewhat similar to those which produce bird's-eye maple, and when cut into veneer, are dyed of various colours, of which French grey seemed to me the most effective ; and made up into furniture which commands a high price.

At St. Petersburg this form of birchwood is known as " Karelsky," being supposed to come from the Karelian peninsula; and is largely used both in the solid and as veneer for furniture making. Though not so elegant as the waved form, or as the bird's-eye maple which it somewhat resembles, it is very quaint and striking in appearance, and can be imported at very reasonable prices.

The bark,[1] when taken off in sheets, is used in Scandinavia for covering the roofs of houses, and remains for many years undecayed between the inner boarding and the outer sod of turf. A strong smelling oil, obtained by destructive distillation from birch wood, is, when mixed with alcohol and rubbed on the skin, the best protection I know of against the swarms of midges and mosquitoes which make life almost unbearable in the short summer of the far north. This oil is used as a preservative, and gives the fragrant odour to Russia leather. Birch bark has no equal for lighting fires, and in the dripping forests of the north I have often had good reason to value it when nothing else would start a fire.           (H. J. E.)

## BETULA DAVURICA

*Betula davurica*, Pallas, *Fl. Ross*, i. 60, t. 39 (1784); Winkler, *Betulaceæ*, 86 (1904).
*Betula Maximowiczii*, Ruprecht, in *Bull. Phys. Math. Acad. Pétersb.* xv. 139 (1856) (not Regel).
*Betula Maackii*, Ruprecht, in *Bull. Phys. Math. Acad. Pétersb.* xvi. 380 (1857).

A tree, attaining 60 or 70 feet in height. Bark purplish brown, separating in small, papery scales, which remain attached, curled and ragged, to the trunk, giving the tree a peculiar appearance. Young branchlets glandular, covered with a minute erect pubescence, interspersed with a few long hairs. Leaves, about 3 inches long and $1\frac{1}{2}$ to 2 inches wide, narrowly ovate or ovate-rhombic, cuneate at the base,

---

[1] Pyrobetulin, obtained by sublimation from the outer bark of birch, is used for depositing films on glass, about to be engraved, and for covering lint with an antiseptic layer. Cf. Wheeler, in *Pharm. Journ.* ix. 494 (1899).

acute or acuminate at the apex; margin ciliate, coarsely and irregularly serrate; nerves, six to eight pairs; upper surface at first pubescent on the midrib and nerves, ultimately glabrescent; lower surface glandular, with scattered pubescence on the midrib and nerves; petiole, ½ inch, slightly pilose.

Fruiting catkins, ¾ to 1 inch long, ovoid-cylindrical, acute at the apex; scales glabrous with scattered glands on the margin and outer surface, middle lobe triangular, lateral lobes broad, rounded and spreading; nutlet obovate, with narrow wings, broadest in their upper part.

This species, which is readily distinguished by its peculiar bark, is widely spread throughout Amurland, Manchuria, Korea, and north China. According to Komarov,[1] it grows throughout the whole of Manchuria, in the drier parts of the valleys and in open places on the mountains, on rocky or sandy soil, occasionally forming small woods, but is never seen in the dense virgin forests.

It is extremely rare in cultivation, the only specimen which we have seen being a small tree in Kew Gardens, about 15 feet high, which was raised from seed sent by Bretschneider from Peking in 1882.                                    (A. H.)

## BETULA CORYLIFOLIA

*Betula corylifolia*, Regel and Maximowicz, in *Bull. Soc. Nat. Mosc.* xxxviii. 417, t. 8 (1865); Regel, in DC. *Prod.* xvi. 2, p. 178 (1868); Winkler, *Betulaceæ*, 59, fig. 17 (1904).

A tree, the dimensions of which are not stated. Young branchlets slightly pubescent. Leaves (Plate 270, Fig. 14), 2½ to 3 inches long, about 2 inches wide, oval; rounded or truncate (rarely cuneate) at the base, acute at the apex; coarsely serrate; nerves, twelve to fourteen pairs, impressed above and very prominent beneath; upper surface pilose on the midrib, elsewhere glabrous; lower surface pale in colour, with conspicuous long silky hairs on the midrib and nerves, elsewhere glabrescent; petiole, ¾ inch, at first pilose, later glabrescent. Fruiting catkins (Plate 270, Fig. 14), 1½ to 2 inches long, ¾ inch wide, cylindrical, often curved; scales large, slightly pubescent, ciliate in margin, deeply three-lobed, lobes linear-oblong, the middle one about twice as long as the lateral lobes; seeds with very narrow wings.

This remarkable species,[2] peculiar in the shape of the leaf and in the stout, long fruiting catkins, was found on the high mountains of the provinces of Senano and Nambu in the main island of Japan, by Tschonoski. Very little is known about it in the wild state, and it has never apparently been introduced into cultivation in Europe.                                    (A. H.)

[1] *Flora Manshuriæ*, ii. 49 (1903).
[2] The Japanese name of this species, according to Matsumura and Goto, is *Urajiro-kamba*.

## BETULA MAXIMOWICZII

*Betula Maximowiczii*,[1] Regel, in *Bull. Soc. Nat. Mosc.* xxxviii. 418 (1865) ; Winkler, *Betulaceæ*, 89 (1904).

*Betula Maximowicziana*, Regel, in DC. *Prod.* xvi. 2, p. 180 (1868) ; Shirasawa, *Icon. Ess. Forest. Japon*, text 45, t. 23, ff. 1-8 (1900) ; Mayr, *Fremdländ. Wald- u. Parkbäume*, 449 (1906).

A tree, attaining in Japan, according to Mayr, 100 feet in height.   Bark grey, smooth, peeling off in thin, papery strips.   Young branchlets, with scattered glands ; glabrous, except for slight pubescence above the insertions of the leaves.   Leaves (Plate 269, Fig. 6) about 6 inches long and 4 inches wide, broadly ovate, deeply and narrowly cordate at the base, acuminate at the apex ; margin non-ciliate, biserrate ; nerves, ten to twelve pairs, each ending in a long-pointed serration ; upper surface at first pubescent with erect hairs, later glabrescent ; lower surface with scattered pubescence throughout, or glabrous, except for pubescence on the midrib and nerves, gland-dotted ; petiole an inch or more in length, pubescent or glabrous.

Fruiting catkins (Plate 269, Fig. 6), two to four in a raceme, about $2\frac{1}{2}$ inches long, and nearly $\frac{1}{2}$ inch in diameter ; scales glabrous, shortly three-lobed, the lateral lobes spreading and shorter than the middle lobe ; nutlets very small, with broad wings.

A variety of the species is in cultivation in Kew Gardens, distinguished by having smaller leaves, with more shortly pointed serrations, and with their under surface (as well as the young branchlets and petioles) covered with long, soft pubescence.

This species is readily distinguished by its large leaves, deeply and narrowly cordate at the base.   In winter the twigs are stout, shining, yellowish, nearly glabrous ; buds about $\frac{3}{8}$ inch long, appressed to the branchlet, curved laterally and ending in a sharp beak, with glabrous scales.

*B. Maximowiczii* occurs in Japan, in the central chain of Hondo, but is more common in Yezo,[2] where, according to Sargent,[3] it is a shapely tree, 80 or 90 feet in height, with a trunk 2 or 3 feet in diameter, covered with pale, smooth, orange-coloured bark.   Towards the base of old trees the bark becomes thick and ashy-grey, separating into long, narrow scales.

The largest that Elwes saw in Japan were growing in a mixed forest of maple, poplar, ash, spruce, and silver fir, on volcanic soil, at about 3000 feet elevation, near the Crater Lake, Shikotsu, in Yezo ; and one measured 90 feet high by 9 feet 9 inches in girth.   The Japanese name [4] of this species is *Udai-kamba*.

---

[1] This species, with *B. luminifera*, Winkler, *B. Bæumkeri*, Winkler, both natives of central China, and *B. alnoides*, Buchanan-Hamilton, distributed throughout the Himalayas and in central and southern China, constitute the section *Betulaster*, distinguished by elongated fruiting catkins and broad-winged nutlets.

[2] Mayr, *op. cit.*, plate 31, gives a picture of this tree growing in a forest in Yezo.

[3] *Forest Flora of Japan*, 62 (1894).

[4] According to Matsumura, in *Shokubutsu Mei-I*, 48 (1895).   The same name is given in Goto's *Forestry of Japan*.

It was introduced into cultivation in England by J. H. Veitch,[1] who sent home seeds from Yezo in 1888; and a large number of seedlings were raised at the Arnold Arboretum, in 1893, from seed received by Sargent[2] from the forestry officers of Yezo.

Sargent has spoken highly of the beauty and value of this tree, which, wherever we have seen it in this country, is thriving. It is one of the most beautiful of the young trees in Messrs. Walpole's lovely garden at Mount Usher, County Wicklow.

A tree at Kew, raised from seed obtained in 1893 from the Arnold Arboretum, was 25 feet high and 13 inches in girth in 1907. At Tortworth, a tree, probably of the same age, 30 feet high and 17 inches in girth, is growing vigorously. At Grayswood, Haslemere, a tree obtained as a small plant from Lemoine in 1894, was 29 feet by 21 inches in 1906, and is very healthy.

This species is common on the Continent in botanical gardens, and is very hardy, having borne without injury the severe winter climate of Grafrath, near Munich; and on this account, and because of its rapid growth, it is recommended by Mayr as worth cultivating as a forest tree.

Shirasawa says that the wood is rather hard, showing no difference in the colour of the sap and heart wood, and is used in Japan for house-building. The fishermen of Yezo make torches out of the bark, as it takes fire easily, even when wet.

(A. H.)

## BETULA ERMANI

*Betula Ermani*, Chamisso, in *Linnæa*, vi. 537, t. vi. f. 8 (1831); Erman, *Reise*, t. 17 (1835); Komarov, *Flora Manshuriæ*, ii. 49 (1903); Winkler, *Betulaceæ*, 66 (1904).

A tree,[3] attaining about 100 feet in height in Manchuria. Bark creamy-white, with raised whitish lenticels, and peeling off in irregular shreds. Young branchlets glabrous, except for a few hairs above the insertions of the leaves, and covered with numerous glands which roughen the shoot in the second year.

Leaves (Plate 270, Fig. 12), about 3 inches long, 2 inches broad, ovate, with a broad truncate or slightly cordate base, acuminate at the apex; margin slightly ciliate at first, coarsely and irregularly serrate; nerves, ten to twelve pairs; upper surface with scattered hairs; lower surface glandular, glabrous except for slight pubescence in the axils and on the midrib and nerves; petiole, $\frac{1}{2}$ inch, glabrescent, glandular.

Fruiting catkins (Plate 270, Fig. 12), ovoid-oblong, about 1 inch long and $\frac{5}{8}$ inch in diameter, sessile or shortly stalked; scales glabrous on the surface, ciliate and glandular in margin, with linear-oblong lobes, the middle lobe longer than the two lateral divergent lobes; nutlets with narrow wings, broadest above.

---

[1] *Hortus Veitchii*, 357 (1906).  [2] *Forest Flora of Japan*, 62 (1894).

[3] According to Matsumura, *Shokubutsu Mei-I*, 47 (1895), var. *nipponica* of this species is known as *Take-kamba*, while the type is called *Ezo-no-take-kamba*, i.e. *B. Ermani* of Yezo.

Var. *nipponica*,[1] Maximowicz, *Mél. Biol.* xii. 923 (1888).   This differs from the type in the branchlets not being so densely glandular, and in the longer, narrower cones, which are cylindrical, 1½ inch long, and ½ inch in diameter.   This variety was found by Maximowicz in the Nikko mountains, and is probably the form of the species occurring in the main island of Japan.

*B. Ermani* is widely distributed in eastern Asia, occurring in Kamtschatka, Manchuria, Korea, Saghalien, Kurile Isles, and Japan.   According to Komarov,[2] it grows in Manchuria, near the sea-coast, from the river Amur to St. Olga Bay, in the Sichote-Alin mountains, in the north-eastern part of the province of Mukden, and in the Korean main range; and is a native of mountain forests at elevations between 2000 and 7000 feet.   Komarov informs us that it is the largest of the birches, which he found in Manchuria, attaining a height of 60 to 100 feet.

According to Sargent,[3] it is the most common birch in the high mountains of the main island of Japan, where it is scattered through the coniferous forests at 4000 to 6000 feet, and is very conspicuous from the white bark of the trunk, and the bright, orange-coloured bark of the principal branches.   It appears to be the birch figured and described by Shirasawa,[4] as *Betula alba*, var. *communis* (not Regel), which he states to be a tree of vigorous growth, attaining 70 feet in height and 3 feet in diameter, occurring in the central chain of Honshu, especially at Shimotsuke, in Nikko, Musachi, in Chichibu, and Kiso, in Shinano.   At Kiso the bark is used as material for writing on, and for envelopes; and also for torches, as it contains a considerable amount of resin.   Elwes saw a birch, which he believes to be *B. Ermani*, growing in great abundance in the forest north of Asahigawa, and also on the volcanic cone near Lake Shikotsu, in Hokkaido.   Here it seemed to be commoner than, but not so large, a tree as *B. Maximowiczii*.

This species was introduced into cultivation through the agency of the St. Petersburg Botanic Garden; and is not uncommon in gardens both on the Continent and in England, where it is often wrongly named *B. ulmifolia*, *B. corylifolia*, etc.   At Kew there are small trees, 20 to 25 feet in height, both of the typical form and of var. *nipponica*.   The former[5] is the first of the birches to come into leaf, the foliage often being fully developed at the end of March; and, in consequence, the trees are usually much injured by spring frosts.   The largest tree we have seen of this species is in the arboretum at Westonbirt, and when measured by Elwes in 1908, was 51 feet by 2½ feet.   There is a good specimen apparently of the typical form at Benmore, in Argyleshire, which is grafted near the ground, and has yellowish scaly bark like that of *B. lutea*.   In 1907 Elwes found it to be about 40 feet by 3 feet.   Var. *nipponica* is later in leafing, and is not usually injured by frost.   Two trees of this variety at Grayswood, Haslemere, are about 30 feet in height.                                                                                                  (A. H.)

---

[1] This variety appears to be identical with a specimen in the Kew herbarium collected by Tschonoski on "high mountains not far from Fuji-yama," which is labelled *B. Bhojpattra*, var. *subcordata*, Regel, in DC. *Prod.* xvi. 2, p. 177 (1868).                                        [2] *Flora Manshuriæ*, ii. 50 (1903).

[3] *Forest Flora of Japan*, 62 (1894).

[4] *Icon. Essences Forest. Japon*, text 42, t. 21 ff. 1-15 (1900).   *Betula alba*, var. *communis*, Regel, is *B. papyrifera*, an American species, which certainly does not occur wild in Japan.

[5] The trees of the typical form were raised from seed sent by the Arnold Arboretum in 1893.

## BETULA ULMIFOLIA

*Betula ulmifolia*, Siebold et Zuccarini, in *Abh. Bayer. Akad. Wiss.* iv. 3, 228 (1846); Winkler, *Betulaceæ*, 62 (1904) (in part); Schneider, *Laubholzkunde*, i. 101 (1904).

A tree, attaining in Japan 70 feet in height and 8 feet in girth. Bark described as greyish-brown, smooth, shining, not separating into thin layers, and resembling that of *Prunus pseudocerasus*. Young branchlets covered with a white, short, somewhat appressed pubescence. Leaves, about 3 inches long, $1\frac{1}{2}$ inch wide, narrowly ovate or ovate-oblong, unequally cordate[1] at the base, acuminate at the apex; margin ciliate, bi-serrate, with falcate serrations; nerves twelve to fourteen pairs; upper surface with appressed, long, brownish hairs both on the midrib and nerves, and in bands between the nerves; lower surface similarly pilose on the midrib and nerves, gland-dotted and glabrous between the nerves; petiole $\frac{1}{2}$ inch, pilose.

Fruiting catkins, on short pilose peduncles, about $\frac{3}{4}$ inch long and $\frac{1}{2}$ inch in diameter, ovoid; scales pubescent, ciliate, strongly veined, with the central lobe oblong and obtuse, nearly twice as long as the ovate rounded lateral lobes; nutlets with narrow wings.

This species is extremely rare in cultivation, the only specimen which we have seen being a tree, about 8 feet high, in Kew Gardens, which was raised from seed, received under the name *B. grossa*,[2] S. et Z., from Tokyo in 1896. It is identical with the type specimen of *B. ulmifolia*, S. et Z., preserved in the Munich herbarium, with which we have compared it.

This species, together with *B. grossa*, S. et Z., *B. carpinifolia*, S. et Z., both natives of Japan, and *B. costata*, Trautvetter, a native of Manchuria, are closely allied; and our knowledge of their exact relationship and distribution is very imperfect. It is possible that *B. grossa* and *B. carpinifolia* are varieties or hybrids of *B. ulmifolia*, while *B. costata* is the continental geographical form of the same species.

Shirasawa,[3] whose figures of *B. grossa* and *B. ulmifolia* do not in either case exactly agree with the type specimen of the latter species, says that both these species are spread throughout the central chain of Hondo, and occur also in Kiushiu and Shikoku. *B. ulmifolia*, which is the representative in Japan of the American *B. lutea*, differs much in bark and other characters from *B. Ermani*, with which it has been confused. (A. H.)

I saw a very fine birch which my guide and companion, Mr. Mochizuki of the Japanese Forest Service, called *B. grossa*, growing in the forest of central Japan, at Ongawa, about 3000 feet above the sea; and measured specimens 80 to 90 feet high.

---

[1] The leaves on the lower part of the branchlet and on the short shoots are markedly cordate; those on the upper part of the branchlet are usually truncate or rounded at the base.

[2] Similarly a dried specimen at Kew, collected in the Etchu province on Mt. Tateyama, and labelled *B. grossa* by the Tokyo University Science College, is identical with *B. ulmifolia*. It bears the Japanese name *Yoguso-minebari*.

[3] *Icon. Ess. Forest. Japon*, text, 42, 43, t. 22 (1900).

The timber, of which I brought home a specimen now at Kew, is a hard wood of a bright pinkish brown colour, and is used for flooring. It seems at least as good as the best American birch timber. It was, however, very difficult to identify the species of birch, of which no less than five[1] are said to be found in this district, and the foresters of Japan were not themselves sure of their scientific names.

(H. J. E.)

## BETULA LUMINIFERA

*Betula luminifera*, Winkler, *Betulaceæ*, 91, fig. 23 (1904).

A tree, the dimensions of which are not known. Young branchlets covered with dense, erect, pale pubescence, non-glandular. Leaves (Plate 270, Fig. 16), about 5 inches long and $3\frac{1}{2}$ inches wide, broadly ovate, sub-cordate or truncate at the broad base, acuminate at the apex; margin ciliate, irregularly serrate, the serrations ending in cartilaginous points; nerves ten to fourteen pairs; covered more or less on both surfaces with white, short pubescence; petiole $\frac{3}{4}$ inch, pubescent. Fruiting catkins (Plate 270, Fig. 16), solitary, cylindrical, elongated, about 3 inches long and $\frac{2}{5}$ inch in diameter; on a peduncle $\frac{3}{4}$ inch long; scales lanceolate, auricled on each side a little below the middle; nutlets pubescent, with broad wings.

This species, which is the representative in central China of *B. Maximowiczii* of Japan, was discovered by Père Farges in the north-eastern mountains of Szechwan, and was introduced into cultivation in 1901 by E. H. Wilson, who sent seeds from the same locality. A young tree in Veitch's nursery at Coombe Wood is now 16 feet high at seven years old, and is very flourishing. This species is remarkable for its fine foliage, and is worthy of cultivation as an ornamental tree.

(A. H.)

## BETULA UTILIS, Himalayan Birch

*Betula utilis*, Don, *Prod. Fl. Nepal.* 58 (1825); Hooker, *Fl. Brit. India*, v. 599 (1888); Gamble, *Indian Timbers*, 668 (1902); Winkler, *Betulaceæ*, 61 (1904); Schneider, *Laubholzkunde*, 102 (1904); Brandis, *Indian Trees*, 622 (1906).
*Betula Bhojpattra*, Wallich, *Pl. As. Rar.* ii. 7 (1832); Brandis, *Forest Flora, N.-W. India*, 457 (1874).
*Betula Jacquemontii*, Spach, in *Ann. Sc. Nat.* sér. 2, xv. 189 (1841).

A tree, attaining in the Himalayas about 60 feet in height, but becoming a shrub at high elevations. Bark on young trees thin, smooth, brownish red, with

[1] Matsumura, in *Shokubutsu Mei-I.* 47 (1895), enumerates eight distinct species of birch as occurring in Japan; but of these, *B. utilis*, as explained in a note under our account of that species, and *B. grossa*, mentioned above, are doubtful. There remain six distinct species, undoubted natives of Japan, viz. :—*B. Maximowiczii, B. corylifolia, B. Ermani, B. globispica, B. ulmifolia*, and *B. verrucosa*, var. *japonica*; all of which are referred to in this account of the genus Betula, and under each species is given its native name.

darker coloured horizontal lenticels, peeling off in transverse rolls; on older trees darkened and thickened at the base. Young branchlets, non-glandular, covered with a dense, greyish tomentum; older branchlets smooth, glabrous. Leaves (Plate 269, Fig. 7) coriaceous, about 3½ inches long and 2¼ inches wide, oval or ovate-oblong, rounded at the base, acuminate at the apex; margin slightly ciliate, irregularly serrate; nerves nine to twelve pairs; upper surface shining, dark green, with scattered pubescence; lower surface yellowish green, glandular, glabrous between the nerves, which are slightly pubescent, and with dense axil-tufts of pubescence; petiole ¾ inch, tomentose.

Fruiting catkins (Plate 269, Fig. 7),[1] cylindrical, 1½ inch long, ⅓ inch in diameter, on tomentose peduncles, variable in length; scales with glabrous, ciliate, spatulate lobes, the central lobe dilated above and obtuse at the apex, and often trifid, about twice as long as the erect or slightly divergent lateral lobes; nutlets with narrow wings.

## VARIETIES

In addition to the type, described above, which occurs in the Himalayas and China, the following varieties can be recognised:—

1. Var. *sinensis*, Franchet, *Journ. de Bot.* xiii. 207 (1899).

*Betula albo-sinensis*, Burkill, in *Journ. Linn. Soc. (Bot.)* xxvi. 497 (1899).

Leaves glabrescent beneath. Fruit-scales glabrous, not ciliate; nutlets smaller than in the type. Discovered in north-eastern Szechwan by Père Farges. Not yet introduced.

2. Var. *Prattii*, Burkill, *Journ. Linn. Soc. (Bot.)* xxvi. 499 (1899). Leaves more pubescent than in the type, the pubescence extending over the whole under surface, and very dense in the axils and along the midrib. Fruit-scales strongly ciliate, with spatulate lobes, the lateral lobes spreading and not erect. This variety occurs in western Szechwan, at high elevations (13,500 feet), and has not yet been introduced.

3. Var. *Jacquemontii*, Regel, in DC. *Prod.* xvi. 2, p. 177 (1868).

*Betula Jacquemontii*, Spach, in *Ann. Sc. Nat.* sér. 2, xv. 189 (1841); Cambessedes, in Jacquemont, *Voyage dans l'Inde, Botanique*, 157, t. 158 (1844); Regel, *op. cit.* 178 (1868).

A tree, with white bark, marked by brownish horizontal lenticels, and peeling off in transverse strips. Young branchlets slightly glandular, and covered with a dense, erect, short pubescence. Leaves (Plate 270, Fig. 15), about 2½ inches long, and 1¾ inch broad, ovate, rounded or slightly cuneate at the base, acuminate at the apex, bi-serrate; nerves seven to nine pairs; upper surface with scattered pubescence or glabrescent; lower surface gland-dotted and glabrous except for long hairs on the midrib and nerves; petiole ¾ inch, glabrescent, glandular. Fruiting catkins (Plate 270, Fig. 15), 1½ inch long, ⅓ inch wide, cylindrical, on long pubescent stalks; scales glabrous, ciliate, with an elongated linear central lobe,

---

[1] In this figure the middle lobe of the scale is represented short and trifid at the apex, as is occasionally the case; but as a rule it is more elongated, and broadened and rounded at the apex.

acute at the apex, and two or three times as long as the divergent short rounded lateral lobes; nutlets with narrow wings.

This variety is represented in Kew Gardens by two trees obtained from St. Petersburg in 1891 and 1894, and 25 feet and 20 feet high respectively. It is very distinct in appearance from the typical form of the species, having white bark; smaller, few-nerved, thinner leaves; and different catkins. Moreover, the branchlets and fruiting peduncles are shortly pubescent in the variety, and tomentose in the type.

This variety,[1] judging from the material in the Kew Herbarium, is common in the Himalayas, and probably constitutes a distinct species, which a careful study in the field may show to occupy a different area of distribution from that of typical *B. utilis*, which is so readily recognisable by its reddish bark and other characters.

### DISTRIBUTION

*B. utilis*[2] is widely distributed in the Himalayas and in China. It occurs in West Tibet, and in the Himalayas from the Kurram valley and Kashmir, to Sikkim and Bhotan, at altitudes usually ranging from 10,000 to 14,000 feet, but descending in the north-west to 7000 feet. It is often gregarious at the upper limit of arborescent vegetation, where it is commonly associated with *Rhododendron campanulatum*. According to Brandis, it attains 50 to 60 feet in height, and sometimes 10 or 12 feet in girth. Gamble's account of the bark includes that of the type and of var. *Jacquemontii*; as he describes it as smooth, shining, reddish white or white, the outer bark consisting of numerous distinct, thin, papery layers, peeling off in broad horizontal rolls; the thicker lower part of the bole becoming rough and dark as in the European birch. He states that the growth is slow, with an average of fifteen rings per inch of radius. The wood is extensively used in the inner arid Himalayan region for building purposes; it is elastic, does not warp, and seasons well. The bark is the most valuable part of the tree, and is used for paper, umbrellas, hookah-tubes, and roofs of houses.

In China it is a moderate-sized tree, growing only at high elevations, between 8000 and 13,500 feet, in the provinces of Szechwan, Hupeh, and Kansu. It was seen in the latter province by Przewalski,[3] who describes the bark as reddish, peeling off and hanging from the tree in long festoons.

This species is very rarely seen in cultivation, and the typical form from the Himalayas, like most of the broad-leaved trees from that region, for some un-explained reason, does not appear to have succeeded in this country. At Grays-

---

[1] Shirai, in *Tokyo Bot. Mag.* viii. 320, ff. 23, 24 (1894), states that this variety occurs in Japan; but the plant figured by him seems to be identical with a specimen gathered by Maximowicz in the province of Shinano, and labelled *B. Bhojpattra*, var. *typica*, Regel, which is certainly not that species, and appears to be *B. ulmifolia*, S. et Z. Shirai's account of the bark of this tree, as not being papery, but greyish brown, smooth, cracking, and falling off in patches, confirms this identification.

[2] There is no evidence that *B. utilis* occurs in Japan, where it is represented by the closely allied species *B. Ermani*. Shirasawa, in *Icon. Ess. Forest. Japon,* text 44, t. 23, ff. 9-22 (1900), figures a tree as *B. Bhojpattra* which is not this species (*B. utilis*), as is confirmed by his account of the bark, as being hard, compact, and falling off in scaly facets. *B. Ermani*, var. *nipponica*, has also been considered erroneously to be a form of *B. utilis*.

[3] Cf. Bretschneider, *Europ. Bot. Discoveries*, 987 (1898).

wood, Haslemere, a tree planted in 1882, which was obtained from a nursery at Newry, was 30 feet high and 2 feet 1 inch in girth in 1906.

There are three trees about 20 feet high in the Botanic Garden of Trinity College, Dublin, which, according to Burbidge,[1] were raised from seed sent by Sir Joseph Hooker in 1881. At Castlewellan there are some young trees, about 8 feet high, which were obtained by grafting branches of the Dublin trees on the common birch. (A. H.)

## BETULA PAPYRIFERA, Paper Birch, Canoe Birch

*Betula papyrifera*, Marshall, *Arbust. Am.* 19 (1785); Sargent, *Silva N. Amer.* ix. 57, t. 451 (1896), and *Trees N. Amer.* 202 (1905); Winkler, *Betulaceæ*, 83 (1904).

*Betula lenta*, Wangenheim, *Nordam. Holz.* 45 (1787) (not Linnæus).

*Betula papyracea*, Aiton, *Hort. Kew.* iii. 337 (1789); Loudon, *Arb. et Frut. Brit.* iii. 1708 (1838).

*Betula grandis*, Schrader, *Ind. Hort. Bot. Goett.* 2 (1833).

*Betula latifolia*, Tausch, in *Flora*, xxi. 2 p. 751 (1838).

*Betula alba*, Linnæus, var. *papyrifera*, Spach, in *Ann. Sc. Nat.* sér. 2, xv. 188 (1841).

*Betula cordifolia*, Regel, *Monog. Betulaceæ*, 86 (1861).

*Betula macrophylla*, Hort., ex. Schneider, *Laubholzkunde*, i. 115 (1904).

A tree, usually attaining in America, in its typical form, 60 or 70 feet in height and 2 or 3 feet in diameter, the variety found on the lower Fraser River in British Columbia being usually much larger in size. Bark thin, smooth, creamy-white, marked with long, narrow, horizontal lenticels, and separating into thin papery layers; becoming on old trunks near the base $\frac{1}{2}$ inch thick, dull brown or blackish, fissured, and scaly. Young branchlets, with scattered long hairs, mostly falling off in summer; in the second year dark brown and glabrous.

Leaves (Plate 269, Fig. 5), 2 to 3 inches long, $1\frac{1}{2}$ to 2 inches wide, ovate; rounded or slightly cordate at the base, acuminate at the apex; margin ciliate and irregularly bi-serrate; nerves six to eight pairs; upper surface dull green, slightly pilose on the nerves; lower surface paler, with numerous minute brown glands, usually glabrous except for dense axil-tufts of pubescence and a few long hairs on the midrib and nerves, occasionally minutely pubescent between the nerves; petiole at first pilose, ultimately glabrescent.

Fruiting catkins (Plate 269, Fig. 5), cylindrical, about $1\frac{1}{2}$ inch long, $\frac{1}{3}$ inch thick, hanging on slender stalks; scales pubescent or glabrous, ciliate, with the middle lobe longer than broad; lateral lobes rounded, erect, spreading or recurved. Nutlet with broad wings.

In winter the twigs usually show a few scattered long hairs; buds, $\frac{3}{8}$ inch long, appressed to the branchlet, ovoid, acute, with glabrous, ciliate scales, glistening with resin.

### Varieties and Hybrids

This species, spread over a vast territory in North America, is very variable in the wild state; and the forms occurring on the Rocky Mountains and in the Pacific

[1] *Proc. Roy. Hort. Soc.* 1901, p. xxxviii.

Coast region have not yet been fully studied in the field.  The following varieties are notworthy :—

1. Var. *cordifolia*, Regel, in *Bull. Soc. Nat. Mosc.* xxxviii. 401 (1865) ; Fernald, in *Rhodora*, iii. 173 (1901) ; Sargent, *Silva N. Amer.* xiv. 55, t. 724 (1902).

Leaves distinctly cordate at the base, smaller than in the type.  An alpine tree, moderate in size, scarcely exceeding 40 feet in height, which occurs on Mount Katahdin in Maine, on the White Mountains in New Hampshire, and in the northern Rocky Mountain region.

2. Var. *kenaica*.

> *Betula kenaica*, Evans, *Bot. Gazette*, xxvii. 481 (1899) ; Sargent, *Silva N. Amer.* xiv. 53, t. 723 (1902), and *Trees N. Amer.* 205 (1905).

A tree in Kew Gardens, about 20 feet high, obtained from Dieck in 1891, and said to be a native of Alaska, has been identified by Sargent with *B. kenaica*. Judging from this specimen and the description given by Sargent, this species is only a small-leaved variety of *B. papyrifera*, from which it cannot be separated by any characters of importance.  The branchlets are minutely pubescent and slightly glandular.  The leaves are about 2 inches long, with five to seven pairs of nerves.  The fruiting catkins are smaller than in the type, about an inch long, with glabrous ciliate scales, the middle lobe of which is narrow, oblong, not much longer than the rounded broad lateral lobes.  The bark is like that of ordinary *B. papyrifera*, but the white colour is slightly tinged with orange.

*Betula kenaica* was discovered in 1897 by Dr. Evans in Alaska, in the Kenai Peninsula, near Cook Inlet ; and was found on Kodiak Island by Coville in 1899. It is described as being a small tree, only reaching 40 feet in height.  It is probable [1] that the variety extends south from Alaska through British Columbia ; and is a form with small leaves growing on poor soil and in mountainous regions, while var. *Lyalliana*, with large leaves, occurs at nearly sea-level in rich alluvial soil.

3. Var. *Lyalliana*,[2] Koehne, in Beissner, Schelle, and Zabel, *Laubholz-Benennung*, 55 (1903) ; Schneider, *Laubholzkunde*, i. 115 (1904).

> *Betula occidentalis*, Lyall, in *Journ. Linn. Soc.* (*Bot.*), vii. 134 (1864) (in part) (not Hooker) ; Sargent, in *Bot. Gazette*, xxxi. 237 (1901), *Silva N. Amer.* xiv. 57, t. 725 (1902), and *Trees N. Amer.* 204 (1905) (not Sargent, in *Silva N. Amer.* ix. 65, t. 453 (1896)).
>
> *Betula Lyalliana*, Koehne, in *Mitt. Deut. Dend. Ges.* viii. 53 (1899).

This variety, which is considered to be a distinct species by Sargent, differs from the type in the greater size of the tree, in the orange tint of the bark, and in the larger leaves, which are thin and membranous in texture, and not so thick and coriaceous as is usual in this species.  The leaves are about 4 inches long and 3 inches broad, with seven to nine pairs of nerves, and are coarsely and doubly serrate, broadly ovate, with a broad, truncate base and a slightly acuminate apex.

---

[1] Specimens just received from trees cultivated in the Arnold Arboretum, U.S., labelled *B. occidentalis*, Sargent, and *B. kenaica*, Evans, are so similar, that they cannot be distinguished even as varieties.  This confirms the opinion that var. *kenaica* and var. *Lyalliana* (*B. occidentalis*, Sargent) are merely geographical forms of *B. papyrifera*, the differences in the size of the leaves and fruiting catkins being due to soil and climate.  Under cultivation in this country they maintain these differences.

[2] This name is preferable to *B. papyracea*, var. *occidentalis*, Dippel, *Laubholzkunde*, ii. 177 (1892), as the plant described there is apparently var. *grandis*, Schneider, a cultivated variety which originated from the eastern form of the species.  Cf. our remarks on *B. occidentalis*, Hooker, given in p. 993.

The young branchlets have a minute, dense, erect pubescence, interspersed with long hairs and a few scattered glands. The fruiting-catkins are like those of the type, with slightly thinner scales, the middle lobe of which is triangular and elongated.

This splendid tree, which attains a height of 100 or 120 feet, and a diameter of 3 or 4 feet, on the alluvial banks of the lower Fraser River, appears to be confined to the lower basin of that river in south-western British Columbia and north-western Washington. It was first collected by Lyall in 1859, " in woods by river banks, on the Sumas and Chilukeveyuh prairies and other low grounds to the westward of the Cascade Mountains "; and his specimens preserved in the Kew Herbarium are identical with my own collected near New Westminster.

Piper[1] recognises the typical variety as occurring in north-western Washington, where he says that it is a tree with dark grey bark, occasionally 3 feet in diameter. He mentions a similar tree, smaller in size and often white-barked, which grows in Stevens County and the Blue Mountains of Washington State. This smaller tree, which also occurs in Idaho, is a connecting link between var. *Lyalliana* and the form of the species which occurs in the Rocky Mountains.

I collected seeds of var. *Lyalliana* on October 20, 1906, from two trees, about 60 feet in height, which I found growing near New Burnaby, on the electric tramline between Vancouver and New Westminster. The virgin forest had been cut down in this neighbourhood, and the few trees which I saw were young and thriving, and growing in open spaces amidst second-growth Thuya and Douglas. I had no time to descend to the alluvial flats of the Fraser River, where Sargent reports the existence of trees of large size. At the large lumber mills of Vancouver and New Westminster, where I made inquiries, the tree is unknown at the present time ; but I was informed that some years ago a small quantity of furniture had been made from large trees cut down near New Westminster.

The seed which I collected has been distributed to various places in Great Britain, and has germinated well. Seedlings raised in a nursery bed at Casewick, Lincolnshire, by Lord Kesteven, are now (August 1908) 12 to 22 inches in height, and for so far have been healthy and vigorous in growth. Some of the seed did not germinate till the following year. At Brocklesby, Mr. Havelock has raised a few plants in a frame, which are 24 to 30 inches high, with fine, large foliage. At Pollokshaws, near Glasgow, Sir John Stirling Maxwell reports that they did better when sown in the open than when grown in a frame, and average 15 inches in height, the tallest being 29 inches. He adds that this variety shows every sign of being a thriving tree.

Elwes has also raised from seeds sent to him from Kaslo, on Lake Kootenay, British Columbia, in 1904, a few young trees which appear to belong to this variety, and others from seed given him by Professor Sargent, and said to be from the lower Fraser Valley. These are growing vigorously at Colesborne.

4. *Betula Andrewsii*, Nelson, in *Bot. Gazette*, xliii. 279, with figure of the tree (1907), is a peculiar form with many branching stems from the base, which has been found in Colorado.

---

[1] *Contrib. U.S. Nat. Herb.* xi. 218 (1906).

5. A hybrid between *B. papyrifera* and *B. populifolia*, found growing wild in New Hampshire and Massachusetts, is described by Sargent in *Garden and Forest*, viii. 356, fig. 50 (1895).

Several varieties and hybrids have originated in cultivation :—

6. Var. *grandis*, Schneider (*B. macrophylla*, Hort.).   Leaves large, cordate, lobulate in margin.   Similar leaves appear on coppice shoots and on lower branches of old trees belonging to the typical form of the species.

7. *B. Koehnei*, Schneider, *Laubholzkunde*, i. 114 (1907), a hybrid between *B. papyrifera* and *B. verrucosa*, is identical with *B. cuspidata* of Späth's nursery.

8. *B. excelsa*, Aiton, *Hort. Kew*, iii. 337 (1789), long supposed to be either a distinct species or a cultivated variety of *B. papyrifera*, is considered by Schneider (*op. cit.* 108) to be a hybrid between this species and *B. pumila*, and differs from the former mainly in the smaller size of the leaves.

### DISTRIBUTION

The paper birch is the most widely distributed species of Betula in North America, the typical form extending northward to Labrador, the southern shores of Hudson's Bay and Great Shore Lake, and southward to Long Island, New York, northern Pennsylvania, central Michigan, central Ohio, northern Nebraska, the Black Hills of Dakota, and northern Montana.   In various forms it also occurs west of the Rocky Mountains to the Pacific Coast, in Alaska, British Columbia, Washington, and Idaho.

It usually grows on rich wooded slopes and on the borders of streams, lakes, and swamps ; and is common in Canada, New York, and northern New England, becoming rarer to the southward and in the Rocky Mountains.          (A. H.)

### CULTIVATION

Notwithstanding the rarity of this tree in cultivation, it seems to grow freely at Colesborne, where I have raised it from seed, and planted it out in situations where it is exposed to cold and damp.   Here it does not suffer from spring frost, and has attained 15 feet in height in seven years.   As an ornamental tree, however, it is not in England superior to the common birch, and has no special merit to justify its being planted except as a curiosity.

The paper birch was introduced into England in 1750, according to Loudon ; but is rarely seen except in botanic gardens, as at Kew, where there are several fair-sized specimens, the largest, a tree with ascending branches, near the Victoria gate, being 45 feet high by 3½ feet in girth.   Close to it is another nearly equal in size, with markedly drooping branches.   In the Cambridge Botanic Garden, a tree, grafted at 1½ foot from the ground, was, in 1906, 47 feet by 4 feet 7 inches.

The largest tree we know of in cultivation is in Mr. Kaufman's garden at White Knights, near Reading, which Henry measured in 1904 as 82 feet by 4 feet 11 inches.   Another tall white-barked tree with a clean stem, grafted on common birch near the ground, grows at Bicton, and measured, in 1906, 75 feet by 7 feet 2 inches (Plate 259). A large tree is growing at Woburn, near the lake on the right of the main entrance from the village. It is on its own roots, and has bark of a brownish-grey colour, quite unlike the trees at White Knights and Bicton. At Arley,[1] a tree measured 41 feet by 4 feet in 1905. There is also an old and sickly tree at Boynton, in Yorkshire, and a young and healthy one on its own roots, about 40 feet high, at Tortworth.

In Scotland and Ireland we have failed to find a single specimen of any size.

The handsomest specimen that I have seen in Europe is at the nursery of Simon-Louis frères at Metz, where on a deep rich loam it has attained 70 feet high by 6 feet 4 inches in girth, and has a fine silvery bark, more beautiful than any that I know in England. (H. J. E.)

## BETULA POPULIFOLIA, Grey Birch

*Betula populifolia*, Marshall, *Arbust. Am.* 19 (1785); Loudon, *Arb. et Frut. Brit.* iii. 1707 (1838); Sargent, *Silva N. Amer.* ix. 55, t. 450 (1896), and *Trees N. Amer.* 200 (1905); Winkler, *Betulaceæ*, 79 (1904).

*Betula excelsa canadensis*, Wangenheim, *Nordam. Holz.* 86 (1787).

*Betula acuminata*, Ehrhart, *Beit. Naturk.* vi. 98 (1791).

*Betula cuspidata*, Schrader, *ex* Regel, in DC. *Prod.* xvi. 2, p. 164 (1868).

*Betula alba*, Linnæus, var. *populifolia*, Spach, in *Ann. Sc. Nat.* sér. 2, xv. 187 (1841).

*Betula alba*, Linnæus, sub-species *populifolia*, Regel, in *Bull. Soc. Nat. Mosc.* xxxviii. 399 (1865).

A tree, attaining in America 30 or 40 feet in height and 18 inches in diameter. Bark similar to that of *B. verrucosa*, but greyish in colour. Young branchlets glabrous, covered with reddish-brown glands, which persist and roughen the shoot in the second year. Leaves (Plate 269, Fig. 4), 2½ to 3 inches long, 1½ to 2 inches wide, deltoid; broadly cuneate or truncate at the base; prolonged into a long caudate-acuminate apex; margin lobulate, irregularly serrate; nerves five or six pairs; both surfaces shining, glabrous, covered with minute brown glands; petiole reddish, long, slender, glandular, glabrous.

Fruiting-catkins (Plate 269, Fig. 4), cylindrical, ¾ inch long, ¼ inch in diameter, pendent or spreading on slender stalks; scales pubescent and ciliate, with short triangular middle lobe and recurved broad lateral lobes; wings broader than the narrow nutlet.

This species is closely allied[2] to *B. verrucosa*; but it is a smaller tree, strikingly different in the colour of the bark, and is readily distinguished by the shape of the leaf, the apex of which is very prolonged, and by the pubescent scales of the fruiting-catkins.

Varieties *laciniata* and *pendula* mentioned by Loudon are not known now in

[1] *Hortus Arleyensis*, 45 (1907).

[2] Sargent, in *Garden and Forest*, ii. 484 (1889), points out the differences between these two species.

cultivation.[1]   Var. *purpurea*, with reddish leaves, was sent out in 1892 by Ellwanger and Barry, of Rochester, New York.   *B. cærulea*, Blanchard, is apparently a form of this species, with dull bluish-green leaves, ovate rather than deltoid in outline, which is common on hills in northern New England and eastern Canada.

This is the smallest of the arborescent birches of America, and grows usually on dry, gravelly or sandy, barren soil, or on the edges of swamps and lakes.   Its area of distribution extends from Nova Scotia and the valley of the St. Lawrence southward to Delaware, and westward through northern New England and New York to the southern shores of Lake Ontario.   It is very abundant in the coast region of New England and the middle states, and springs up in abundance after forest fires or on abandoned farm lands.

It was first cultivated in England by Archibald, Duke of Argyll, at Whitton, near Hounslow, in 1750, and is rarely met with now except in botanical gardens. At Kew, the largest specimen, growing near the end of the rhododendron dell, is 35 feet high and about 6 inches in diameter.   It produces fruit regularly.   Loudon mentions a birch supposed to be of this species at Dodington Park in Gloucestershire, 60 feet high in 1838, but no such tree now survives there.          (A. H.)

## BETULA NIGRA, RED BIRCH

*Betula nigra*, Linnæus, *Sp. Pl.* 982 (1753); Loudon, *Arb. et Frut. Brit.* iii. 1710 (1838); Sargent, *Silva N. Amer.* ix. 61, t. 452 (1896), and *Trees N. Amer.* 198 (1905); Winkler, *Betulaceæ*, 58 (1904).
*Betula lanulosa*, Michaux, *Fl. Bor. Am.* ii. 181 (1803).
*Betula rubra*, Michaux f., *Hist. Arb. Am.* ii. 142 (1812).

A tree, attaining in America 80 or 90 feet in height, with a trunk occasionally 5 feet in diameter, and usually divided into two or three diverging limbs at 15 or 20 feet above the ground.   Bark at first smooth, reddish brown; with age, separating into successive layers, which curl up and persist on the trunk as thin papery scales of various tints of red and brown; ultimately turning black and becoming an inch thick and deeply furrowed at the base of old trunks.   Young branchlets tomentose, with numerous glands; older branchlets glabrous and roughened with the remains of the glands.   Leaves (Plate 270, Fig. 13), 1½ to 3 inches long, 1 to 2 inches wide, deltoid-ovate, with cuneate base and acute apex; margin non-ciliate, coarsely and irregularly bi-serrate, and often lobulate; nerves seven or eight pairs; upper surface shining, with fine pubescence mainly on the nerves; lower surface greyish, with pubescence chiefly on the midrib and nerves, and with numerous white glands; petiole tomentose and glandular.

Fruiting-catkins (Plate 270, Fig. 13), cylindrical, 1 to 1½ inch long, ½ inch in diameter, erect, on stout tomentose peduncles, about ½ inch long: scales pubescent

---

[1] There is a var. *laciniata*, and also a var. *purpurea* assigned to *B. verrucosa* in Späth's and Simon-Louis's nurseries, which may be what Loudon referred to.

and ciliate. The fruit ripens in May or June, and Sargent has called attention to the fact that the early ripening of the seeds of this and other trees, as the red and silver maples, growing beside rivers assures their germination, as they fall on the banks at the season of low water, immediately germinate, and grow speedily.

*Betula nigra* is readily distinguishable by its peculiar bark, the only other species in cultivation which at all resembles it in this respect being *B. dahurica*. It is also very distinct in the greyish colour of the leaves beneath, which are cuneate at the base, acute and not acuminate at the apex, and usually lobulate in margin with sharp double serrations.

In winter the twigs are brown, glandular, and almost glabrous; buds minute, $\frac{1}{8}$ inch long, appressed to the branchlet, with a sharp beak directed inwards; scales ciliate and pubescent.

This species, which is known as the red or water birch,[1] grows usually on the banks of streams and ponds or in swamps, in deep rich soil, liable to inundation. It occurs from Massachusetts, southwards to Florida, east of the Alleghany Mountains; through the Gulf States to Trinity River, Texas; and throughout the Mississippi valley to Indian territory, Kansas, Nebraska, Minnesota, Wisconsin, and Ohio. It attains its largest size in the damp lowlands of Florida, Louisiana, and Texas, being the only birch tree of these warm regions. (A. H.)

Sargent says[2] that its distribution is peculiar, as though it grows abundantly and luxuriantly on the banks of the Merrimac and Spicket rivers in north-east Massachusetts, it occurs nowhere else in New England, and only becomes common in the south of New Jersey, extending from thence to Iowa in the west, and to Florida and Texas in the south, growing in the south on the banks of almost every stream which has a gravelly bed, and of which the banks are not marshy, and attaining a height of 80 or 90 feet with a trunk 3 to 4 feet in diameter; but in North Carolina, according to Ashe, its average size is 40 to 60 feet high by 1 to 2 feet in diameter.

It was introduced into cultivation in England by Peter Collinson in 1736; but is rarely met with, though it ought to be more suitable for cultivation[3] in the warmer parts of England than the more northern species. There are several good specimens in Kew Gardens, the largest, near the rhododendron dell, being 57 feet high and 5 feet in girth. This tree divides at 7 feet from the ground into two main stems; and one or two smaller trees in the collection of birches branch similarly near the ground into two or three stems. A tree near the Victoria gate has a single stem, 48 feet in height and $4\frac{1}{2}$ feet in girth. Some of these trees bore ripe fruit in June 1908.

Though Sargent says that large specimens may be seen in some of the older European parks, neither Pardé nor Correvon mention any trees of this species; but Bean[4] saw one $7\frac{1}{2}$ feet in girth at Herrenhausen, Hanover. (H. J. E.)

---

[1] It is also known as the river birch, and though known to botanists as *B. nigra*, it is very seldom called in America black birch, the latter name being very commonly applied to *B. lenta*. [2] *Garden and Forest*, ii. 591.

[3] In an article on this tree, in *Gard. Chron.* xxv. 21 (1899), Mr. Bean recommends it for ornamental planting on low islands in lakes, and beside water-courses.

[4] *Kew Bull.*, 1908, p. 392.

## BETULA LUTEA, Yellow Birch

*Betula lutea*, Michaux f., *Hist. Arb. Am.* ii. 152 (1812); Sargent, *Silva N. Amer.* ix. 53, t. 449
  (1896), and *Trees N. Amer.* 197 (1905); Winkler, *Betulaceæ*, 65 (1904).
*Betula lenta*, Linnæus, var. *lutea*, Regel, in DC. *Prod.* xvi. 2, p. 179 (1868).
*Betula lenta*, Linnæus, var. *genuina*, Regel, in *Mém. Soc. Nat. Mosc.* xiii. 126 (1861).
*Betula excelsa*, Pursh, *Fl. Am. Sept.* ii. 621 (1814) (not Aiton); Loudon, *Arb. et Frut. Brit.* iii.
  1711 (1838).

A tree, attaining in America 100 feet in height and 3 or 4 feet in diameter. Bark smooth, shining, silvery or golden grey, breaking into ribbon-like strips and curls, which long remain attached; in old trees, ½ inch thick, reddish brown, and fissured.   Young branchlets covered with long pale hairs; in the second year smooth, brown, and usually glabrous.

Leaves (Plate 270, Fig. 11), 3 to 4½ inches long, 1½ to 2 inches wide, ovate-oblong, rounded at the base, acute or slightly acuminate at the apex; margin finely and sharply serrate, ciliate between the teeth; nerves nine to twelve pairs; both surfaces with long silky hairs mainly on the midrib and nerves; pale beneath; petiole pilose.

Fruiting-catkins (Plate 270, Fig. 11) erect, sessile or sub-sessile, ovoid-oblong, 1 to 1½ inch long, ¾ inch in diameter; scale lobes nearly equal, ciliate, pubescent.

In winter the slender twigs are more or less pilose; buds fusiform, ¼ inch long, rather blunt at the apex, with minutely pubescent, ciliate scales.

This species occurs in Newfoundland and along the northern shores of the Gulf of St. Lawrence, to the valley of the Rainy River, extending southwards to Delaware and Minnesota, and along the Alleghany Mountains to the high peaks of North Carolina and Tennessee.   It is one of the largest broad-leaved trees of the eastern provinces of Canada and New England, where it is abundant, usually growing in rich soil on moist uplands, in company with the beech, sugar and red maples, black and white ash, and white elm.   The leaves turn a bright yellow in autumn.                                                                    (A. H.)

The yellow birch was introduced into England, according to Loudon, about 1767, but has never become common.

The finest we know of in England grows in a shrubbery near the kitchen garden at Tortworth, and measured in 1907 about 50 feet by 4 feet.   It has borne fruit, from which Lord Ducie has raised seedlings.   There are small specimens in Kew Gardens.

At Auchendrane, Ayrshire, Mr. Renwick measured a tree in 1907, 57 feet high, with a bole of 15 feet girthing 5 feet 2 inches.

A very fine tree is growing at Oriel Temple, Co. Louth, the seat of Lord Masserene, which was mentioned by Loudon under the name *B. lenta*.   When I saw it in July 1908, it was in perfect health, and measured 58 feet high by 7 feet 4 inches in girth (Plate 260).   Loudon states that in his time it was about fifty years planted, and 50 feet high with a diameter of 1 foot 9 inches.   (H. J. E.)

## BETULA LENTA, Cherry Birch, Black Birch

*Betula lenta*, Linnæus, *Sp. Pl.* 983 (1753); Loudon, *Arb. et Frut. Brit.* iii. 1713 (1838); Sargent, *Silva N. Amer.* ix. 50, t. 448 (1896), and *Trees N. Amer.* 196 (1905); Winkler, *Betulaceæ*, 64 (1904).
*Betula nigra*, Du Roi, *Obs.* 30 (1771) (not Linnæus).
*Betula carpinifolia*, Ehrhart, *Beit. Naturk.* vi. 99 (1791) (not Siebold and Zuccarini).

A tree, attaining in America 80 feet in height and 2 to 5 feet in diameter. Bark smooth, close, dark brown, with pale, elongated, horizontal lenticels, peeling off transversely in thin strips; on old trunks deeply fissured and broken into large, irregular scaly plates. Young branchlets glabrous, except for a few hairs above the leaf-insertions, slightly glandular.

Leaves (Plate 269, Fig. 3), $2\frac{1}{2}$ to 5 inches long, $1\frac{1}{2}$ to 3 inches wide, ovate-oblong, rounded or slightly cordate at the base, acuminate at the apex; margin non-ciliate, bi-serrate; nerves nine to thirteen pairs; upper surface dark green with a few long hairs confined to the midrib and nerves or scattered throughout; lower surface lighter in colour, with silky hairs on the midrib and nerves, forming axil-tufts; petiole pilose at first, ultimately glabrescent.

Fruiting-catkins (Plate 269, Fig. 3) erect, sessile, ovoid-oblong, 1 to $1\frac{1}{2}$ inch long, $\frac{1}{2}$ inch in diameter; scales glabrous, with nearly equal lobes, the lateral lobes being divergent.

This species is characterised by the pleasant aromatic flavour and fragrance of the leaves, twigs, and inner bark; and on that account is sometimes named sweet birch in America. It is, however, more often called, on account of the colour of the bark, cherry birch or black birch. In winter the twigs are shining and almost glabrous; buds fusiform, $\frac{5}{16}$ inch long, ending in a sharp beak, brownish, viscid, shining.

Var. *laciniata*, Rehder, in *Rhodora*, ix. 111 (1907). Leaves with six to nine pairs of sharply serrated lobes. A single tree of this variety, which resembles *B. verrucosa*, var. *dalecarlica*, in the form of the leaves, was found in 1901 at New Boston in New Hampshire.

A hybrid between this species and *B. pumila*, L., has been described by Sargent.[1]                                                                              (A. H.)

This species is an inhabitant of Newfoundland and Canada from Nova Scotia to Lake Superior, growing in its greatest perfection in central Ontario, Algoma, and Parry Sound, where Macoun says it is often more than 4 feet in diameter. In the United States it extends west to Iowa, and along the Alleghany Mountains to Kentucky and Tennessee, attaining a large size in the valleys of North Carolina. Ashe figures on plate 12 of his work a splendid tree 80 feet in height, with a clean bole 5 feet in diameter. He says the bark is reddish brown and rough on old trees, while on young trees and branches it is smooth and dark, resembling that

[1] *Garden and Forest*, viii. 243, fig. 36 (1895).

of *Prunus serotina.* It produces seed once in three or four years, and does not bear shade well, though young trees will sprout from the stump. The timber is heavy, hard, and strong, dark brown in colour, and takes a good polish.

I saw a very fine tree of this species in a garden at Lancaster, Massachusetts, 62 feet by 10 feet, dividing at 3 feet into four stems, which were covered with a very pretty ragged yellow and grey bark.

*Betula lenta* was introduced into England in 1759, according to Loudon. We are not aware that it has anywhere attained a large size, except at Oakly Park, near Ludlow, the property of the Earl of Plymouth, where on a rich sheltered flat on the banks of the Teme, I found a tree of considerable age, which in August 1908 measured about 60 feet by 4 feet 9 inches. The trees in Kew Gardens are about 20 feet in height. A specimen at High Canons, Herts, measured 36 feet high by 4 feet 2 inches in girth, and bore fruit in 1907. Another at Bicton, 38 feet by 3 feet 5 inches, is growing in the Arboretum walk, near the Paper Birch.

Beer is sometimes obtained in America by fermenting the sugary sap of this tree. Oil of birch, which is made on a considerable scale in Pennsylvania, is a more important product. This is obtained by distilling the wood,[1] one ton of which yields about 4 lbs. of oil. This oil is nearly identical, both in chemical and physical properties, with oil of winter-green, which is manufactured in the same district; and commercial oil of winter-green is a mixture of the two oils in varying proportions. (H. J. E.)

## BETULA FONTINALIS

*Betula fontinalis,* Sargent, in *Bot. Gazette,* xxxi. 239 (1901), and *Trees N. Amer.* 207 (1905).
*Betula occidentalis,* Sargent, *Silva N. Amer.* ix. 65, t. 453 (1896) (not Hooker); Winkler, *Betulaceæ,* 86 (1904); Schneider, *Laubholzkunde,* i. 114 (1904).

A tree, occasionally attaining 40 feet in height and 3 or 4 feet in girth, more commonly shrubby, with many branching stems. Bark about $\frac{1}{4}$ inch thick, dark brown, shining, not separating into thin layers, marked by pale brown horizontal lenticels. Young branchlets viscid, densely covered with resinous glands, interspersed with long, pale hairs; older branchlets dark in colour and roughened with the persistent glands. Leaves about $1\frac{1}{2}$ inch long, and 1 to $1\frac{1}{2}$ inch broad, thin in texture, broadly or narrowly ovate; rounded, truncate or subcordate, and often unequal at the base; acute at the apex; margin ciliate, sharply and doubly serrate; nerves six to eight pairs; both surfaces glandular with scattered long hairs, at first paler beneath, becoming glabrescent; petiole, $\frac{1}{2}$ inch, glandular, glabrescent.

Fruiting-catkins, about 1 inch long, $\frac{1}{4}$ inch in diameter, cylindrical, on slender glandular stalks; scales pubescent, ciliate, with the three lobes triangular and nearly equal in size, the lateral lobes divergent; nutlets with broad wings.

This species is readily distinguished by its conspicuously glandular branchlets and its small, thin leaves, which are variable in width, and in the form of the base.

[1] Cf. article by H. Trimble in *Garden and Forest,* viii. 303 (1895), where the process is described. The oil of birch is contained in the inner bark only; and on this account the wood used in distillation is obtained from small trees, usually coppice shoots.

*B. fontinalis* × *B. papyrifera.* There are two small trees in Kew Gardens, with wide-spreading pendulous branches, which were obtained from Dieck in 1891, and were said to be *B. occidentalis* from Alaska. One of these trees has firm, dark-brown, shining bark like that of *B. fontinalis*; while the other has white bark, peeling off in shreds, indistinguishable from that of *B. papyrifera.* The branchlets are exactly similar to those of *B. fontinalis.* The leaves (Plate 269, Fig. 8) are very variable and not precisely the same on both trees; resembling those of *B. fontinalis* in colour, but much larger and much thicker in texture; 2 to 3 inches or more in length, broadly ovate; rounded, truncate, or cuneate at the base, acute at the apex, coarsely serrate or toothed in margin, pubescent and gland-dotted on both surfaces. The fruiting-catkins (Plate 269, Fig. 8) and the scales are as large as those of *B. papyrifera*; but the scales are more like those of *B. fontinalis* in shape, the three lobes being almost triangular, glabrous, and ciliate. In all probability these two trees, with such variable characters in the bark and foliage, are hybrid between *B. fontinalis* and *B. papyrifera*, which occur in the same region.

The small-leaved birch, described above under the name *B. fontinalis*, was identified by Nuttall[1] with *B. occidentalis*, W. J. Hooker, *Fl. Bor. Amer.* ii. 155 (1839); and most botanists, including Sargent[2] in 1896, Winkler, and Schneider, have followed Nuttall. Sargent,[3] however, in 1901 advanced the opinion that the tall, large-leaved birch of the lower Fraser River, considered by us to be *B. papyrifera*, var. *Lyalliana*, is a distinct species, which he identified with Hooker's *B. occidentalis*; and he proposed the name *B. fontinalis* for the small-leaved birch.

The material[4] in the Kew Herbarium, on which Hooker founded his species, includes no less than three distinct birches, none of which, however, is the large-leaved variety of *B. papyrifera*; and as his description is confused and not confined to a single species, the name *B. occidentalis*, Hooker, must be entirely abandoned; and, in consequence, *B. fontinalis* is rightly adopted for the small-leaved birch, as being the first valid name for this species.

*B. fontinalis* is a small tree or spreading shrub, widely distributed in western North America, where it usually grows on moist soil near the banks of streams in mountain valleys. It extends from the basin of the upper Fraser and Peace rivers in British Columbia, Alberta, and the valley of the Saskatchewan, southwards to Mount Shasta and the northern Sierra Nevada in California, and through the Rocky Mountains and the interior ranges to Nevada, Utah, and northern New Mexico; extending eastwards in the United States to the Black Hills of Dakota and north-western Nebraska. Mr. M'Innes, of the Canadian Geological Survey, has recently discovered this species in the district north of Lake Superior.[5]

[1] *N. Amer. Sylva*, i. 22 (1842).     [2] *Silva N. Amer.* ix. 65 (1896).     [3] *Bot. Gazette*, xxxi. 237 (1901).

[4] This material includes :—

I. One specimen collected by Dr. Scouler, labelled "De Fuca Straits"; another, collected by Dr. Tolmie, "N.-W. Coast"; and a third collected by Douglas "west of the Rocky Mountains." These three specimens are a small-leaved variety of *B. papyrifera*, identical with *B. kenaica*, Evans.

II. Two specimens, collected by Dr. Richardson, labelled "Arctic Sea-Coast," one of which is *B. fontinalis*, Sargent; and the other, *B. alaskana*, Sargent.

III. A specimen, with young foliage, collected by Drummond, near Edmonton, which is probably *B. fontinalis*.

[5] *Canadian Forestry Journal*, 1905, 175.

It was introduced into cultivation in the Arnold Arboretum in 1874; but is rarely met with in Europe, except in Botanic Gardens. Four or five specimens, in the lower nursery at Kew, which were received from Sargent in 1903, are now 8 to 10 feet in height, and are vigorous in growth, promising to become trees of considerable size. Another specimen in the collection at Kew, raised from seed obtained from the Pinehurst Nursery in 1897, is about 10 feet in height and is more shrubby in appearance. (A. H.)

### TIMBER OF THE AMERICAN BIRCHES [1]

All American authors agree in saying that the tree which supplies the best wood of commerce is that of the *Betula lenta*, known in New England as black birch, in Canada as cherry birch. Macoun says that the yellow birch is scarcely distinguished in commerce, and judging from the specimens in Hough's *American Woods*, it would be difficult for any one but an expert to do so. Michaux and Macoun both say that the timber of the white birch is less valuable than either of the above; and Sargent says that the wood of the red birch has lately been found suitable for furniture of the best quality. Mr. Weale tells me that large quantities of red and yellow birch are imported into Liverpool, chiefly in the form of hewn logs of 15 to 18 inches square, which are valued in proportion to their size at from 1s. 6d. to 2s. a foot and upwards for logs showing figure. The wood often shows a beautiful undulation called "roll figure," which, when cut into veneers, was a few years ago very fashionable for bedroom furniture, and is sometimes rather difficult to distinguish from satin wood, though its colour is rather pinkish than yellowish and the undulations larger and more open. Mr. C. L. Willey, of Chicago, tells me that this fine "curly birch" is principally found at altitudes of 3000 to 4000 feet in North Carolina, and is very light in colour, having a yellowish tint; whilst other trees produce wood of a reddish colour, resembling that of cherry (*Prunus serotina*).

A large quantity of American birch is also imported in the form of planks 2 inches to 5 inches thick, and averaging 8 inches wide, which are sold in Liverpool at about 1s. per cubic foot, and consumed for chair-making at High Wycombe and elsewhere. At this low price it is the most formidable competitor to the native beech, and in some of the factories which I visited, seemed to be the more popular wood of the two.

The bark of the Paper Birch, and to a less extent that of other species, is as important to the inhabitants of Canada as that of the common birch is to those of northern Europe. Canoes and lodges are covered with large sheets of bark; it is placed on shingled roofs under the shingles to prevent the water from coming through; and very ornamental boxes, baskets, and other articles are made from it by the Indians. It also serves as a writing material, and I have a clearly written letter from Prof. Elrod, sent me by him, when during an expedition in Montana he ran out of paper. (H. J. E.)

[1] Emerson distinguishes the five common birches of New England as follows :—
1. *B. lenta*, black birch; bark dark coloured.
2. *B. lutea*, yellow birch; bark yellowish, with a silvery lustre.
3. *B. nigra*, red birch; bark reddish or chocolate coloured, very much broken and ragged.
4. *B. papyrifera*, canoe birch; bark white with a pearly lustre.
5. *B. populifolia*, grey or white birch; bark chalky-white, dotted with black.

# DIOSPYROS

*Diospyros*, Linnæus, *Gen. Pl.* 143 (1737); Hiern in *Trans. Camb. Phil. Soc.* xii. 1, 144 (1873); Bentham et Hooker, *Gen. Pl.* ii. 665 (1876); Gürke in Engler u. Prantl, *Pflanzenfam.* iv. 1, 161 (1890).

*Cargillia*, R. Brown, *Prod. Fl. Nov. Holl.* 526 (1810).

*Leucoxylum*, Blume, *Bijdr. Fl. Ned. Ind.* 1169 (1825).

*Noltia*, Schumacher, *Dansk. Vidensk. Selsk. Skrift.* iii. 189 (1828).

*Rospidios*, A. de Candolle, *Prodr.* viii. 220 (1844).

Trees or shrubs, belonging to the order Ebenaceæ. Leaves alternate or rarely sub-opposite, deciduous or persistent, simple, entire, without stipules.

Flowers diœcious or rarely polygamous, monœcious or perfect,[1] in cymes or solitary from the axils of the leaves of the current year, or in a few species arising from the old wood. Calyx, three- to seven-, usually four-lobed, pubescent, and accrescent under the fruit. Corolla urn-shaped, campanulate, tubular, or salver-shaped, three- to seven-, usually four- to five-lobed, pubescent. Male flowers small, usually in cymes; stamens four to sixteen, inserted on the base of the corolla or hypogynous; filaments slender and often united by pairs, forming an outer and inner series; anthers opening longitudinally or by apical pores; ovary aborted or wanting. Female flowers often solitary; staminodes four to eight, sometimes wanting, occasionally with fertile anthers; ovary with four to sixteen cells, which are double the number of the styles and one-ovuled, or rarely of the same number as the styles and two-ovuled. Fruit a berry, with the enlarged and persistent calyx at its base, containing one to ten or more oblong seeds, which have a copious albumen.

The alternate, simple, stalked, entire leaves, without stipules, and the shoots without true terminal buds and with two persistent bud-scales at their base, are distinguishing marks of the genus.

About 180 species of Diospyros are known, mostly confined to the subtropical and tropical regions of both hemispheres. The wood is usually hard and close-grained, the heartwood black, the sapwood soft, thick and yellow.

Only three species are in cultivation in this country; and of these *Diospyros Kaki*, Linnæus f., the Chinese Persimmon, a shrub or small tree, usually only met with in England in greenhouses or trained against a wall, does not come within the scope of our work. It ripens its fruit in warm summers in England.     (A. H.)

---

[1] In many species the sexes are unstable; cf. Wright, *Ann. R. Bot. Gard. Peradeniya*, ii. pt. i. 1, 133 (1904).

## DIOSPYROS VIRGINIANA, American Persimmon

*Diospyros virginiana*, Linnæus, *Sp. Pl.* 1057 (1753); Loudon, *Arb. et Frut. Brit.* ii. 1195 (1838)
    Sargent, *Silva N. Amer.* vi. 7, tt. 252, 253 (1894), and *Trees N. Amer.* 749 (1905).
*Diospyros guajacana*, Romans, *Nat. Hist. Florida*, 20 (1775).
*Diospyros concolor*, Moench, *Meth.* 471 (1794).
*Diospyros pubescens*, Pursh, *Fl. Am. Sept.* i. 265 (1814); Loudon, *Arb. et Frut. Brit.* ii. 1196
    (1838).
*Diospyros caroliniana*, Rafinesque, *Fl. Ludovic.* 139 (1817).
*Diospyros Persimmon*, Wikström, *Jahr. Schwed.* 1830, p. 92 (1834).

A deciduous tree, attaining occasionally in America 115 feet in height and 6 feet in girth, but usually smaller. Bark[1] deeply divided into square corky plates. Young shoots with a minute dense erect pubescence, persistent usually in the second year. Leaves (Plate 199, Fig. 3) oblong or elliptical; rounded and unequal or broadly cuneate at the base; shortly acuminate at the apex; margin entire and ciliate; upper surface dull, light green, and glabrous except for some pubescence on the midrib at the base; lower surface pale, glabrous; veins pinnate, arcuate, and looping near the margin; petiole pubescent, $\frac{1}{2}$ to 1 inch long.

Flowers appearing, when the leaves are more than half-grown, on the current year's shoot, dioecious. Staminate flowers in two- to three-flowered pubescent pedunculate cymes; calyx with four broadly ovate acute ciliate lobes; corolla tubular, slightly contracted below the very short acute reflexed lobes; stamens sixteen, in two series, with pubescent filaments. Pistillate flowers, solitary, on short recurved peduncles; stamens eight, usually with aborted anthers; ovary pilose towards the apex, eight-celled; styles four, two-lobed at the apex, pubescent at the base.

Fruit solitary, on short woody peduncles, persistent on the branches during winter; depressed, globose; surrounded at the base by the persistent calyx, which has four broadly ovate pointed recurved lobes. The fruit is variable in size, from that of a small cherry to a large plum; and its flavour is very different in different localities and even on trees growing close together—sometimes sweet without the action of frost, or ripening after frost, or at other times acid and never edible. Seeds oblong, flattened, $\frac{1}{2}$ inch long. Seedless forms occur, and experiments are being made in America with these and other good varieties.

The leaves on trees, growing in the Southern States, are strongly pubescent beneath; and this variety, which we have not seen in cultivation in England, is scarcely to be distinguished by the foliage alone from *D. Lotus*, which has pubescent leaves, and differs in this respect from the ordinary form of *D. virginiana*, with

---

[1] The bark is well figured in *Gard. Chron.* iv. 504, fig. 71 (1888).

glabrous leaves. The buds and leaf-scars of the two species are very different, and constitute the best marks of distinction.

In winter, the American Persimmon (Plate 200, Fig. 4) shows the following characters :—Twigs slender, covered with a minute dense erect pubescence, with two broadly ovate scales (of the previous season's bud) persisting at the base of the shoot. Leaf-scars, oblique on prominent pulvini, small, semicircular, with a deep transverse lunate depression, showing indistinctly the coalesced cicatrices of the vascular bundles. True terminal bud not formed, the tip of the branchlet dying off in summer and leaving at the apex of the twig a short glabrous stump with a terminal scar, which subtends the uppermost axillary bud. Buds ovoid, slightly compressed, small, brown, shining, glabrous, usually with a minute curved beak tipped by a few hairs; outer scales two, imbricate, ovate, acute, non-ciliate, concave interiorly, pubescent at the tip, glabrous elsewhere.

This species is widely distributed in the United States, its most northerly point being at Newhaven in Connecticut. It is not uncommon in Long Island, and extends southwards to Alabama and Florida, and westwards through Ohio and Iowa to Missouri, Arkansas, Louisiana, eastern Kansas, Indian Territory, and the valley of the Colorado River in Texas. It usually grows on light, sandy, well-drained soil, but attains its largest size in the deep alluvial lands of the Mississippi basin, where it sometimes reaches a height of more than 100 feet, with a slender trunk free from branches for 70 or 80 feet. It is exceedingly common in the south Atlantic and Gulf States, often covering with its suckers abandoned fields, and springing up by the sides of roads and fences. Sargent[1] gives a figure of a tree with wide-spreading branches, not unlike the specimen at Kew in size and appearance, which is growing in an old corn-field near Auburn in Alabama.

Elwes saw a fine tree of this species in a damp river bottom near Mount Carmel, Illinois, in 1904, which measured 100 feet high by 6 feet in girth, with a clean, straight trunk 60 feet high; but the late Dr. Schneck, who showed it to him, measured one as much as 115 feet high, 80 feet to the first limb, and only $5\frac{1}{2}$ feet in girth at the base. When growing in open fields or along road-sides, where it is most frequently seen, it forms a more spreading tree, usually 30 to 40, and rarely more than 60 feet high. (A. H.)

## CULTIVATION

This tree is easy to raise from seed, and perfectly hardy in England, but requires a warm, dry soil, and a much hotter summer than usual to make it thrive. The seedlings which I have raised grow very slowly and do not root freely in my soil. Judging from the extreme rarity of the tree in cultivation, it is hardly likely to be worth planting generally, and, so far as we know, has never borne fruit in England. Even in the climate of central France it fruits, according to Pardé,[2] very rarely, and grows slowly, having only attained about 20 feet in height at Les Barres. Neither

---

[1] *Garden and Forest*, viii. 262, fig. 38 (1895).     [2] *Arboretum Nat. des Barres*, 215 (1906).

Mouillefert nor Mayr mention this tree; and I have seen none in Europe worth mentioning, except a fine tree in the Botanic Garden at Padua, which appears to be the pubescent variety of this species, although it is labelled *D. Lotus*. According to Prof. Saccardo, it was planted in 1760, and measured 30 metres high by 2 metres in girth in 1887.

The American Persimmon was introduced into England some time before 1629, when an account of a cultivated tree appeared in Parkinson's *Paradisus* published in that year.

It is extremely rare in this country at the present day, and Loudon in 1838 only mentioned six trees, two of which still survive. One of these, which is a staminate tree, growing in Kew Gardens, now measures 64 feet in height by 5 feet 3 inches in girth, and, according to Sargent,[1] is apparently as thriving as if it were in its native habitat. It is one of the denizens of the original Kew Arboretum, which was laid out by W. Aiton, and in all probability was one of the numerous trees presented in 1762 to the mother of George III. by Archibald, Duke of Argyll, who was a 'great introducer and cultivator of rare trees at Whitton, near Hounslow.[2] (Plate 261.)

Another, mentioned by Loudon as being 24 years planted and 18 feet high in 1838, is growing at the Wilderness, White Knights, near Reading, and is now 45 feet high by 4 feet 1 inch in girth. At Barton, Suffolk, another is 40 feet high by 2 feet 1 inch in girth. At Bushey Lodge a tree with a broken top was, in 1904, 30 feet high by 5 feet 8 inches in girth. Suckers are growing from its roots as far away as 50 feet, and one of these, 10 feet high, is said to be about ten years old.

### TIMBER

The wood is very hard and heavy, of a pale yellowish-white colour, with black heartwood, which, however, usually shows only in old trees. Hough states that he felled one 14 inches in diameter for the specimens in his work,[3] but though there were over sixty rings of sapwood, only two or three in the heart were black. It is used in America for shoe-lasts and shuttles, for which latter purpose it is imported to a small extent to Liverpool. Michaux[4] states that it was used at Charleston for shafts, and preferred for that purpose to ash or any wood except lancewood, but the quantity available is too small to give this timber much economic importance. The fruit is little valued as human food, though eaten by animals.

(H. J. E.)

[1] *Garden and Forest, loc. cit.*

[2] Cf. J. Smith, *Records of Kew Gardens*, 258; and Nicholson, in *Gard. Chron.* iv. 504 (1888), in which is given a good picture of the tree (fig. 72). Cf. also *Kew Bulletin*, 1891, p. 292.

[3] *American Woods*, iii. No. 61.

[4] Mich. *fil., N. Am. Sylva*, ii. 222.

## DIOSPYROS LOTUS, Date-Plum

*Diospyros Lotus*, Linnæus, *Sp. Pl.* 1057 (1753); Loudon, *Arb. et Frut. Brit.* ii. 1194 (1838); C. B. Clarke, in Hooker, *Fl. Brit. India*, iii. 555 (1882); Hemsley, in *Journ. Linn. Soc. (Bot.)* xxvi. 70 (1889); Shirasawa, *Icon. Ess. Forest. Japon*, 123, t. 79 (1900).
*Diospyros microcarpa*, Siebold, in *Ann. Soc. Hort. Pays.-Bas.* 1844, p. 28.
*Diospyros japonica*, Siebold et Zuccarini in *Abh. Bayer. Acad.* iv. 3, p. 136 (1846).

A tree attaining 60 feet in height and 6 feet in girth. Bark remaining a long time smooth, finally rough and with plate-like scales. Young shoots with a moderately long dense pubescence, often persistent in the second year. Leaves (Plate 199, Fig. 4) oblong or elliptical, base rounded or broadly cuneate, apex acuminate or acute; margin entire, ciliate; upper surface dark green, shining, usually becoming glabrous except at the base of the midrib, but often with scattered minute hairs on the veins and veinlets; lower surface pale and pubescent throughout; veins pinnate and looping towards the margin; petiole, $\frac{1}{4}$ to $\frac{1}{2}$ inch, pubescent.

Flowers diœcious. Staminate flowers, two to three together in subsessile cymes; calyx with four short ovate acute ciliate lobes; corolla urceolate, with four short obtuse lobes; stamens sixteen, in pairs in two series; filaments glabrous. Pistillate flowers, solitary, subsessile; staminodes, eight; ovary eight-celled, one ovule in each cell; styles, four. Fruit subsessile, almost globose, yellow or blackish, $\frac{1}{2}$ to $\frac{3}{4}$ inch in diameter; fruiting-calyx spreading, with a ring of short dense silky hairs on the inside beneath the fruit. The fruit varies considerably in size, and is astringent in flavour.

The leaves pubescent beneath, and the different buds and leaf-scars distinguish this species in summer from *D. virginiana*. In winter the following characters (Plate 200, Fig. 3) are available :—Twigs slender, usually with scattered long hairs, occasionally glabrous; two long acuminate scales of the previous season's bud persist at the base of the shoot. Leaf-scars small, nearly parallel to the twig on prominent pulvini, semicircular, marked with a raised transverse crescentic ridge, composed of the coalesced bundle cicatrices. True terminal bud absent, stump at the apex of the twig pubescent. Buds long, ovoid, acuminate, blackish, pubescent; outer scales two, imbricate, long, acuminate, pubescent, ciliate, concave interiorly.

This species has been long in cultivation, and its exact distribution in the wild state is difficult to define. It appears to be indigenous in Asia Minor; in the Caucasus,[1] where it occurs wild throughout the whole territory between sea-level and 3500 feet; in Afghanistan; in the north-west Himalaya[2] at 2000 to 6000 feet in Hazara and Kashmir; and in central and northern China.

It has been cultivated for centuries in the countries bordering on the Mediterranean, and has become naturalised in many places, as in the south of France and in Dalmatia. It is not wild in Japan, but is often planted there, either for its own

---

[1] Radde, *Pflanzenverb. Kaukasusländ.* 181 (1889).  [2] Gamble, *Indian Timbers*, 455 (1902).

fruit or as a stock on which to graft the persimmon. In China it is largely cultivated, and the fruits, known as *hei-tsao* or "black dates," are an article of commerce. (A. H.)

The date-plum was early introduced into England, being cultivated by Gerard[1] in 1633 or earlier. It grows easily from seed, and, according to Loudon, at the rate of 12 to 18 inches annually, if planted in rich soil,[2] but requires a warmer climate than ours and never attains a large size, so far as we know, in England; though trees of 20 feet or so in height are sometimes seen in botanic gardens and in parks. It ripens fruit usually every year at Kew.

Mayr[3] figures the wood, which is very remarkable on account of the contrast in colour between the black heart-wood and the pale sap-wood. Judging from this and the wood of the Japanese kaki, this wood if procurable would be valuable for cabinet-making; but, so far as I know, it is nowhere common enough to have acquired any recognised commercial value. (H. J. E.)

[1] *Herball*, Johnson's edition, 1495 (1633).

[2] The seedlings which I have raised from seed collected in France, seem liable to injury by frost, and do not ripen their shoots when young.

[3] *Fremdländ. Wald- u. Parkbäume*, 464, t. xvii. fig. 29 (1906).

END OF VOL. IV

*Printed by* R. & R. CLARK, LIMITED, *Edinburgh.*

PLATE 208.

SILVER FIR AT COWDRAY

PLATE 200.

PLATE 210.

SILVER FIR AT ROSENEATH

PLATE 211.

SILVER FIR AT TULLYMORE

PLATE 212.

SPANISH FIR IN ANDALUSIA

SPANISH FIR AT LONGLEAT

PLATE 213.

PLATE 214.

GREEK FIR AT BARTON

HIMALAYAN FIR IN SIKKIM

PLATE 215.

PLATE 216.

JAPANESE FIR IN JAPAN

GIANT FIR AT EASTNOR CASTLE

PLATE 217.

PLATE 218.

GIANT FIR IN VANCOUVER'S ISLAND

CALIFORNIAN FIR AT LINTON

PLATE 219.

LOVELY FIR IN BRITISH COLUMBIA

PLATE 221.

NOBLE FIR IN OREGON

PLATE 222.

RED OR SHASTA FIR AT BAYFORDBURY

PLATE 223.

RED OR SHASTA FIR AT BONSKEID

PLATE 224.

BRISTLE-CONE FIR AT EASTNOR CASTLE

PLATE 225.

ROCKY MOUNTAIN FIR IN MONTANA

PLATE 226.

MEXICAN FIR AT FOTA

PLATE 227.

DOUGLAS FIR ON BARKLEY'S FARM

PLATE 228.

DOUGLAS FIR FOREST IN VANCOUVER'S ISLAND

PLATE 229.

DOUGLAS FIR AT EGGESFORD

PLATE 230.

DOUGLAS FIR AT LYNEDOCH

PLATE 231.

DOUGLAS FIR AT TORTWORTH

SPANISH CHESTNUT GROVE AT BICTON

PLATE 232.

SPANISH CHESTNUT AT ALTHORP

PLATE 233.

SPANISH CHESTNUT AT THORESBY

PLATE 234.

SPANISH CHESTNUT AT RYDAL

PLATE 235.

PLATE 336.

SPANISH CHESTNUT AT ROSSANAGH

PLATE 237.

JAPANESE CHESTNUT AT ATERA, JAPAN

PLATE 238.

WEEPING ASH AT ELVASTON CASTLE

TALL ASH AT COBHAM PARK

PLATE 239.

TWISTED ASH AT COBHAM PARK

PLATE 240.

PLATE 241.

TALL ASH AT ASHRIDGE

PLATE 242.

ASH AT WOODSTOCK, KILKENNY

PLATE 243.

DISEASED ASH AT COLESBORNE

DEFORMED ASH AT CIRENCESTER

PLATE 244.

PLATE 245.

NARROW-LEAVED ASH, ROUGHAM HALL

WHITE ASH AT KEW

PLATE 246.

BILTMORE ASH AT FAWLEY COURT

PLATE 247.

PLATE 248.

ZELKOVA CRENATA AT WARDOUR CASTLE

ZELKOVA CRENATA AT GLASNEVIN

PLATE 249.

PLATE 250.

ZELKOVA ACUMINATA AT CARLSRUHE

CELTIS OCCIDENTALIS AT WEST DEAN PARK

PLATE 251.

PLATE 252.

ALDERS AT LILFORD

PLATE 253.

ALDERS AT KILMACURRAGH

PLATE 254.

ITALIAN ALDER AT TOTTENHAM HOUSE, SAVERNAKE

BIRCH AT SAVERNAKE FOREST

PLATE 255.

PLATE 256.

BIRCH AT MERTON HALL

PLATE 257.

BIRCH IN SHERWOOD FOREST

GNARLED BIRCHES IN GLENMORE

PLATE 258.

PLATE 259.

PAPER BIRCH AT BICTON

PLATE 260.

YELLOW BIRCH AT ORIEL TEMPLE

DIOSPYROS VIRGINIANA AT KEW

PLATE 261.

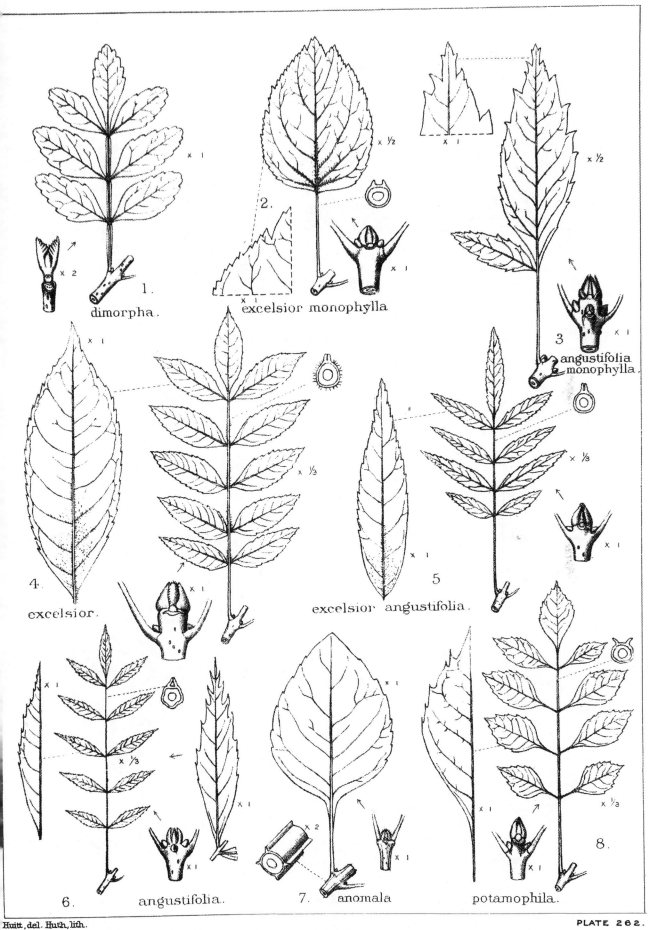

1. dimorpha.

2. excelsior monophylla

3. angustifolia monophylla.

4. excelsior.

5. excelsior angustifolia.

6. angustifolia.

7. anomala

8. potamophila.

PLATE 262.

FRAXINUS.

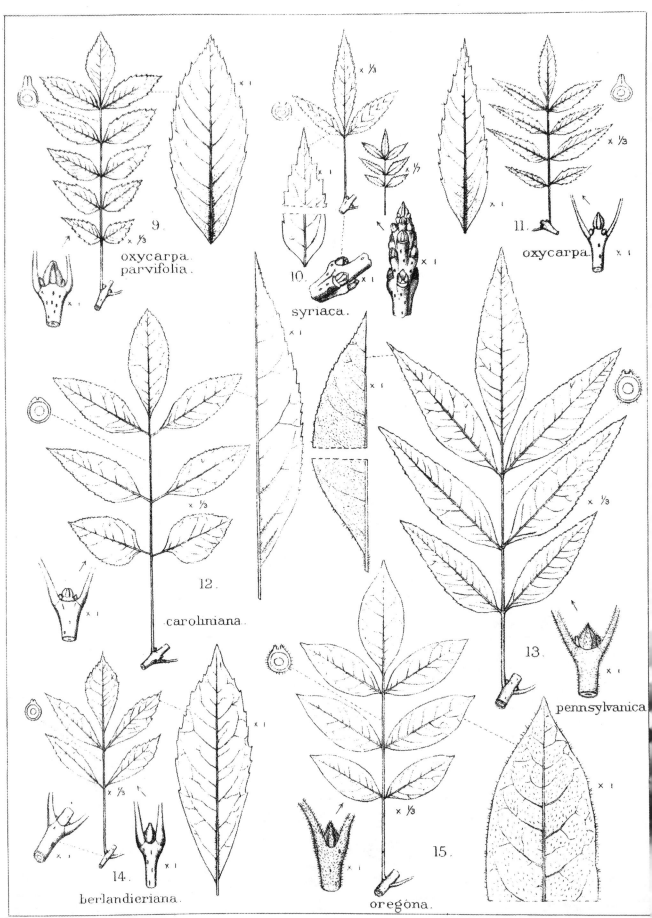

9. oxycarpa parvifolia.

10. syriaca.

11. oxycarpa.

12. caroliniana.

13. pennsylvanica.

14. berlandieriana.

15. oregona.

Hurtt, del. Huth, lith.

PLATE 263

FRAXINUS.

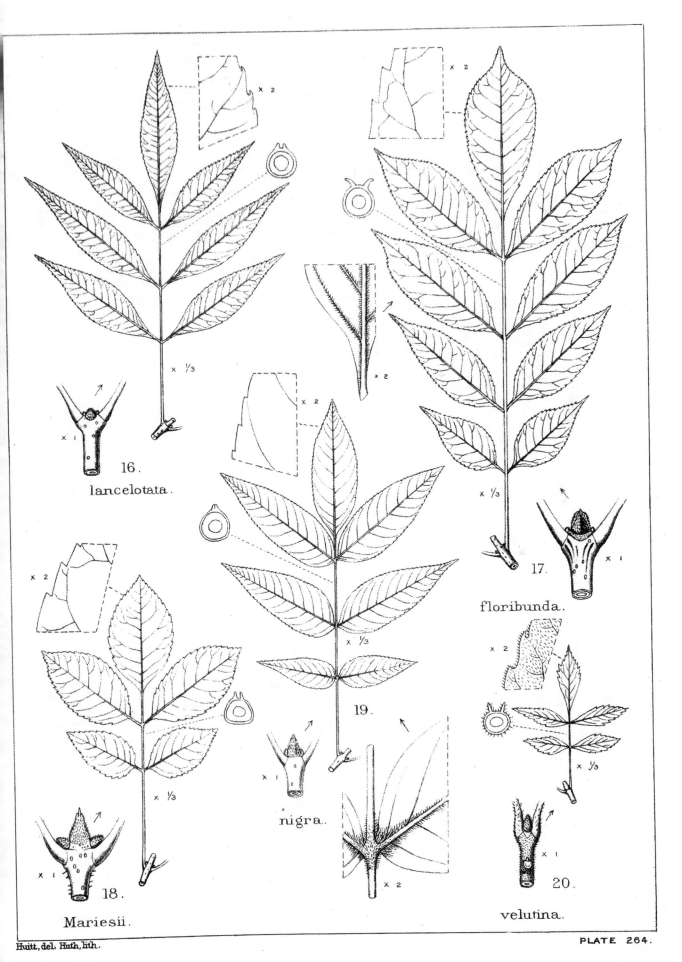

16.

lancelotata.

17.

floribunda.

18.

Mariesii.

19.

nigra.

20.

velutina.

PLATE 264.

FRAXINUS.

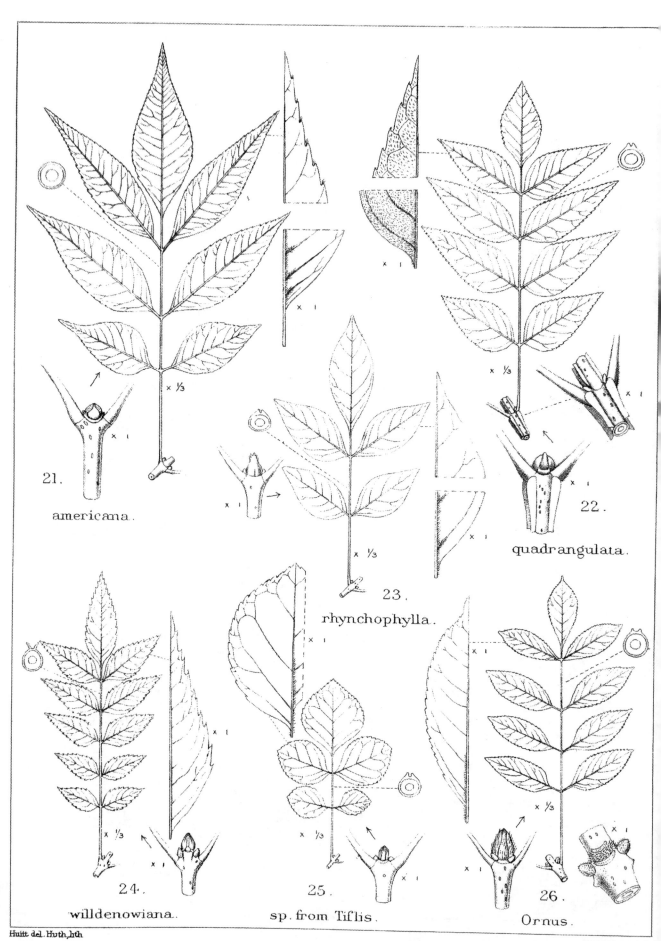

21.

americana.

23.

rhynchophylla.

22.

quadrangulata.

24.

willdenowiana.

25.

sp. from Tiflis.

26.

Ornus.

Huitt del. Huth, lith

PLATE 265.

FRAXINUS.

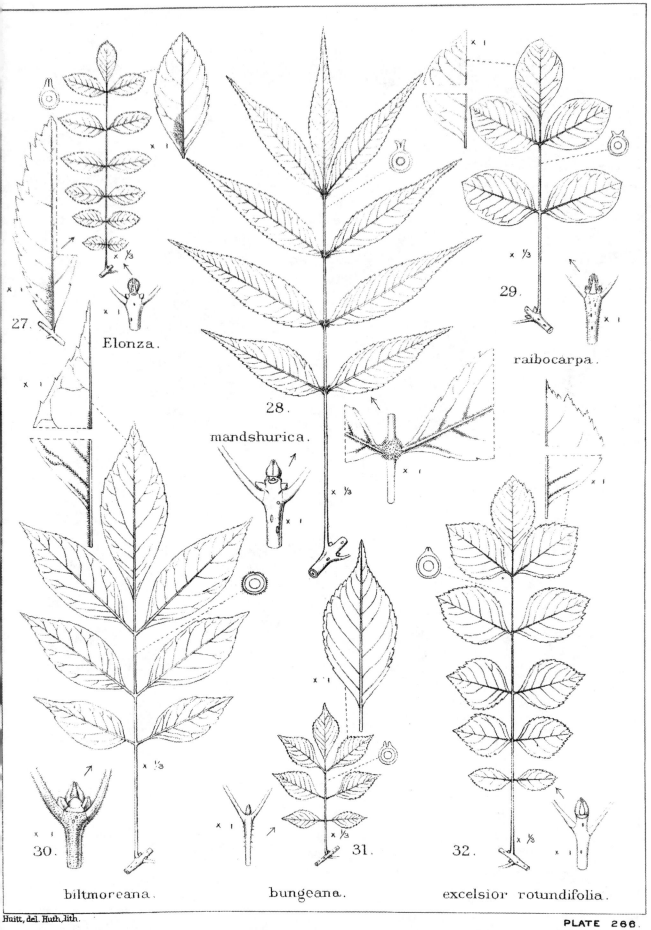

27. Elonza.

28. mandshurica.

29. raibocarpa.

30. biltmoreana.

31. bungeana.

32. excelsior rotundifolia.

PLATE 266.

FRAXINUS.

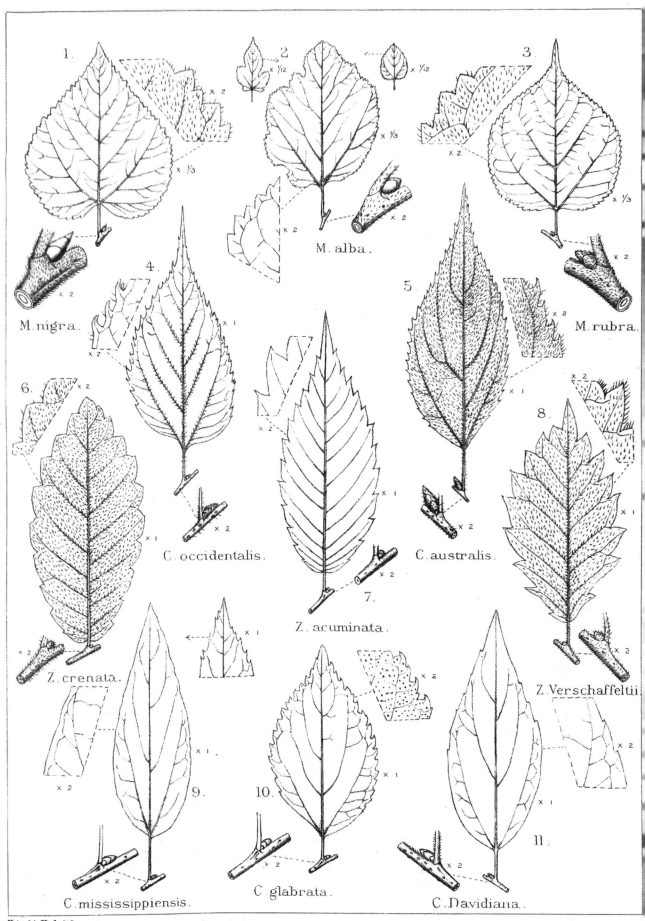

1. M. nigra.

2. M. alba.

3. M. rubra.

4. C. occidentalis.

5. C. australis.

6. Z. crenata.

7. Z. acuminata.

8. Z. Verschaffeltii.

9. C. mississippiensis.

10. C glabrata.

11. C. Davidiana.

Hmitt.del. Huth.lith.

PLATE 267.

MORUS, CELTIS, AND ZELKOVA.

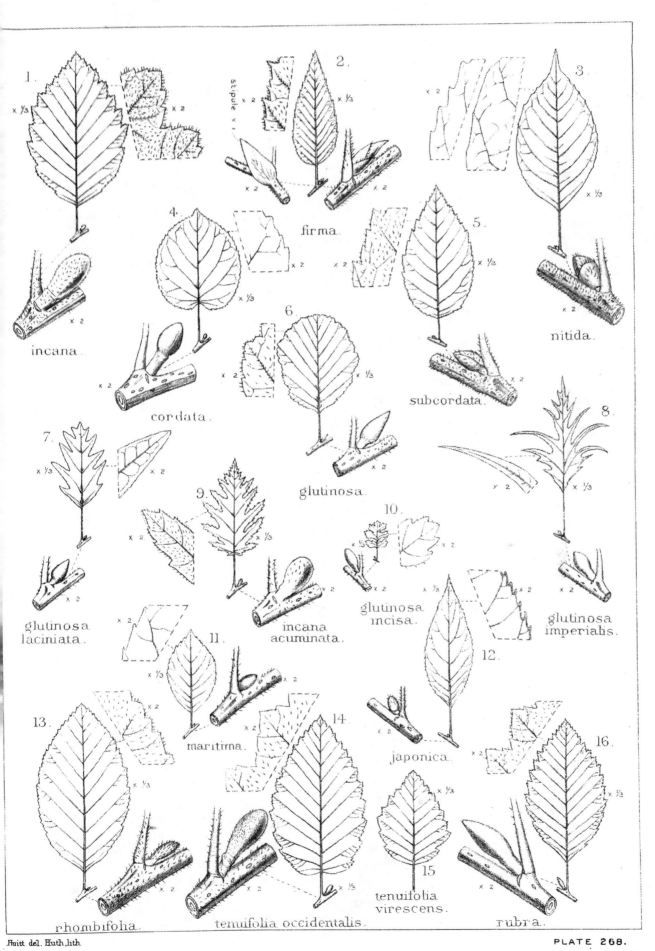

1.
x ⅓
x 2

incana.

2.
stipule x 1
x 2
x ⅓
x 2

firma.

3.
x ⅓
x 2

nitida.

4.
x ⅓
x 2

cordata.

5.
x ⅓
x 2

subcordata.

6.
x 2
x ⅓
x 2

glutinosa.

7.
x ⅓
x 2
x 2

glutinosa
laciniata.

8.
x 2
x ⅓
x 2

9.
x 2
x ⅓
x 2

incana
acuminata.

10.
x 5
x 2
x 2

glutinosa
incisa.

glutinosa
imperialis.

11.
x ⅓
x 2

maritima.

12.
x 2
x ⅓

japonica.

13.
x ⅓
x 2

rhombifolia.

14.
x ⅓
x 2

temuifolia occidentalis.

15.
x ⅓

tenuifolia
virescens.

16.
x ⅓
x 2

rubra.

Hoitt del. Huth, lith

PLATE 268.

A L N U S .

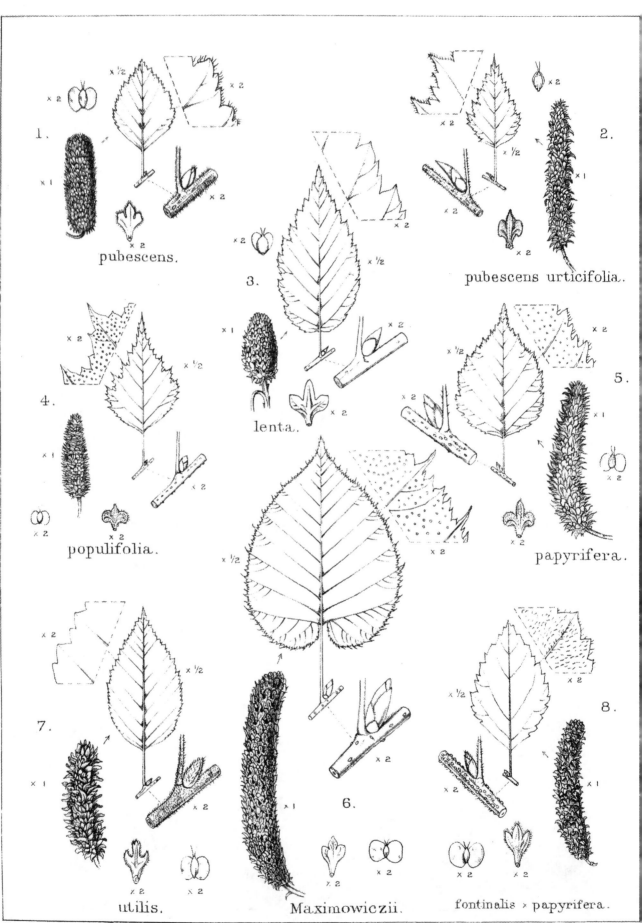

1. pubescens.

2. pubescens urticifolia.

3. lenta.

4. populifolia.

5. papyrifera.

6. Maximowiczii.

7. utilis.

8. fontinalis × papyrifera.

Huitt, del. Huth, lith.

PLATE 269.

BETULA.

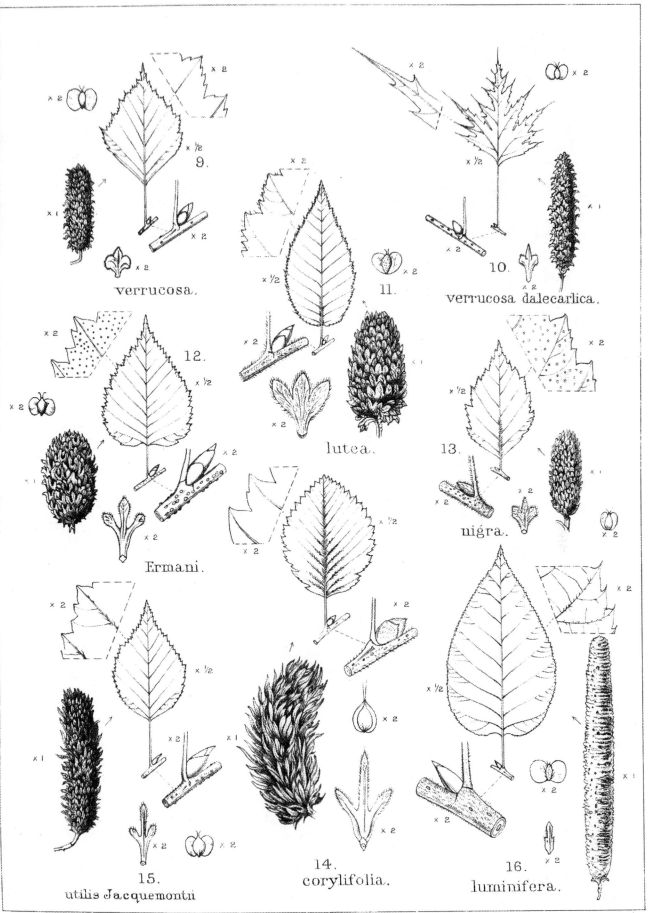

9.

verrucosa.

11.

10.

verrucosa dalecarlica.

12.

lutea.

13.

Ermani.

nigra.

15.

utilis Jacquemontii

14.

corylifolia.

16.

luminifera.

PLATE 270.

BETULA.

Printed in the United States
By Bookmasters